Reading Benedict / Reading Mead

New Studies
in American Intellectual
and Cultural History

Howard Brick, *series editor*

Reading Benedict / Reading Mead

Feminism, Race, and Imperial Visions

Edited by Dolores Janiewski and Lois W. Banner

The Johns Hopkins University Press / Baltimore and London

For the two Marys,
Mary Catherine Bateson
and
Mary Wolkskill,
with gratitude

Printed in the United States of America on acid-free paper
9 8 7 6 5 4 3 2 1

The Johns Hopkins University Press
2715 North Charles Street
Baltimore, Maryland 21218-4363
www.press.jhu.edu

Library of Congress Cataloging-in-Publication Data

Reading Benedict/reading Mead : feminism, race, and imperial visions / edited by Dolores Janiewski and Lois W. Banner.
p. cm.—(New studies in American intellectual and cultural history)
Includes bibliographical references and index.
ISBN 0-8018-7974-4 (hardcover : alk. paper) — ISBN 0-8018-7975-2 (pbk. : alk. paper)
1. Benedict, Ruth, 1887–1948. 2. Mead, Margaret, 1901–1978. 3. Women anthropologists—Biography. 4. Women anthropologists—Attitudes. 5. Feminist anthropology. 6. Racism in anthropology. 7. Anthropologists' writings, American—History and criticism. 8. American literature—Women authors—History and criticism. 9. Women and literature—United States. I. Janiewski, Dolores E., 1948– II. Banner, Lois W. III. Series.
GN20.R43 2004
301′.092′2—dc22 2004004571

A catalog record for this book is available from the British Library.

Contents

Introduction

Being and Becoming Ruth Benedict and Margaret Mead

> Ruth said to Naomi "Entreat me not to leave thee . . . for whither thou goest, I will go."
>
> Ruth 1:16–17

The partnership between Ruth Fulton Benedict and Margaret Mead continues to exert a fascination more than a half-century after Benedict's death in 1948. Within their home discipline of anthropology, Mead and Benedict produced books that compete for the status of being that discipline's Urtexts. At the same time, their reformist aims and their decision to write for a popular audience continue to elicit both praise and criticism. Controversies over the validity of their interpretations and revelations about their emotional, intellectual, and sexual intimacy render them particularly interesting to scholars concerned about the history of ideas, gender, sexuality, and identity. As demonstrated by this collection of essays, their writings continue to engage historians, anthropologists, political theorists, literary specialists, sociologists, and practitioners in cross-disciplinary fields.

They met in 1922, in a class in anthropology at Barnard College. Mead, then twenty-one, was a student in the class, and Benedict, fifteen years older and a Ph.D. candidate in anthropology at Columbia University, was the teaching assistant. Their relationship—as teacher and student, fictive mother and daughter, professional colleagues, and lovers—continued until Benedict's death twenty-five years later, when she was sixty-one. During those years both became anthropologists and public intellectuals of note. Mead lived another forty-three years, dying in 1978 at the age of seventy-seven. During those four decades, she became one of the nation's most famous public figures, crisscrossing the nation and the world to speak on a wide variety of intellectual and public matters while publishing a stream of articles and books.

As public intellectuals, Benedict and Mead navigated the boundaries between the private space of intimacy and the public glare of publicity, presenting a carefully constructed respectability while risking the danger of scandal because of their personal involvements. Granted a full professorship only after a quarter-century of service to Columbia University, Benedict never ceased her efforts to educate the public to accept different cultures, sexualities, races, and temperaments. Her antifascist and antiracist activities would expose her to allegations of Communist sympathies, while her lesbianism jeopardized the camouflage of married name and marital title.

Often treated with scholarly condescension as a popularizer, Mead attracted antagonism for her continual efforts to communicate with a "general and educated audience." Mead's regular appearances in the mass media culminated in a monthly column in *Redbook,* co-written with Rhoda Metraux. Determined to communicate with American housewives, Mead continued the campaign begun with Benedict to mediate between the academy and the public in a career often overlooked in laments about the disappearance of the public intellectual.[1] Selective concealment of aspects of their sexuality was a cost Benedict and Mead paid to step "into public view" to warn about the dangers of racism, promote tolerance, and advocate flexibility in gender and sexual practices.

The writings of Benedict and Mead occupy a place in the popular imagination as interpretations of cultural issues that translated disciplinary insights into compelling and accessible prose. Benedict's *Patterns of Culture* (1934) persuaded the reading public that anthropology could reveal useful truths about their own and other people's society. Mead's *Coming of Age in Samoa: A Psychological Study of Primitive Youth for Western Civilization* (1928) helped to perpetuate the sexual liberalism of the 1920s. Her *Sex and Temperament in Three Primitive Societies* (1935) offered a feminist critique of normative gender and sexual relations that prepared the way for the feminist revival of the 1960s. Benedict's *The Chrysanthemum and the Sword: Patterns of Japanese Culture* (1947) seared itself into Japanese consciousness, initiating a debate about the correct portrayal of their culture that continues into the present—even as it sought to persuade American readers to relinquish their conceptions of the Japanese as a "race enemy." Having earned themselves a place in a feminist pantheon as scholars, activists, and sexual rebels, Benedict and, especially, Mead would become the targets of conservative opponents to radical changes in gender and sexuality and of leftist critics who saw them either as not consistently feminist or not sufficiently critical of racial and colonial dominance within the United States or the islands within imperial systems.

This collection pays attention to their writings, the outcome of their intense and ongoing conversations, letters, poems, and debates. As Mead acknowledged in *Blackberry Winter: My Earlier Years* (1972), referring to Benedict, "We read and reread each other's work, wrote poems in answer to poems, shared our hopes and worries." Had Benedict written more publicly about Mead, she might have emphasized similar aspects of their relationship, while expressing frustration at having to take on editorial or other chores while Mead went on field trips and lecture tours. Benedict, however, suffered from the loss of hearing in one ear; aside from a few field trips among North American Indians, she left extensive ethnography up to Mead, who specialized in the tribal societies of the South Pacific. Poetry allowed them greater freedom to discuss their feelings during the first years of their relationship in the 1920s and early 1930s. Their correspondence played a crucial role in continuing their emotional and intellectual collaboration when marriages, field trips, journeys, and love affairs kept them apart. Summing up their relationship after Benedict's death, Mead wrote, "I had read everything she had ever written and she had read everything I had ever written. No one else had, and no one else has."[2]

The biographical details of Benedict's life are well known to readers of the published biographies. A new interpretation of Benedict's and Mead's lives by Lois Banner provides the most fully researched analysis, based on recently opened Margaret Mead papers and an extensive search of other archives.[3] These biographies detail Benedict's life from her birth in 1887. The poetry and letters of Benedict, married in 1914 to Stanley Benedict, reveal a series of romances with younger women (and several men) to whom she acted as lover, mentor, friend, and confidante, even as she dealt with an increasingly troubled marriage. Following a separation in 1930, Benedict began to receive formal academic recognition when Franz Boas, whom she had served for a decade, secured her appointment as an assistant professor at Columbia.

By 1933 Benedict had become "lesbian," although she and Stanley never divorced. The favorable reception of *Patterns of Culture* in 1934 established her scholarly reputation. Popular writing such as *Race: Science and Politics,* published in 1940 and republished in Britain as *Race and Racism,* helped to introduce both the word and the critical concept of racism into the English language as a contribution to the battle against racial hatred and anti-Semitism.[4] During World War II, as a research specialist for the Office of War Information (OWI), Benedict applied her anthropological skills to the service of the war effort by studying Allied and Axis cultures at a distance.

This effort culminated in the postwar publication of *The Chrysanthemum and the Sword* to foster better understanding between the former enemies, Japan and the United States. Written to persuade American readers to think of the Japanese not as racially different but as having a different culture, Benedict's last significant publication became a key text in Japan's cultural adaptation to a world dominated by the United States.[5] The wartime work led her to launch, with Margaret Mead, a project in the study of contemporary cultures located at Columbia University. Benedict's death in October 1948 spared her from the uncongenial constraints that Cold War culture would undoubtedly have imposed on a woman who questioned cultural norms.

The main contours of Mead's life are documented in her autobiography, *Blackberry Winter*, in Mary Catherine Bateson's *With a Daughter's Eye: A Memoir of Margaret Mead and Gregory Bateson*, and in several popular biographies.[6] The Banner biography reveals more about the hidden aspects that have been only partially revealed by previous biographers. In contrast to Benedict's early acquaintance with loss, the portrayal of Mead's early life generally emphasized her essentially happy childhood, beginning with her birth in 1901 as the first and favored child of Edward and Emily Fogg Mead. Closer reading of the evidence suggests a childhood sometimes troubled by the conflicts between a frustrated mother whose desire for intimacy had been better satisfied by her college classmates and feminist friends, and a domineering father who found other female companionship. In Luther Cressman she found an early suitor whose docile temperament differed markedly from her father's.

The young Mead entered Barnard in 1920 determined to find a career and professional success. Having chosen to take a course with Boas during her senior year, Mead discovered anthropology shortly after Benedict had entered the field. She quickly became emotionally intrigued with Ruth Benedict, Boas's shy assistant. Sometime between 1922 and 1925, the relationship between Benedict and Mead became a love affair, nourished by poetry, conversation, and a triangular involvement with Edward Sapir, the eminent linguist and anthropologist trained by Boas at Columbia. As Mead would acknowledge in an unpublished portion of *Blackberry Winter*, she wrote *An Anthropologist at Work: Writings of Ruth Benedict* as a four-part conversation among Boas, Benedict, Sapir, and herself, without fully explaining to her readers the emotional and sexual connections between the last three.

Editing *Blackberry Winter* to its published length erased almost all traces of Sapir's significance, while also eliminating Benedict and Mead's other partners except for their husbands. The coverage of personal life largely ended in the mid-1940s, when divorce ended Mead's last marriage, to concentrate on her subsequent career as a public intellectual. The significance of her thirty-year personal and intellectual partnership with Rhoda Metraux disappeared from this highly selective version of her life. As published,

Blackberry Winter produced a public persona for Mead as a confident, self-controlled, but feminine woman who lived as a "normal" wife, mother, and delighted grandmother.

Mead was married three times: in 1923 to Luther Cressman, an Episcopal priest and student in anthropology and sociology at Columbia; to Reo Fortune, an anthropologist from New Zealand, whom she married in 1928; and in 1936 to English anthropologist Gregory Bateson, whom she divorced in 1950. She had one child, Mary Catherine Bateson, in 1939. She also had numerous lovers, both male and female.

Beginning her career with a famous study of adolescence on Samoa, Mead continued until World War II to study cultures in the South Seas, including Manus, three cultures in New Guinea, and Bali—with a summer in 1930 studying the Omaha tribe in Nebraska. During World War II she served as the administrator for the Committee on Food Habits that launched her into her career as a member of executive boards and as a networker among government officials, scholars, and intellectuals. A curator at the American Museum of Natural History from 1926 until her death in 1978, she lectured throughout the nation; revisited Manus five times; became involved in the founding of the United Nations; wrote numerous books and articles, including the column for *Redbook* magazine; and was welcomed as a media personality, especially on television.

Readers of this collection will discover lives and texts that are far more complicated and tangled than the smoother public versions previously published. The texts require closer scrutiny to reveal the biographical elements opaque to conventional modes of interpretation. As Bateson tellingly observed in *With a Daughter's Eye,* interpreting her mother's life required the "need to deal with the fact of concealment." Letters recently opened to scholars reveal a more emotionally overwrought and insecure young woman and anthropologist than the self-assured person who appears in *Blackberry Winter.* Frequently assuring Benedict of her love and longing in letters written from Samoa, Mead desperately tried to find the key to explain Samoan culture in order to meet the expectations of her mentors. Finally discovering what she saw as the vital clue, Mead continued to struggle with competing emotional commitments. When she embarked on a voyage to Europe via Australia, her correspondence ceased for a period, to Benedict's perplexity. Only later, when Mead began writing after she had arrived in Europe, did it emerge that a shipboard romance with Reo Fortune had added yet another person to the emotional tangle. Benedict and Mead, despite the pain of that interrupted correspondence and the emergence of an emotional rival, would resume their intellectual collaboration when they returned to New York. Out of that crisis would come Mead's second marriage, to Fortune, a future partner for fieldwork, and her first book, *Coming of Age in Samoa.*

Apparently convinced that Mead had developed a stable, sexually exclusive marital relationship with Fortune, Benedict sought to help the couple by finding research support and assisting with the publication of their work. In 1930 economic necessity and Benedict's ability to get funding to address a gap in the scholarly literature led to a field trip to the Omaha reservation for Mead and Fortune. Assigned responsibility for investigating the "woman question" while Fortune looked for evidence of Omaha visions, Mead undertook "the unrewarding task of discussing a long history of mistakes in American policy toward the Indians." Conscientiously, Mead completed her analysis of "the cultural shipwreck of the Indian tribe," published in 1932 as *The Changing Culture of an Indian Tribe.*[7]

Having secured more funding, Mead and Fortune returned to the Pacific in the early 1930s. During their time in New Guinea, Benedict performed editorial duties for the absent couple, wrote positive reviews of their work, and worked her way out of a frustrating marriage into a home with a female partner. She would use Fortune's research on Dobu as one of the three cultures compared in *Patterns of Culture.* The manuscript arrived in New Guinea in the midst of another emerging emotional tangle as Gregory Bateson became Mead's potential new research partner and Fortune grew increasingly estranged. Enlisting Australian friends, Fortune's brother in New Zealand, and Benedict in New York, Mead managed to secure a second divorce, complete *Sex and Temperament,* organize a research trip to Bali, and marry Bateson without creating a public scandal that would harm her own or her new husband's careers.

In contrast to Fortune's insistence on sexual fidelity, Bateson disavowed any sense of jealousy or competition with Benedict. Neither requiring an exclusive relationship nor conforming to one, the marriage survived the challenge of other attachments until war separated the couple. After the war, neither psychoanalysis nor Mead's tolerance for his other partners could keep the marriage together. Once again, Benedict lent emotional support as a distraught Mead traveled to Europe during the summer of 1947 with her daughter, Mary Catherine. Mead would endure another devastating blow a year later as Benedict's death ended an experience of personal and intellectual closeness that had endured for twenty-five years. Repeating the pattern that had produced *Sex and Temperament* in 1935, Mead rapidly published *Male and Female: A Study of the Sexes in a Changing World* before the news of a divorce could undermine her claims to expertise about relations between men and women.

Looking back at the events in the late 1940s as a crucial turning point, Mead divided her life into two relatively equal segments, split apart by the bomb that exploded over Hiroshima and by personal losses. In *Blackberry Winter,* Mead wrote that Hiroshima had ended her existence as a "collaborating wife, trying to combine intensive field work and an intense personal life." No longer a partner in a marriage "which once occupied so much of my time and attention," Mead lost a sense of "wholeness," now understanding herself as "fated for a life which is no longer sharable." The intimacy experienced with Bateson and Benedict was replaced with "a multitude of special relationships, collaborations, slight gaieties, and partial intensities" in a fast-paced, peripatetic existence.[8]

Mead's conviction that the "new age" that dawned at Hiroshima required her active engagement as a public intellectual meant that she felt the necessity to operate within the constraints of the emerging Cold War. Published in 1949, *Male and Female* acknowledged "the basic difference" between men and women. Mead attacked the idea of "crusades based on the rights of women." Competition with men was "dangerous." At the same time, she warned that excessively rigid sexual segregation would "push a large part of society towards celibacy or homosexuality," appearing to suggest that either would be an unfortunate outcome. *Male and Female*'s conclusion noted the painful paradox of "a tangled situation" created by two sexes having "different gifts" in a world that desperately needed to use the "gifts of the whole of humanity."[9] Ideological incoherence showed the difficulties Mead now faced in adapting her prewar views and public persona to the postwar context.

Making another contribution to the Cold War consensus defining the United States as the ideal society, Mead published *New Lives for Old: Cultural Transformation—Manus 1928–1953.* She applauded the Manus people's "active choice" to enter "into our way of life." Such commentary, like the antifeminist statements in *Male and Female,* would subject Mead to later criticisms, but it resembled the unabashed patriotism she had displayed during wartime writings when she had similarly glorified the United States.[10] Mead's writings lacked the ironic tone that had distinguished Benedict's commentary about the differences between the United States and other cultures.

Subsequent research suggests that Mead's characterization of a dramatic split in her life between an intense personal life that ended at Hiroshima and an active public life that began in the late 1940s may have been a part of the "concealment" to which she resorted to protect her public persona. A relationship with Metraux began in the late 1940s and continued until Mead's death in 1978. Metraux edited Mead's writings, conducted research, and planned future publications, sometimes writing in Mead's place although only occasionally being acknowledged for her contributions. So far as it can be determined, Metraux never sought such recognition, keeping silent about her contributions to Mead's productivity during her last three decades.[11] Close reading of Mead's writings after Benedict's death must always take into consideration that Mead rarely wrote entirely alone.

There would be periods of significant change in Mead's life which *Blackberry Winter* did not discuss. The radicalism of the late 1960s and early 1970s gave Mead scope during the colder parts of the Cold War era. Praised by Betty Friedan in her feminist classic *The Feminine Mystique* as the "most powerful influence on modern women," Mead seized the opening made by the feminist revival to address questions that she had treated ambiguously in *Male and Female.* Aided by Metraux, Mead encouraged her readers to confront the dangers of "nuclear holocaust"; participate in politics; rethink marriage, family, and heterosexuality; and take action on their own behalf in a popular form of feminism that extended the movement into the mainstream. The moon landing in July 1969, with its accompanying pictures of a vulnerable earth suspended in space, took its place beside Hiroshima as a turning point in Mead's life. Entering her eighth decade in the 1970s, Mead published *A Rap on Race, World Enough,* and *Twentieth Century Faith.* Mead debated race relations with James Baldwin, discussed the fragility of the earth and its environment, insisted on the need to correct "colonial bias" in scholarship, and urged her readers to take responsibility for restraining the awesome power that could destroy "every living thing on this planet." She criticized the "extravagant hopes that high technology" and American-style democracy could liberate the world's peoples as outmoded ideas, implicitly apologizing for her contributions to the Cold War consensus.[12] These writings, particularly *World Enough,* served as a coda to her public career.

Hindered by secrets, camouflage, and gaps in the documentary record, interpreters of Benedict's and Mead's writings must display sensitivity to historical context, recognize the aims of the authors, and attend to issues of popular reception. In view of the enduring appeal of texts like *Coming of Age in Samoa, Patterns of Culture, Sex and Temperament in Three Primitive Societies,* and *The Chrysanthemum and the Sword,* assessing both context and reception are daunting but necessary tasks. The interpretation of these texts, written by founding mothers of anthropology and commentators on controversial issues, is a complicated enterprise. As our readers will discover, the contributors to this collection do not always agree as to their meaning and significance. Reflecting diverse disciplinary frameworks, the analyses also reveal different emotional, intellectual, and political responses. Some scholars draw biographical insights from analyzing Benedict's and Mead's writings. Others focus on disciplinary controversies. Some locate these texts within an ideological or cultural context that obviously changed over their lifetimes and our own. The question of whether both Benedict and Mead should be described as complicit with colonialism inside and outside the United States, referred to in our title as "imperial visions," will be answered in different ways in some of the essays that follow.

The presence of such conflicts or interpretative differences testifies to the vitality of Benedict's and Mead's writings and to the multiple vantage points from which they can be seen. As provocative intellectuals who raised troubling questions in their own writings and sometimes provided inadequate but always interesting answers, Benedict and Mead deserve the respect of critical essays that seek to explore their contradictions, silences, and commitments.

I / Becoming Benedict, Becoming Mead

Tall, cerebral, and eminently respected in her profession, Ruth Benedict appeared to many observers to differ markedly from the short, exuberant Mead, whom many considered prolific rather than profound. Indeed, as some of the contributors to this section will show, such comparisons have often designated Benedict as the "good" woman and Mead as the "bad." Mead's willingness to engage with the public led to disparagement from members of her profession who saw her as courting popularity by issuing glib statements.

Beneath the surface, however, biographers have found that these contrasts fail to consider the commonalities that enabled Benedict and Mead to work together for twenty-five years. Despite her seeming exuberance, Mead could be emotionally needy, while Benedict's cool surface hid her emotional turmoil from all but her intimates. They borrowed ideas from each other that would shape each other's work. Their unpublished letters display their need for each other's support as they confronted the problems of a fledging discipline, hostility to women's intellectual ambitions, and the danger of scandal if their private behavior was found out. Whether overtly discussed in personal communications or covertly expressed in coded messages, the personal experiences of Benedict and Mead as women, anthropologists, and public intellectuals shaped their words, giving their writings the energy that still engages readers.

As the chapters in this section suggest, the "public" Benedict and Mead never completely separated themselves from their private intimacy. Their writings, as argued by Dolores Janiewski, contain clues to their relationship that can be deciphered for autobiographical evidence. Lois Banner explores one of Benedict's early unpublished writings, a short story called "The Bo-Cu Plant," to understand the nature of her feminism and the conflicts over gender in her life and her career. Maureen Molloy explores the context in which Mead developed a public image that, along with her first book, *Coming of Age in Samoa,* would etch her into the public imagination as a "girl scientist." Perhaps Benedict's combining "high culture" elements from literature and poetry with her anthropology in *Patterns of Culture,* while Mead associated herself with popular culture in *Coming of Age in Samoa* may help explain their differing reputations as public intellectuals.

1 Woven Lives, Raveled Texts

Benedict, Mead, and Representational Doubleness

Dolores E. Janiewski

Mead described her relationship with Benedict as a twenty-five-year collaboration in which "we read and reread each other's work" in a pattern of intellectual closeness that she would never enjoy with anyone else. Benedict's death was one of the crucial events that profoundly changed her life in the late 1940s by foreclosing the possibility of an "intense personal life." Afterward, she would devote her energies to a public career.[1] Like Mead's other writings, her depiction of her personal relationship with Benedict concealed crucial aspects of their emotional, intellectual, and sexual partnership and of her life after Benedict's death.

Writing to Mead after her sister's death, Marjorie Fulton Freeman sought to comfort the bereft partner. Her words suggested that Benedict also felt a special bond with Mead: "One of the deepest satisfactions of her life has been the privilege of stirring up your interest and then watching you carry the torch into fields where she could never go." The desire for a surrogate could be traced back to the early days when Benedict and Mead first met. Benedict's diary had recorded her acute need for "a companion in harness," which Mead would become. In choosing to pursue anthropology as her career, Mead understood her choice as including "a closer relationship to Ruth." Benedict, for her part, found a disciple who could go places where the partially deaf Benedict could not venture. The written record that began in 1922 offers clues about the interweaving of two lives by two women who enjoyed "the pleasures of similarity and the pleasures of contrast," while concealing from prying eyes the extent of their emotional and physical intimacy.[2]

In 1984 Mary Catherine Bateson partially opened the textual closet that had hidden her mother's intimate partnerships. She wrote about a "double pattern of intimacy" that included the relationship with Benedict, while revealing her painful sense of betrayal as a result of this discovery. Publishing her memoir of her mother and father, Bateson wrote about one of the secrets that Mead had chosen to conceal. Seeking to exonerate her mother from charges of deception, the daughter claimed that the meanings in "Margaret's words" might be missed by literal-minded readers who limited themselves to "a conventional mode of interpretation," but they could be decoded by more sophisticated interpreters. As Bateson suggested, reading beneath the surface of both Benedict's and Mead's writings offers insights hidden within the "epistemology of the closet" that Bateson did not completely open.[3]

The alert reader of Benedict's and Mead's writings must act like a detective, interpreting verbal clues about a complex relationship. Making use of the critical insights associated with the "politics of the closet," one must investigate their texts "for rearrangement, ambiguity, and representational doubleness." Acutely aware of the penalties imposed for "appearing gay," Benedict and Mead may have manipulated "the identity they attach[ed] to themselves, both in the secrecy of their own minds and on the public stage," as Bateson ruefully concluded. Seeking to avoid the scandalous exposure of the conflicts between "their personal desires, sexual behavior, subjective identity, and their public identity," they developed coded ways to represent themselves that would enable them to speak without being easily understood. Even without access to the private letters only recently available to researchers, decoding their poems, anthropological writings, and other publications may disclose secrets they shielded beneath a verbal disguise. By continuing to use the honorific "Mrs." even after her separation, Benedict, like Mead, tried to maintain the protective camouflage of respectability as a "presumptive" heterosexual even as she developed emotional and physical connections with female partners, including Mead.[4]

During the decade that she became an anthropologist and became involved with Mead, Benedict devised ways to write while disguising the autobiographical elements. Anthropological texts like *Coming of Age in Samoa* provide clues about their long-lived partnership. Benedict's *Patterns of Cultures* and *The Chrysanthemum and the Sword* would become her most notable contributions to their jointly conceived oeuvre. After *Coming of Age in Samoa,* Mead's *Sex and Temperament in Three Primitive Societies* most powerfully reflected Benedict's influence. Together, Benedict and Mead developed an interdisciplinary synthesis to explain the interactions between personality and culture while presenting their findings in a literary form that could convince a popular audience of the value of their insights.[5]

Poetry became a favorite mode of theirs. It was made even more discreet by Benedict's use of the pseudonym Anne Singleton, who expressed Benedict's emotions and desires while protecting her from the risks of public exposure. Closely read, her poems contain frequent shifts from masculine to feminine pronouns that would make the resolutely heterosexual reader uneasy. Mead also wrote poetry, luxuriating in "romantic unhappiness" as her relationship with Benedict increased in intensity. They delighted in the freedom to write "poems in answer to poems," while also sharing their work with Edward Sapir, who formed the third side of a volatile triangle of professional ambition and sexual ambiguity. Their collaboration grew out of "hours telling each other stories about people whom the other had never met," as the relationship grew more intimate. They discussed anthropological concepts such as deviance. Benedict responded to the "zest of youth" and to the admiration that came from her new disciple. Mead believed that Benedict had released her from the anxiety of a "child awed by articulate elders." On the journey to Samoa in 1925, Mead "took into the field all the questions about deviance she [Benedict] had raised." She also took a collection of verses that Benedict hoped would allow Mead to feel "me speaking them to you." Words and comments made Benedict's presence palpable as Mead embarked on this first research trip. Later Benedict sent her annotated copy of Nietzsche's *Thus Spake Zarathustra* to Mead because it offered the "right, poetical way" to understand culture.[6]

When the communications ceased during the time that Mead was sailing toward Europe after her departure from Australia, Benedict felt perplexed and anxious. Her eventual discovery of Mead's shipboard romance with Reo Fortune that put him "in the foreground of your life" changed their relationship, but silence would not come between them again. Poetry expressed the emotional dilemma. Although Benedict was willing to take the risk that the adoption of a lesbian identity would create, Mead was not. Plaintively one poem asked, "Why did you shout your misery instead of whispering to me?" Another poem's reference to desire as a knife "severing the twisted threads that bind our lives so closely together" alluded to the conflict between desire and respectability. Mead, as her daughter would later explain, disliked "all forced choices." In these circumstances, however, she chose to marry Fortune. "Is this the end of love?" one poem asked—as another spoke of "what was one" becoming "two again." Mead agonized as Benedict learned to accept the decision.[7]

When Mead returned from Samoa, the two scholars began a spirited "discussion that continued for many years," as Mead focused on how to study the interaction between culture and personality, while Benedict began to work on her "configurational approach to culture." When Benedict spent the summer of 1927 among the Pima, the contrast between the Pima and the Zuni struck her with the force of a revelation. From Nietzsche she took the concepts of Dionysian emotionalism and Apollonian rationality to describe the Pima "culture of painful and dangerous experiences" prone to "emotional and psychic excesses" and the Zuni "pursuit of sobriety, of measure," with its rejection of "excess and orgy." Bringing those insights back with her to New York, she and Mead spent the winter of 1927–28 discussing her findings, bringing together concepts drawn from Freud, Jung, Sapir, and others as they sought to combine insights from anthropology and psychology.[8]

The first fruits of that collaboration would soon appear in print as *Coming of Age in Samoa* and *The Social Organization of Manu'a,* in which Mead wrote about "dominant cultural attitudes." Mead planned a trip to Manus to demonstrate that the minds of "children, neurotics, and primitives" shared common characteristics. Living together the summer of 1928, Benedict and Mead wove together their personal concerns and their intellectual desires to understand "human cultures as organic and functioning wholes." Writing to Benedict on the journey that would take her to Manus and marriage with Fortune, Mead talked about having developed a "common mind" with Benedict "in everything from science to love." Mead wrote about "the pattern which winds woven and beautiful all about me, woven by our four hands in the last six years."[9] Although Benedict would have to accept the "double pattern of intimacy," Mead reassured her that their intellectual partnership would endure.

A testy exchange with Edward Sapir reveals Benedict's deep personal investment in the work that the other had published. Taking public vengeance against Mead, Sapir published "Observations on the Sex Problem in America" in 1928, denouncing the "modern woman," homosexuality, and "sex freedom." Sapir's remarks about "books about pleasure-loving Samoans" and the "smart and trivial analysis of sex by intellectuals" were clearly aimed at Mead. It was Benedict, however, who accused Sapir of attacking her. His assurances that "you were never once in my thoughts when I wrote the paper on sex" failed to placate Benedict. They revealed, moreover, his ignorance of their collaboration.[10] The relationship with Sapir never recovered, but that with Mead would continue despite recurrent emotional upheavals.

Benedict's paper "Psychological Types in the Cultures of the Southwest" grew out of the collaboration to produce *Patterns of Culture.* Arguing that each culture selects certain traits and temperaments, Benedict sympathized with the "misfit whose disposition is not capitalized by his culture." In an autobiographical reference she wrote, "The person who has an ineradicable drive to face the facts and avoid hypocrisy may be the outlaw of a culture." Remembering that Sapir had classified her in Jungian terms as a "rational"—that is, a feeling-thinking introvert—Benedict may have recognized the personal significance of her terms describing the differences between cultures. Publicly appearing calm and restrained like the Zuni, but inwardly turbulent and passionate like the Pima, Benedict no doubt found that this Nietzschean dichotomy served both as an ethnographic purpose and as a self-referential subtext.[11]

Benedict's decision to live with a female partner in the early 1930s partially opened the door to her sexual closet, lending greater intensity to her need to change popular attitudes toward those defined as deviant. Her essay "Anthropology and the Abnormal" asked whether the categories of normal and abnormal are "culturally determined" or "absolute," that is, universal. Beginning with discussions of trances, Benedict delicately approached her personal concern. A trait like homosexuality "exposes an individual to all the conflicts to which all aberrants are always exposed" in some cultures, but others gave homosexuality an "honored place." Answering her own question, she insisted that "normality," like "morality," was a culturally derived term "for socially approved habits." An ironic commentary directed attention to the admiration for "unbridled and arrogant egoists" in her own culture. Referring to such men as "mentally warped to a greater degree than many inmates of our institutions," she asked her readers to agree that "our normality is man-made and is of our own seeking." Turning anthropological knowledge to political purpose, Benedict coolly discussed what she felt most passionately. Her Apollonian voice articulated the right to Dionysian pleasures despite assuming the "hair shirt of discipline" to communicate her message.[12]

In the throes of writing *Patterns of Culture,* Benedict asked Mead and Fortune for their help. She told Fortune that she wished "you and Margaret" could read the manuscript "and hand it back blue penciled." She asked to use his material on Dobu to broaden the scope of analysis beyond North America. The arrival of the manuscript in the midst of a volatile triangle between Mead, Fortune, and Gregory Bateson rendered Benedict's ideas about patterns, personalities, and cultural configurations even more influential than they might otherwise have been. Eventually, Mead sent criticisms about Benedict's style as "intimate, sometimes heavily formal, sometimes colloquial or journalese, sometimes in the jargon of anthropology and sometimes in the phrases of good literature." In her view, Benedict was trying too hard to communicate with the "intelligent man in the street," academics, and students while trying to teach an overly complicated message about tolerance, the public uses of anthropology, and cultural configuration.[13]

Contrary to Mead's advice, the published version of *Patterns of Culture* included a plea for the acceptance of those defined as deviants in the terms of their culture. Benedict sought to convince her readers to see their culture as one among many possible ways of living. Arguing against differentiating humankind into the "Chosen People" and "dangerous aliens," Benedict warned against "the gospel of the pure race." She appealed for sympathy for homosexuals in a culture that defined them as abnormal. *Patterns of Culture* would help prepare the intellectual climate for critiques of racism.[14] It would take much longer for her plea for sexual tolerance to achieve the same influence.

Mead's *Sex and Temperament* reflected Benedict's influence but also her own autobiographical impulse, stemming from concerns about the "basis of my own apparent deviance from the accepted style of the career-minded women that I met." Mead, moreover, admitted the subjective nature of her discoveries as she sought to explain the reasons for the differences between her reactions and those of Reo Fortune to the people of New Guinea. Their responses "had nearly as much to do with us, as individuals, as it had to do with the nature of the cultures we studied." Following Benedict's example in *Patterns of Culture,* Mead focused on three New Guinea cultures to illustrate her belief that "human nature is almost unbelievably malleable."[15]

Cautioned when she returned to New York against providing ideological support for beliefs in biologically determined cultural differences, Mead avoided discussing biological factors as shaping temperament. Instead, she emphasized cultural selection from the "arc of human potentialities" as helping to explain both sex and temperament. The result was the most explicitly feminist work that she ever published. She criticized "artificial distinctions, the most striking of which is sex" in a passage which indirectly referred to race. *Sex and Temperament* culminated a decade of collaboration emphasizing the "particularistic/pluralistic" version of culture and personality as Benedict's and Mead's preferred alternative to "natural" or biological explanations of sex and race.[16]

After her return from New Guinea and the hurried completion of *Sex and Temperament,* Mead's writings began to show new influences stemming from collaborations with psychologists and social scientists and her new intimate partnership with the biologically trained Bateson. Her adoption of a "modified Freudian theoretical orientation" contributed to a more biologically inflected interpretation of culture and personality. As reflected in her 1937 publication, *Cooperation and Competition among Primitive Peoples,* the new paradigm emphasized the importance of early childhood, including toilet training, nursing, and weaning, as shaping the personalities of children, and ultimately, the culture in which they came of age. A book jointly authored with Bateson on Bali interpreted culture through a psychoanalytic lens. The birth of Mary Catherine provided a personal motive for an increasing emphasis on the maternal/child relationship as crucial to personal and cultural formation.[17]

Mead's enjoyment of motherhood and marriage may have made it easier to accept the constraints of the epistemological closet. An ambivalence about homosexuality, reinforced by her Freudian-inspired scholarly network, frequently appeared in her writings through a tendency to associate homosexuality with "maladjustment." Fearful that Benedict might endanger her own academic career by revealing her sexual orientation, Mead upbraided her friend for taking unnecessary risks. Now safely enclosed within normative heterosexuality, Mead taught her daughter that "proper behavior was important because it allowed choices and opened doors." Read positively, Mead "affirmed and respected" conventional ways of doing things "even when she did not herself conform." Interpreted as a strategic decision, Mead preferred the "enabling privacy" of the closet. Benedict, living with a female partner, demonstrated less commitment to the norms of respectability as she continued to criticize sexual repression in her writings.[18]

Although differences between Benedict and Mead were becoming more apparent in the late 1930s, they continued to share common commitments to "sane and scientific direction" by enlightened, tolerant experts. In the same year that World War II began, Benedict wrote *Race: Science and Politics* in "payment of the moral debt which anthropologists owed a world threatened by Nazism." Republished as *Races of Mankind* in 1943, Benedict's critique of race relations in the United States attracted the attention of anti-Communist conservatives, who condemned it as "subversive." Benedict's tendency to focus on the darker side of American culture even during wartime contrasted with Mead's propensity for giving American culture "the benefit of the doubt." Returning from Bali in 1939, Mead shared Benedict's alarm about the dangers of "totalitarian disordering or authoritarian control of culturally patterned behavior." As one contribution to the war effort, she wrote *And Keep Your Powder Dry*, an idealized portrait of American culture that banished "the bogies of racism" by defining the southern United States as not a part of mainstream American culture. The United States could instruct other nations in how to live in "a world of plenty, of great expansion, of room for everybody to make a contribution and succeed." Benedict shared Mead's belief in the need to ensure victory for "democratic ways of life," but she qualified that goal by asserting that the "world must be made safe for differences." Benedict asked policymakers to take "into account different habits and customs of other parts of the world" while contributing her expertise to psychological warfare and to reconstruction after the war.[19]

Temperamentally better adapted to cope with tragic ironies, Benedict found a way to respond to "a murky era of moral ambiguity and uncertainty" that began with the dropping of the atomic bombs on Hiroshima and Nagasaki. On August 15, 1945, she wrote to her wartime colleague,

Robert Hashima, a former occupant of one of the internment camps, to express her joy "that the slaughter had been stopped." Referring to Japan, she told him that "no Western nation has ever shown such dignity and virtue in defeat" and asked Hashima to "help me say it." Referring to the occupation of Japan, Benedict hoped "that America will play her part with restraint and dignity." A month later she began writing *The Chrysanthemum and the Sword,* based on wartime research. She chose not to refer directly to the climactic event that had ended the war with Japan, an event that Mary McCarthy called "a kind of hole in human history." Instead, Benedict hoped to influence the ways the American public, the military, and policymakers would treat the Japanese during occupation and reconstruction.

Inverting the usual association of softness and liberalism, Benedict told her readers that the "tough-minded respect differences." She advocated a world in which the United States could "be American to the hilt without threatening the peace of the world, and France may be France, and Japan may be Japan on the same conditions." The Japanese needed to develop an economy "on the profits of peace" to "buy their passage back to a respected place among peaceful nations," while the United States must avoid turning the postwar world into an "armed camp." Using the only weapon she possessed, Benedict wanted her readers to understand that "no foreigner can decree for a people who have not his habits and assumptions a manner of life after his own image."[20] Benedict offered a vision of "one world" that Mead felt unable to write.

Having chosen to avoid discussing the bomb in *The Chrysanthemum and the Sword,* Benedict presented her own views in a review of John Hersey's *Hiroshima.* She applauded the "calmness of the narrative" and the simplicity that enabled the reader to understand "the nightmare magnitude of the destructive power the brains of man have brought into being." She spoke about the "painful guilt" that some Americans have felt "at the thought that they belonged to the nation which catapulted this horror into the houses and streets of a city." Other Americans, no doubt including herself, "have read it as a portent of things to come." The Japanese had demonstrated their capacity to "bear the unbearable and to tolerate the intolerable" with "patient fortitude." She hoped the book would make its readers understand the "necessity for outlawing war." In speaking about Hiroshima, Benedict also revealed her own intentions in writing *The Chrysanthemum and the Sword.*[21]

During the last two years of her life, Benedict sought to restrain the nationalistic fervor being whipped up in the early days of the Cold War. Writing about the need to recognize "cultural diversities in the post-war world," she urged Americans to resist the temptation to assume that "given the slimmest chance all peoples will pattern themselves upon our model." She warned against the "arrogance of race prejudice." Reiterating the moral of *The Chrysanthemum and the Sword,* she insisted, "We shall succeed better if we respect" other people's values because "we shall sow only bitterness if we try to impose our own by force." She warned against the temptation to turn other people into "faithful replicas of themselves." Benedict's death spared her from the full fury of the Cold War. She would also avoid the need to defend herself against critical comments about national character studies and that her service to the U.S. war effort had helped to lead the discipline astray.[22]

Less temperamentally attuned to dealing with moral ambiguities, Mead faced the need to adapt to Cold War constraints without the support of the person most able to help her negotiate a complex ideological and moral terrain better suited to those with a penchant for irony and pessimism. This era of "half-light" narrowed the scope for debate, making it difficult for Mead to find the right tone to take with a popular audience she wished to influence. Her new status as a divorced single mother and a career woman made her fearful that she might lose her credibility at a time when a book like *Modern Woman: The Lost Sex* could become a best seller. Published in 1949, *Male and Female* represented Mead's initial attempt by accentuating "the basic difference" between men and women and warning against "crusades based on the rights of women." Competition with men was "dangerous." At the same time, she argued that rigid sexual segregation would "push a large part of society towards celibacy or homosexuality." Seeking to accommodate her prewar penchant for feminism and sexual liberalism to the stridency of postwar anti-feminism, *Male and Female* displayed ideological incoherence. At its conclusion, Mead described the painfulness of dealing with "a tangled situation" created by two sexes having "different gifts" that must be used for the benefit "of the whole of humanity."[23] Sometimes confused and contradictory, *Male and Female* textually marked a period in which the author's life was becoming fragmented as divorce, death, and a divided world made it difficult to regain her accustomed optimism.

Disappointed readers accused Mead of contradicting herself. When she prepared a British edition of *Sex and Temperament* in 1950, Mead responded to such criticisms by proclaiming the right to "not only have it both ways, but many more than both ways." Using the same ethnographic data to both criticize and endorse notions of biologically determined sexual differences, Mead tried to keep faith with the intellectual legacy she had created with Benedict while maintaining a place for herself as a public intellectual after Benedict's death. In 1963 Betty Friedan criticized her in *The Feminine Mystique* as someone who had abandoned her earlier "revolutionary vision of women finally free to realize their full capabilities" to become a proponent of the "feminine mystique."[24] Benedict might have appreciated the irony of an author, shielding her own leftist background, attacking Mead for yielding to the compulsion to deny, at least in part, her feminism in response to Cold War pressures. Despite their common need to hide unacceptable beliefs and behaviors in a Cold War closet, Friedan and Mead found themselves ideological adversaries, if only briefly.

As another part of her accommodation to cultural imperatives, Mead concentrated on research that could serve the Cold War agenda. The study of cultures at a distance would help researchers who, for "exigent political reasons," wanted to know about a culture that was inaccessible. Funded initially by the U.S. Navy and then by the Rand Corporation, research in contemporary cultures focused on the Soviet Union, Eastern Europe, China, and countries in Western Europe susceptible to electing left-leaning parties. Mead's 1951 study, *Soviet Attitudes toward Authority: An Interdisciplinary Approach to Problems of Soviet Character*, continued the studies of the enemy begun during wartime. The targets of John F. Embree's criticisms in the *American Anthropologist* in 1951 included Mead as helping to impose "American notions of the 'good' state on other peoples." Mead's Soviet analysis distanced author and reader from the culture being dissected.[25] Benedict had been able to maintain an ironic distance from American assumptions of cultural superiority, but Mead found herself submerged within the ideological currents of a period that demanded social and political conformity from those who wished to participate in public discussion.

Mead's return to Manus in 1953 provided her a brief escape from domestic U.S. politics but not from the Cold War consensus. Defining the United States as the ideal society for so-called backward societies to emulate, Mead published *New Lives for Old*. In contrast to her pre-war writings, utopia had moved from the South Seas to the postwar United States, the land of "production and plenty." Mead applauded the Manus people's "active choice" to enter "into our way of life." Change should be speeded up. The "very primitive and some of the simplest peasant peoples of the world" had to learn a "whole new set of habits" in order to "be fit to live in the modern world." Perhaps responding to anti-imperialist critics like Embree, Mead accused them of attempting to deny "full participation in the culture which we claim to value so highly." Americans should welcome the new converts to the "fresh bright dreams of a system based on a belief that all men are created equal" for which she had become an ardent supporter. Benedict's warnings about the arrogance of power disappeared from Mead's depiction of "this strange emergence of a group of erstwhile savages" upon "the world stage."[26] Whether unwillingly or enthusiastically embracing her part, Mead welcomed the people of Manus onto the twentieth-century stage constructed by a benevolent United States.

By the mid-1950s, Mead began to examine the personal consequences of a life confined to an epistemological closet. A letter drafted in 1955 reveals that she was becoming "increasingly conscious of the extent" to which her life had become fragmented since Benedict's death and her divorce from Bateson. She blamed the "exigencies of the mid-twentieth century" for requiring that she keep family and friends "in ignorance of some part of my life." That same year she moved in with Rhoda Metraux, a decision that may have led to her concern about the dangers of exposure and the need to conceal aspects of her life. Four years later she revisited her relationship with Benedict in *Anthropologist at Work.* In the "counterpoint between almost forgotten words," she invited her readers to seek new meanings as she read herself back into the prose and poetry created by two people writing answers to each other's questions.[27]

Inquiring readers can unravel the distracting threads that partially obscure the central pattern of *Anthropologist at Work,* an examination of the relationship between Benedict and Mead within a biographical disguise. Published in 1959, the biography contained autobiographical clues about its author. Grammatical shifts in the introduction from third-person singular to first-person singular to first-person plural conveyed through their instability that author and subject co-existed but could sometimes merge into a single persona. The "she" who "saw her work in a context of three other anthropologists" denoted Benedict in the literal reading of the sentence but just as plausibly applied to Mead. Describing Benedict as living "two lives," Mead exhibited the same propensity in this autobiography manqué.[28]

Denying that she was writing a detective story, Mead began *Anthropologist at Work* with verbal clues to the reader prepared to read her words unconventionally. Benedict's poem "The Eucharist" began the biography with lines claiming that "light" the "more given is the more denied." Unconventionally, Mead began the biography neither with Benedict's birth nor her marriage. Instead she chose to begin with the decade in which Benedict discovered anthropology and her relationship with Mead began. The segment containing Benedict's diary entries for 1923 opened with the poem, "New Year," which began "I shall lie once with beauty, breast to breast," reversing the positions of poet and lover. Reading *Anthropologist at Work* after her mother's death, Bateson understood why Mead had quoted "Ruth's lines." As Bateson recognized, the biography must be read by someone attentive to silences, coded references, and double meanings to avoid being mislead. A difficult personal history of "alienation and unmet need" that culminated in Benedict's discovery of a "distinctive identity and creativity,"

Anthropologist at Work was an implicitly feminist work written in an unfeminist era that spoke about lesbian desire and women's need for professional fulfilment.[29]

The middle sections described their collaboration as spending "hours discussing how a given temperamental approach to living could come so to dominate a culture." Out of their conversations emerged Benedict's theory that "culture could be seen as 'personality writ large.'" Benedict's *Patterns of Culture* grew out of their mutual search for the "key" that would unlock cultures. Contained within this discussion was a letter that called into question the value of their joint enterprise. Warning that anthropologists' "infernal curiosity" and "thirst for scientific data" could destroy the people they studied, Jaime de Angulo called Benedict to account. "Don't you understand the psychological value of secrecy at a certain level of culture?" Reprinting the letter without commenting on its relevance, Mead left its accusations unanswered.[30] Perhaps de Angulo's appreciation of the value of secrecy was something that Mead felt needed to be said in a biography about two lives that contained many secrets.

The section dealing with the brief postwar period of Benedict's life carried the title, "The Gathered Threads," a phrase repeated from the letter Mead had written "to those I love" in 1955 and later used in the epilogue to *Blackberry Winter.* Benedict's loss meant the dissolution of that "net of common knowledge" that they had woven together, but Mead's writing offered a way to reweave it. Once again breaking with biographical convention, however, Mead did not end the biography with Benedict's death. Rather, she concluded with the writing over which she had exerted no influence, Benedict's unpublished biography of Mary Wollstonecraft, written in the 1910s.[31] Another instance of the representational doubleness that shaped *Anthropologist at Work,* this biographical fragment described a feminist heroine who had asserted "that women are more than men's playthings, that they have lives and understandings of their own and that anything short of a full development of their powers is a duty left undone." Serving as a feminist affirmation of the choices made by both author and subject, Benedict's analysis of the young Wollstonecraft's romantic attachment to an older girl retold the love story of Benedict and Mead. The biography revealed the same taste "for rearrangement, ambiguity, and representational doubleness" that Mead displayed. As Mead apparently recognized, Benedict had written an autobiography in biographical form.[32]

Rhoda Metraux, who would take up many of the tasks for Mead that Benedict had performed, displayed her ability to understand her predecessor. In her foreword to the 1966 edition of *Anthropologist at Work,* Metraux described it as "both biography and autobiography." She advised the readers that this was a "record of a friendship" in which apparent difference was underlaid by a "symmetry that allowed them to build on each other's work." Only partially hinting at her own editorial contributions to the biography of Benedict, Metraux would later edit Mead's autobiography, *Blackberry Winter,* significantly reducing the role that both she and Benedict had played in Mead's personal and professional life. Together, Metraux and Mead created an autobiography far less revealing than *Anthropologist at Work* to a discerning reader.[33] Despite the advent of the so-called sexual revolution, neither Mead nor Metraux felt that they could entirely abandon the "enabling privacy of the closet."

Mead published another biography of Benedict in 1974, the same year that she wrote sympathetically about bisexuality in a *Redbook* column. Writing in *Redbook,* Mead described bisexuals as people for whom "the differences between men and women and the differences between individuals in temperament and gift were the basis of love affairs of great intensity and meaning." Writing in the third person, however, neither Mead nor her collaborator, Metraux, referred to themselves. The new biography of Benedict emphasized Benedict's feminism but still remain enclosed in the "epistemology of the closet" in regards to sexuality. Mead's and Metraux's portrayals of their relationships with their silences, coded references, and double meanings suggest that they continued to fear that public exposure might destroy Mead's credibility as a public intellectual.[34]

After her mother's death, Bateson revealed her anger at discovering the "fact of concealment." She wrote about the sexual aspects of Mead's relationship with Benedict but followed the example of Metraux in not discussing the other long-enduring partnership with Metraux that ended only with Mead's death in 1978. No doubt justifiably fearing the consequences of complete openness, Mead, Benedict, and Metraux chose to conceal aspects of their lives. At the same time, their writings helped to make it possible for their heirs to avoid the same need for concealment. Speaking at Benedict's memorial, Mead expressed the hope that "others of their kind might somewhere, sometime, be at home."[35] Recognizing that Mead sometimes used the third person to avoid the public exposure that would come with a more forthright use of the first person, it is possible to read these words as describing the cause of sexual variance to which they dedicated their lives. Refusing to be exiles, they made a place for themselves in public life, sometimes using unconventional modes of representation to accomplish their purpose.

2 "The Bo-Cu Plant"

Ruth Benedict and Gender

Lois W. Banner

In 1916, two years after her marriage to Stanley Benedict, Ruth Benedict drew on his knowledge of plants and poisons to write a mystery/horror story titled "The Bo-Cu Plant." The unpublished story is melodramatic and overwritten. Not surprisingly, her biographers have dismissed it as a minor collaboration with Stanley, but that judgment needs to be reconsidered. Stanley helped with the story's plot and the choice of the pseudonym Edgar Stanhope for its author, referring to Edgar Allan Poe and to Ruth's hope for a career as a writer. Yet the story reflects both the mode of Benedict's writing and themes in her life.[1] Set on the Amazon River in South America, it anticipates her interest in anthropology and in race. Dealing with men and women at odds, it reveals an early feminist phase in her thought and foreshadows her later gender strategies in the male-dominated profession of anthropology. It relates to her quest to determine her gender orientation by understanding the male and female sides of herself. Containing a story within a story, it hints at the complex levels of her literary intent—and of her self. Benedict's recently opened letters to Mead at the Library of Congress reveal her as complex and passionate, with ambivalent attitudes toward men—characteristics that the story portrays.

The "Bo-Cu Plant" begins one stormy November night in a mansion in a fashionable area of London. Five male explorers are smoking "oriental" tobacco and sharing tales of their adventures. Among them are Ah Sing, a Chinaman from India, "who has visited all the secret places from Tibet to the Red Sea"; Lord Dunstan, an Englishman who knows "the Eastern mind better than any other man living"; Bollo, "an inky black prince from the Upper Congo" who holds degrees from "two leading British universities"; an unnamed wealthy Cambridge University researcher; and Dr. Moran, a physician and expatriate American, who tells them the story of the Bo-Cu plant.

A lumber company, Moran begins, hired him as the doctor for a group of surveyors the company sent to the Amazon River region in South America to assess cutting down a stand of mahogany for market. In the forest near the surveyors' camp where Moran has his dispensary, is a lush black vine with beautiful—and deadly—white flowers. That vine is the Bo-Cu plant. Its scent, Moran discovers, can cause hallucinations: after smelling it, he sees nymphs running through the forest.

Moran's story now turns to his assistant, Carl, "a blond, Viking type," and to a local Spanish woman, the exotic Nita. Carl and Nita fall in love, while Moran learns that before Nita's birth, her mother, Dolores, would sneak into the forest late at night to embrace a giant tree. Then, during the birth, Dolores died in a "hell of torture." Before dying, she cursed Nita for being born. She swore that if Nita married she would die. Nita agrees to marry Carl. On their wedding day the scent of the Bo-Cu plant fills the air. Dr. Moran, the rational scientist, tries to persuade them to postpone their marriage, but they refuse. They enter their cabin on the wedding night, anticipating sexual consummation. Instead, Nita turns into a hideous monster. Looking into her eyes, Carl sees "the devil sitting there grinning at me." In response, he bludgeons Nita to death with an axe.

In the first instance, this alarming story, culminating in an axe murder, reflected the Benedicts' troubled relationship. After Ruth married Stanley in a rush of passion, they moved to the suburbs of New York, where he was a research chemist at Cornell Medical College in uptown Manhattan. By the time she wrote "The Bo-Cu Plant," their marriage was in crisis. Controlled and unemotional, Stanley wanted a domestic wife, whereas Ruth, prone to depressions and mood swings, disliked suburban life. When it seemed impossible for Ruth to conceive a child, she decided on a career, despite Stanley's objections. The battle between Carl and Nita on their wedding night in "The Bo-Cu Plant" reflected the conflict between Ruth and Stanley in real life.

The story reflected more than Benedict's troubled marriage, however. With her bucolic Amazon region breaking into violence, it is the opposite of the peaceful, fantasyland of the daydream she had constructed as a child and continued to retreat to in her imagination even as an adult. In that daydream she lived with her dead father—reconfigured as a feminized Christ—in calm and peaceful "delectable mountains," a term she borrowed from John Bunyan's *Pilgrim's Progress*, where it is the name of an earthly Eden on the road to the Celestial City that the pilgrim travels. Moreover, as Dolores embraces a phallic tree and Nita turns into a monster, each woman comes to resemble the female demons of turn-of-the-twentieth-century literature and art by male authors, which often linked dark forests and poisonous flowers to the perceived threat to masculinity from the women's rights movement and the "new" emancipated woman of that age. Like the feminist writers discussed by Nina Auerbach in her book on these representations, however, Benedict reshaped those female demons into agents of subversion of male power.[2]

Indeed, "The Bo-Cu Plant" can be read as a feminist tale. Dr. Moran's story is set on the Amazon River, a setting that evokes ancient myths of female Amazons battling men. Like those legendary female warriors, Dolores and Nita fight back against the male explorers in London who are consuming both "oriental" tobacco and their eroticized story. Dolores dies in childbirth; Nita turns into an avenger instead of a willing bride. Moreover, the men in London are a unified group. That each has a distinct racial or national identity underscores their unity as upper-class men engaging in the male rituals of smoking tobacco together and watching a sexualized female performance. (Even Bollo from the Congo is a prince with an elite education.) The class camaraderie of these men suggests the existence of a world of elite males differentiated from others, not by race or ethnicity, but by economic standing. They share an imperialist mentality that could turn them from telling tales of adventure to each other into engaging together in conquest and war. Even the surveyors in Dr. Moran's story participate in an imperialist project: they are assessing cutting down the forest, which is identified with Dolores embracing a tree and the nymphs running through the forest. And the "blond, Viking Carl" is a prototypical Nordic male conqueror, while Dolores and Nita, Spanish women, are darker-skinned.

Benedict further criticized the men in her story by employing the dualism between Apollo (rationality) and Dionysus (emotionality) that she drew from Nietzsche's *Birth of Tragedy* and used in 1934 in *Patterns of Culture* as typologies to characterize certain cultures. That dualism appears in "The Bo-Cu Plant" in the contrast between the white flowers and the black vine of the plant, and between Dr. Moran, the rational narrator, and the violent Dolores and Nita. Yet beneath Apollonian rationality as positive (and male) and Dionysian emotionality as negative (and female) lies another reality revealed at the end of the story. It turns out that the devil, the ultimate Western representation of evil in a male form, controls Dolores, Nita, and the Bo-Cu plant, as he grins at Carl through Nita's eyes. Thus beneath Dolores's and Nita's rage lies an ever-present patriarchy, entirely in control.

That patriarchy was in evidence with the advent of World War I. By 1916 the war had been stalemated for two years. Both armies were deployed in huge trenches along the western front, while exploding shells, poisonous gases, deadly rounds of bullets from the new machine guns, and bombs from the new airplanes caused death and horrible wounds. That year massive offensives on the part of both the Allies and the Germans resulted in staggering casualties and no gains for either side. The war came to the Benedict household when Stanley was accidentally gassed while conducting experiments in his laboratory for the government on poisonous gases, causing his health to deteriorate.[3]

The male aggressiveness of the war appalled Benedict. "How useless to attempt anything but a steady day-by-day living with this tornado of world-horror over our heads," she wrote in her journal. She compared World War I with Thomas Hobbes's state of nature as a "war of all against all" in his seventeenth-century *Leviathan*—which she had read in a college class at Vassar in 1909. At that time, convinced by the doctrine of progress that "all the crucial problems of the world had been solved," she had seen it as irrelevant. Before World War I, Benedict wrote, "we did not think of ourselves as dwarfed by a Leviathan." The war allowed Leviathan, Behemoth, and the other beasts of the Battle of Armageddon in the biblical book of Revelation and the missionary rhetoric of her Baptist childhood to re-emerge. Those beasts would appear in metaphors about the horrors of war in her poetry, as in her poem "Resurgam"—a word for resurrection—when she wrote, "This is the season when importunate rains rutting the graves unearth slim skeletons."

The poetry that Benedict continued to write in the 1920s also implicitly criticized men. Like the well-known poets Léonie Adams and Louise Bogan—her close friends—she adopted the metaphysical, symbolist style of the male modernists and sometimes took on a male persona, but she wrote in the lyric mode typical of female poets since Sappho. Benedict and her friends employed the emotionalism identified with women, but they controlled it with a modernist emphasis on irony and symbolism and an elegiac point of view. As Gloria Bowles has written: "They absorbed the male modernists and then used them to elaborate female themes of love and madness."[4]

Benedict's brief biography of Mary Wollstonecraft revealed her feelings about men. She interpreted Wollstonecraft as driven by anger against men, beginning with her alcoholic father. His physical (and perhaps sexual) abuse of Wollstonecraft, her mother, and her sister led her to become a feminist and a radical. As a young adult she rescued her sister from an abusive husband. Then came her passionate encounter with the adventurer Gilbert Imlay, who, after fathering her child, abandoned her for another woman. By that point, according to Benedict, Wollstonecraft viewed heterosexual love as "a stalking mischief" and male power as a "Juggernaut," referring to the followers of a Hindu god whose ecstatic followers supposedly flung themselves to death under the wheels of the cart bearing his image.[5] Benedict applied the word to Wollstonecraft's critique of patriarchy, implying that women too often embraced self-destruction as romantic liberation.

When Houghton Mifflin turned down "Mary Wollstonecraft," Benedict abandoned it, noting in her journal that the war had made such feminist projects irrelevant. Reforming humanity was now more important to her than agitating for women's rights. "We can hardly drag back from oblivion," she wrote, "the vital questions that were life and death to us in the early summer of 1914"—questions that concerned suffrage and women's rights.[6] In 1919, after several years as a social worker, she took classes in anthropology at the New School for Social Research, and in 1921 she entered the graduate program in anthropology at Columbia University, beginning her lifelong career.

She continued to challenge men, both in her personal life and in her writings. By entering anthropology, she defied Stanley, who identified with his profession and did not want his wife as a competitor in science. Moreover, Benedict did not follow the other female anthropologists at Columbia by studying the Pueblo tribes, which were matrilineal and matrifocal (although she would do so later). For her first subject of study in anthropology, like most of the male anthropologists trained by Boas, she chose the "masculine" Plains Indian tribes. Just as characteristically, she symbolically absorbed these male anthropologists into her library study of the vision quest among the North American Indian tribes, in which she integrated the ethnographic work of the male Boasians on individual cultures into a new synthesis which stressed differences between cultures and their institutions, while hinting at the thesis of cultural wholeness that she would present in *Patterns of Culture.*

Her emphasis on male aggressiveness is apparent in her 1922 article on the vision quest in the *American Anthropologist,* her first important publication. What stands out in the article are the masochistic cruelties males inflicted on themselves to attain their vision. Blackfoot males cut off finger joints. Crow males cut off strips of skin. Cheyenne males were attached to poles by wooden pins driven through their flesh. Benedict connected that male violence to male warfare. Among the Plains Indians, she wrote, male adulthood meant warfare, and their "reward" for their self-torture was "enhanced prowess in deeds of warfare."[7]

Her most forceful early writing criticizing aggressive masculinity and connecting it to war was "The Uses of Cannibalism," a brief Swiftian satire that she wrote in the mid-1920s. Obviously referring to World War I, she described mankind's need for national glory and personal violence as resulting in "the death, in great numbers and with distressing tortures, of young men in sound health and vigor." In contrast, she noted sarcastically, some tribal cultures dealt with these needs through ritualized cannibalism. The head-hunters of the Malay Archipelago, for example, battled each other "with the fierceness of males in breeding season" in pre-arranged battles. After one side killed an opponent, the victors feasted on the body of the slain warrior. To satisfy a community's need for violence, Benedict wrote sarcastically, "nothing could be more harmless" than consuming only "one useless body per year."[8]

Trying to establish herself in the male academic world, Benedict did not publish her early anti-male writings. In this era many academic males felt under siege, threatened by the appearance of women in academia, who were representatives in the flesh of the "new woman" that had troubled turn-of-the-century male artists and writers. In response, the men in disciplines like psychology and sociology downplayed humanistic and social service approaches to emphasize techniques like testing and quantification that were perceived as more rational and thus male; while the women turned to their own specialties, such as social work.[9] Benedict took a more daring path by abandoning social work for anthropology. Moreover, unlike most of the other women in the discipline, she did not confine herself to observing behavior, doing interviews, and writing ethnographies of individual cultures. Like the men, she turned to theory, the ultimate arena of male rationality.

Her confrontation with the men in her discipline was subtle. She rarely challenged them directly. In personal dealings she was always "intensely sympathetic and kindly." Tall in stature, with a low voice, she had a haunting, regal beauty and a remoteness that could be off-putting. She herself described the distant expression she usually wore in public as a "mask" she put on and off. A flickering "Mona Lisa" smile on her face made her intriguing. The men at Columbia found her sexually appealing, but she also seemed witchlike and threatening.

Intrigued by "the Benedictine enigma," male anthropologists fell under Benedict's spell. According to Mead, she had a fragility that appealed to the "chivalry and solicitude of those with whom she worked." Alexander Goldenweiser spent many hours counseling her when he taught her at the New School. Robert Lowie supervised her dissertation. Alfred Kroeber helped her with her first fieldwork, while she was still a graduate student, among the Serrano Indians of Southern California. Edward Sapir became her major intellectual supporter during her early years in the discipline. So successful was she in joining their circle that they provided her with contacts to major intellectual journals like *The New Republic* and included her in their informal "press committee" publicizing their discipline. Lowie named her, along with Sapir, Goldenweiser, and Paul Radin, as the "superintelligentsia" of anthropology. No other woman in anthropology achieved that sort of intellectual respect, in spite of the men's need for women to study tribal societies in order to secure full data about kinship, work, and family since tribal societies often would not permit the men to interact with their women.[10]

According to Mead, however, these men did not extend an "unequivocal" welcome to women. The racism and anti-Semitism of this era made those of Jewish or non-Wasp descent likely to develop "uncertainties" about themselves. The assumptions of turn-of-the-century masculinism and its fascination with bodybuilding and violent sports exposed intellectuals to the charge of effeminacy. According to Sander Gilman, Jews were especially vulnerable to these charges, which were often lodged against ethnic men in this era. In his autobiography, Lowie called himself a "marginal man," referring to both his ethnicity and his masculinity. He remembered that in his high school on the lower East Side of New York he was interested in his studies, while the other boys were interested in who was the world heavyweight boxing champion.[11]

Among the male anthropologists trained by Boas, Kroeber was controlled and paternal. He was best known for his highly rational theory that a "superorganic" force controlled human history, a "majestic order" that pervaded all civilization. Theodora Kroeber, his wife, described him as appreciating independent women but as also having a problem with women with loud voices or those who lionized famous men. To prove the scientific regularity of history and the existence of his "superorganic," Kroeber first studied the cycles of fashions in women's dress. In this work he reduced women's bodies to objects that he measured to produce mathematical statistics he could quantify. Benedict found his criticism mostly nuanced and not offensive, but she distrusted his attitude toward women in the profession and expressed astonishment when he telegraphed her in 1928, asking her to teach a summer session course at Berkeley, where he chaired the department. When he taught at Columbia during the spring semester of 1932, Benedict found him dismissive of her work.[12]

In contrast to Kroeber, Alexander Goldenweiser was flamboyant and outspoken. In his early work he praised women's creativity in tribal cultures, while noting that most of those cultures were "androcentric," or male dominated. Yet he was a womanizer; Benedict found love letters to him from women "hidden" in a desk drawer in the anthropology seminar room at Columbia. He was professionally disgraced when his extramarital affair with an Iroquois woman from a tribe he was studying became public. By 1929 his career was floundering. In that year he drew from Freud's misogynist writings on women to describe them as "an enigma, a menace as well as a joy." The next year he wrote that women were by nature inferior to men in rationality; thus, they couldn't be real intellectuals. By 1936 he referred to them as "peculiar creature[s] with a distracting and at times repulsive periodicity in [their] life cycles . . . and a fascinating but always excessive and always disturbing influence on men via sex."[13]

Like Goldenweiser, Radin also had a difficult career, especially after he was fired from his job at the Bureau of Indian Ethnology in Washington, D.C., in the early 1920s because of suspicions that he had embezzled funds. He was brilliant and charming, but even the male Boasians found him difficult. Lowie and Sapir called him a "trickster" and a "scamp," although eventually he produced distinguished work. Teaching at Columbia as a visiting professor in the late 1920s, he charmed graduate students away from Benedict. He privately expressed distaste that women were in the professions at all.[14]

More than most of these men, Lowie became directly involved with feminist reasoning when he co-authored an article with the renowned feminist psychologist Leta Hollingworth, a professor at Columbia Teachers' College. The article repeated Hollingworth's rebuttal to the scientific arguments for male superiority along with Lowie's response to the feminist argument that matriarchal societies preceded patriarchal ones. In writing about Indian religion, Lowie went so far as to chide Benedict in her work on the vision quest for underplaying women's participation. Like Kroeber, however, he preferred the masculine Plains Indians to the matrifocal Pueblos who, he contended, "do not evince the manly, upstanding incisiveness of the Indians of the Plains, their directness in personal intercourse, the interesting play of individuality." Lowie confessed to his sister that only six-foot-tall, powerfully built women attracted him sexually. He called his preference his "complex." In 1920 he wrote to Radin, then a visiting professor at Berkeley, that he looked forward to getting together so that they could discuss luscious figs and eucalyptus trees and "the perversity of women, that most inexhaustible of subjects."[15]

As for Sapir, Benedict was attracted to this man whom the male Boasians considered the brightest among them. Yet seemingly oblivious to her interest, he called her "a well-articulated rational type" and became romantically involved with Mead. Ruth Bunzel, who studied with him, found him "abnormally innocent" regarding women. He made fun of woman suffrage and wrote poetry in which, like many male writers at the turn of the twentieth century, he depicted women as sirens luring men to destruction. He described himself as a "dainty man" with a strong aesthetic drive, but he also described himself as a "hungry man" craving the "crassness of life." After Mead rejected him as a lover and revealed her free love views to him, he turned against her in fury, attacking her ability as an anthropologist in two articles that he wrote after she published *Coming of Age in Samoa* in 1928—

one in the popular *American Mercury* and the other in the scholarly *American Journal of Psychiatry.* In 1929 he attacked her by name in a review of a book by Franz Boas he wrote for the *New Republic.* That same year he wrote to Benedict that Mead was a "loathsome bitch," a "malodorous symbol of everything he hated in American culture."[16] A conservative in matters of marriage and morals, Sapir hated the sexual freedom of the 1920s that Mead canonized in *Coming of Age in Samoa.* He privately shared his version of his affair with Mead as well as his dismissal of her anthropology with his male friends in anthropology.

In 1932 Mead wrote to Benedict that Sapir was trying to do them in because of wounded vanity, while Kroeber would do anything to please him. Four years later Benedict wrote Mead that Bunzel had learned all the "gory details" of the affair from an anthropologist at Yale who had heard them from Radin. Benedict surmised that Radin had told every man with whom he had ever gone drinking about the affair. Over the years Benedict and Mead speculated that Sapir's animus lay behind the many negative reviews of Mead's work in professional journals. As late as 1937, Mead complained to Benedict that she was not being consulted in the profession because of the old canards spread by Sapir.[17]

Although Sapir and the others usually did not include Benedict in their indictments of Mead, she had her own problems with them. In 1931, after Boas became ill, she took over running the Columbia anthropology department, and her difficulties with men increased. The department was housed administratively in the Faculty of Political Science in the graduate school at Columbia, which functioned as "a very exclusive gentlemen's club." In a letter to Mead in 1932, she described Kroeber, then visiting at Columbia, as arrogant and insulting. She was appalled by his dismissal of her work and his approach to personality and culture in his lectures to his students in which he examined "half a dozen or more cultures for whether Superman could have been bred in them," and "gave the laurels to the Plains." His insensitivity, according to Mead, prompted Benedict to begin writing *Patterns of Culture.*[18]

In *Patterns of Culture* Benedict put her own scholarship on the Zuni and the Plains Indians together with that of Reo Fortune on the Dobus and Franz Boas on the Kwakiutl to create a theoretical framework for anthropology—and a best-seller. She captured the public market all these men wanted to reach while becoming a major contender to succeed Boas as the discipline's intellectual leader. Benedict's use of Fortune's and Boas's work was similar to the strategy of her doctoral dissertation, in which she had taken over the men's research to create a new synthesis. Fortune had taken Mead away from her. Benedict challenged him by absorbing his research.

Indeed, Benedict's Dobu are even more paranoid than Fortune's. Benedict deeply respected Boas, who wrote a laudatory introduction to *Patterns of Culture*, yet her characterization of the Kwakiutl culture as "megalomaniac" described Boas as well as the Northwestern tribe, for his efforts to create and dominate a new anthropology had a "megalomaniac" cast. Boas eventually criticized Benedict's argument in *Patterns of Culture* for overlooking the Kwakiutl's gentle family life and making them too harsh.[19] In doing so, he included himself, for he prided himself on his gentleness at home.

By the time she wrote *Patterns of Culture*, Benedict had begun to absorb Boas. She also delivered a major attack on male aggressiveness in the book, defining the deviants in U.S. society as "family men," "officers of the law," and "business men," and characterizing them as arrogant egoists akin to a Dobu out to kill his neighbor or a Kwakiutl holding an extravagant, destructive potlatch. Members of these male prestige groups in the United States, like the seventeenth-century Puritan leaders who put women to death as witches, she said, could be described as psychopathic: "Like the behavior of Puritan divines, their courses of action are more often asocial than those of the inmates of penitentiaries."[20]

Male critics occasionally criticized her attacks on men. Evaluating *Patterns of Culture*, literary critic Richard Chase, a former student of hers, chastised her for wanting a "too bland social order in which masculine aggression is stilled." In toppling the culture's major male figures "one by one," according to Chase, she displayed only her feminism, not an "independent mind." Benedict saw it differently. In her 1935 Introduction to *Zuni Mythology*, she drew from the work of psychologist Alfred Adler to claim that powerful women engendered a "masculine protest" from men. She found such male behavior rare among the matrifocal Zuni, but it was "so common an element in our civilization." "This business of masculine protest being reinforced by being in the same field," she commented, "is terrible."[21]

Benedict also had difficulty in her intimate relations with men. She had problems with Stanley; Sapir rejected her for Margaret Mead; the young Tom Mount, with whom she fell in love in 1929 after Mead married Fortune, rejected her for a younger woman. In 1932, a year after the younger Natalie Raymond had become her partner and Benedict had opted for lesbianism as her sexual orientation, she had a conversation with Sapir that she found especially pointless. She wrote to Mead that she was "just giving thanks to God that there was no man living whose whims and egotisms I had to take seriously."

Despite such comments, Benedict was not hostile to all men. Like present-day men's studies theorists, she realized that there are different types of masculinities, that men can be gentle as well as aggressive. She had difficulty with Stanley, Sapir, and Mount when she was involved with them, but she remained friends with each of them after their intimate relationship ended. She never divorced Stanley, and she gave Mount money after they broke up to help him support his wife and child. In the spring of 1932 in a letter to Mead she excoriated Kroeber, but she also noted that she gossiped with him "amiably" and "quite a lot."[22] On the wall of her office at Columbia hung photos of Franz Boas and an aged Crow chief, men who had gained wisdom through life experience and whose personalities had been softened by the process of aging.

In her writings, Benedict fulsomely cited male philosophers such as Nietzsche and Santayana, but she never cited feminist authors such as Charlotte Perkins Gilman or Ellen Key, even though before World War I she had praised their work. She may have wanted to establish a male genealogy to enhance her acceptance as an intellectual by the male anthropologists. She carried this emphasis on male authors to an extreme in her presidential address before the American Anthropological Association in 1948 some months before her death. In that address she called for a new humanistic direction in anthropology of the sort she had learned in college at Vassar. In describing that college learning, however, she did not cite her professors Laura Wylie and Gertrude Buck, modernist literary critics who had trained her. As an example of her college reading, she cited works on Shakespeare written by male critics.[23]

Benedict could be critical of women. She formed a strong bond with Mead, but she never got along with Elsie Clews Parsons, the folklorist and wealthy patron of Columbia anthropology who was close to Boas. She did not like separate organizations of women, preferring those that included men. After all, the mother and daughter in "The Bo-Cu Plant" are at odds, with Dolores's curse at Nita's birth causing the problems in her marriage. As a child Benedict had disliked her mother. Since the name "Dolores" of the mother in the story means 'woman of sorrow,' the choice may refer to Benedict's mother, who mourned for her dead husband for many years, while "Nita" might be a loose anagram of "Benedict."

In 1936, after Mead published *Sex and Temperament,* Benedict wrote that the question of sex differences had been a "passion" with her throughout her life, but she did not regard women as superior to men. "I have raged at times at the virago and the pin-headed female," she asserted. As an analyst for the federal Office of War Information during World War II, she wrote an analysis of gender roles in Thailand in which she concluded that, although men seemed dominant, women were in control behind the scenes.[24]

Until her mid-30s Benedict searched for a way to express her "masculine" side and to reconcile it with her femininity. Given her culture's definition of homosexuality as "inversion," doing so pointed her toward lesbianism. For high school she attended St. Margaret's Episcopal School for Girls in Buffalo between 1903 and 1905. In her "Story of My Life" she referred to a classmate at St. Margaret's as the first of a number of young women to be devoted to her, and she described her schoolmate Mabel Ganson—later the famed memorialist and Greenwich Village saloniere Mabel Dodge Luhan—as living for something that Benedict recognized, "something different from those things for which most people around me lived."[25]

As a student at single-sex Vassar, she participated in the culture of "smashing"—of love relationships between students—that flourished at such colleges at the turn of the century. It was part of the Victorian sex system, which interdicted sex before marriage and regarded romantic friendships between girls as both natural and a way to avoid sex relations with boys and possible premarital pregnancies. In that student culture, with males absent, girls mimicked—or mocked—the dating culture with boys. Older students dressed as males to escort younger students to dances and parties. In the caption under Benedict's photograph in her graduation yearbook, she is described as "a salad: for in him we see / Oil, vinegar, pepper and saltness agree." Thus, she was a mixture of opposites—oil and vinegar, pepper and salt. She was both male and female, called "him" in the caption. In marrying, she opted for her female side, but she regretted giving up the gender crossing of her college experience. In her journal she wrote: "We thought once, in college perhaps, that we were the artificers of our own lives. We planned our usefulness in social work, in laboratories, in schools. And all the time we did not yet know we were women—we did not know that there lay as certain a moral hindrance to our man-modeled careers as to a man-modeled costume of shingled hair and trousers."[26]

What was the "moral hindrance" to wearing shingled hair and trousers and to having a male career? It was that sexologists—and even some feminists—connected the college dress of shingled hair and trousers to the "mannish lesbian," a figure who haunted the early-twentieth-century imagination. With her single-sex identity, she had presumably so strongly inverted to the male side of herself that she had, to all intents and purposes, become a man.

That every individual is composed of male and female elements was a major belief about gender in the early twentieth century. That belief is present in the work of sexologists such as Edward Carpenter and Havelock Ellis, of feminists such as Ellen Key, of Jungians such as Beatrice Hinkle. Still, these writers also believed that male and female remained separate, even in the individual self: the male/female binary was fundamental to this age of anxiety about the "new woman."[27] It is central to Virginia Woolf's brief for androgyny in *A Room of One's Own*. "The normal and comfortable state of being," she wrote, "is when the two live in harmony together, spiritually cooperating. If one is a man, still the woman part of the brain must have effect, and a woman must have intercourse with the man in her. Coleridge perhaps meant this when he said that a great mind is androgynous."[28]

Her assumption of the name "Benedict" when she married, of Edgar Stanhope as her pseudonym in writing mystery stories, and of Anne Singleton as her pseudonym for the author of her poetry suggests an identification with masculinity. Esther Schiff Goldfrank, a graduate student in anthropology at Columbia, observed Benedict closely. She thought that Ruth had taken her husband's name when they married partly because "Benedict" could be the first name for a man. As for Anne Singleton, Anne rhymes with Stan, and Singleton merges the genders into a "single tone." Elizabeth Stassinos has speculated that Benedict derived that pseudonym from Joseph Conrad's "old Singleton" in his *Nigger of the Narcissus*. Singleton is a "learned and savage" patriarch on a ship of sailors, "tattooed like a cannibal chief," with a powerful chest and enormous biceps, who reads Bulwer Lytton's *Pelham*, a novel about an effeminate dandy. No direct evidence supports this hypothesis, but it is plausible.[29]

In marrying Stanley, Benedict opted for her female side—although her taking his last name might be seen as a rebellious gesture. Within several years, however, the marriage was faltering, and Benedict was not satisfied with her heterosexual solution. By the time she met Mead in 1922, she had decided to "lie with beauty breast to breast," as she stated it in her diary on New Year's Day 1923. She had also decided that personal contentment lay in fulfilling both the heterosexual and homosexual sides of the self by having committed relationships with both a woman and a man. For her part, as Mead expressed it, meeting and loving Benedict had replaced her "foolish one-sided preference"—for a heterosexuality implicit in her marriage to Luther Cressman—with a "great glowing affirmation of reciprocity."[30]

Still, Benedict's "bisexual" solution did not satisfy Benedict over time, even though Mead remained satisfied with it. What she perceived as the masculine side of herself remained strong. One of the most revealing statements she made about herself is in the accounts of two dreams that she gave Mead in 1927 for the book that Mead intended to write on dreaming—a book she never completed. In the first, titled "The Remodeled Farm House," Benedict is in her grandfather's farmhouse, where she spent a good deal of her childhood. It is decrepit and disordered, and she decides that she must remodel it. In the second dream, which she titled "The Chosen Twin," she is in a tower with Franz Boas's twin grandsons. In that dream the first grandson is named Robert and the second X. Both boys jump into a pool of water hundreds of feet below: X survives, but Robert is killed.

In her analysis of these dreams, Ruth interpreted the dream about her grandfather's farmhouse as meaning that she had "rebuilt her life according to the most uncongenial pattern." In other words, in marrying Stanley she had rejected the same-sex orientation of her adolescence to become heterosexual, but that orientation was not working for her. In the second dream she identified with "Robert" because Robert was the name her parents had intended to give her had she been a boy. She thought that the fall into the water symbolized her baptism at the age of eleven, when she had been spiritually reborn. She continued: "I presume I identified with a self I might have been." That self was male; in her dream, Benedict wanted to undergo a rebirth to that self.[31]

Mead was a satisfying intellectual comrade, but as a personal partner she left much to be desired. Benedict wanted an exclusive relationship, but Mead kept falling in love with men: first Sapir, then Fortune. Benedict was willing to tolerate Mead's staying with her first husband, a believer in free love. Fortune, however, wanted a monogamous commitment. Having decided to give up men, Benedict opted out of her intimate relationship with Mead for a number of years after Mead chose to marry Fortune. She formed a monogamous partnership with Natalie Raymond, finding a new stability as a lesbian. She had a healing vision of the Egyptian sphinx, a bisexual, masculinized figure, representative in Egypt of royal power. "I was alone on a great desert that was dominated by a magnificent Egyptian sphinx. Nothing can describe the wisdom and iron of that sphinx's face, and I went to it and buried my face in its paws and wept and wept—happily and with confidence." She felt as though she was living permanently in her fantasy world in her "delectable mountains" with her father. It's a little bit of eternal life," she enthused.[32]

Benedict saw her adoption of lesbianism as coming to terms with her masculine side, but she did not adopt the "mannish" behavior that the sexologists identified with lesbians. She wore tailored attire—dresses and fitted suits in soft shades of gray, blue, and green—with a frilly blouse, so that she looked both professional and feminine. Benedict's graduate students remembered her as having "the gracious manners and tone of a well-bred woman."[33] She lived quietly with Raymond, but the relationship fell apart in a few years. In the late 1930s she had a brief affair with Ruth Bunzel, an anthropologist at Columbia University, before settling on psychologist Ruth Valentine, who remained her partner until her death in 1948.

In choosing lesbianism as her orientation, Benedict found a sense of contentment and a control over her depressions that enabled her to write *Patterns of Culture, Race and Racism,* and *The Chrysanthemum and the Sword.* At the same time she served as the de facto chair of the Columbia anthropology department from 1931 to 1937; maintained her relationship with Mead; dealt with the continued hostility of men in the profession; and began a career as an involved public intellectual, speaker, and organizer that would be cut off tragically with her death.

How did Benedict's becoming lesbian affect her stance on gender? *Patterns of Culture* contains a discussion of the inferior status of women in most cultures in addition to an attack on aggressive men, but those themes were muted in her later work. Mead published *Sex and Temperament in Three Primitive Societies* in 1935, a year after Benedict published *Patterns of Culture.* After that, when the issue of women's position arose, she deferred to Mead's work on the subject. In 1938 Mead wrote Benedict that she was thinking of writing a book about sex differences. Benedict replied that she had no ideas on the subject—either from personal experience or from her knowledge as a scholar: "I'm certain they're there but I don't have anything to say." In 1943 Benedict received a letter from a Dr. Oliver Cope of Vassar College pointing out that although she had written extensively about race, she had never written about "sex prejudice." Benedict replied that she was "shocked" to realize that he was correct.[34]

After completing *Patterns of Culture,* Benedict searched for new concepts to organize anthropological data around humanistic values, but she did not choose feminist ones. She identified cultures she thought secure and positive, and those she thought nasty and surly. She tried race, geography, gender, climate, size, wealth, and complexity as possible variables to explain the differences, but none of them worked. Nor did gender, which she applied in terms of whether or not family structures in her sample of societies were matrilineal or patrilineal. Societies that she disliked, like the Dobu, were matrilineal; societies that she admired, like Mead's Arapesh, were patrilineal. In the end, she came up with the abstract concept of *synergy,* which she derived from the field of medicine, where it meant an action of chemicals and cells that produces a benign result.[35]

Moreover, in Benedict's major work on race, *Race: Science and Politics,* published in 1940, she didn't employ gender as a category of analysis. Rather, she used generalized, nongendered categories like "people," "humanity," and "the human race." She castigated late-nineteenth-century imperialism as an ideology, not as a gendered concept. Dr. Moran and the London explorers of "The Bo-Cu Plant" telling their tales of masculine adventure are not present in this work. Nations and class, not gender, are her villains. In *The Chrysanthemum and the Sword* Benedict subordinated gender to an emphasis on institutions and hierarchies. Richard Handler notes that "the narrative [in *The Chrysanthemum and the Sword*] is presided over by an apparently objective persona. Indeed, that persona is more than objective: it is detached, its existence grounded . . . in the universal comprehension of science."[36]

What impact did Benedict's lesbianism have on her professional relationships with men? Her sexual orientation may have caused the considerable animosity between her and Ralph Linton after Linton became chair of the department in 1937 when Boas retired. Linton was a highly masculine man. He identified so strongly with male "feats of daring" that he often recounted with admiration how, when he explored the Guatemalan backcountry during college, one of his Mayan bearers, bit on the hand by a poisonous snake, cut off that hand with a machete.

Benedict's letters to Mead during these years are filled with complaints about Linton. In 1938 she wrote to Mead, "It's perfectly possible that I may have to wallop Linton hard in public someday. He is a swine." For his part, Linton became "near psychotic" in his dislike of Benedict; in fact, he wrote detective stories under a pseudonym in which a female anthropologist tries to kill a male one. Remembering Linton as a colleague several years after Benedict's death, Sidney Mintz recollected that he claimed he had killed her through witchcraft. "He produced for me, in a small leather pouch," Mintz recounted, "the Tanala material he said he had used to kill her."[37] Certainly the depth of animosity suggests more powerful sources than mere academic rivalry.

Soon after he came to Columbia, Linton became friendly with psychoanalyst Abram Kardiner. He brought Kardiner to Columbia to conduct a faculty seminar on the intersections between psychoanalysis and anthropology. Kardiner did not like Benedict any more than Linton did. In an oral interview he gave toward the end of his life, he praised her physical beauty—while describing her as not intelligent and her work as unsubstantial. Perhaps Kardiner's homophobia shaped this view. By 1945 he joined the crusade against homosexuality that was widespread in U.S. culture. His book *Sex and Morality* contended that homosexuality was a grave social menace that had to be destroyed.[38]

When she wrote "The Bo-Cu Plant," Benedict articulated themes that foreshadowed her later positions on gender. She criticized male aggression and imperialism, but she also saw flaws in women. She juxtaposed the narrator of her story, the rational Dr. Moran, against the malevolent devil, who controlled the hysterical women and the destructive Bo-Cu Plant. In the final analysis, if anyone was the hero of the story, it was Moran: the rational male who tried to persuade Nita and Carl to postpone their marriage while the scent of the plant was in the air and thus to stop the horror of the axe murder from occurring. Benedict validated the rational, male side of herself by becoming an anthropologist and a theorist. She combined the male and female sides of herself in her bisexuality and then in her lesbianism. In 1936 she declared to Mead that over the course of her life she had turned herself into an androgyne. In Virginia Woolf's terms she had brought the two genders inside of herself into harmony, cooperating both spiritually and sexually, with her rational masculine side in the preeminent position.

3 Margaret Mead, the Samoan Girl and the Flapper

Geographies of Selfhood in Coming of Age in Samoa

Maureen Molloy

A headline in the *New York Sun Times* announced "Scientist Goes on Jungle Flapper Hunt," as it introduced Margaret Mead to its readers while she was in the midst of her first field trip in Samoa. On her way to becoming a cultural icon, Mead would establish anthropology as a popular science and entrench an unsuspecting Pacific Island people into American popular culture as the epitome of sexual permissiveness. *Coming of Age in Samoa,* the product of this "jungle hunt," would become uniquely durable among anthropological texts in its ability to capture the popular imagination. If, as Micaela di Leonardo suggests, Margaret Mead became the public's Ur-anthropologist, *Coming of Age in Samoa* surely became its Ur-text.[1]

Despite, or because of, its contentious scholarly status, *Coming of Age* retains a vivid place in the popular imaginary. The reasons for this durability, however, have never been explored, perhaps because it has been assumed that this was due to its sexual subject matter. While it is tempting to argue that its popularity stemmed from a combination of sexual prurience and professional and public gullibility, that interpretation runs the risk of glibness. It too readily dismisses the possibility that the book reached more deeply into cultural meanings, and it fails to address the tenacity of the book's hold on both the public and the academic imaginations. This chapter explores some of the social meanings that *Coming of Age in Samoa* both tapped and fed in order to understand the book's reception and its enduring place in popular culture.

The *Sun Times* headline encapsulated many of the contrasts and contradictions of mainstream American culture in the 1920s—faith in science and progress, fascination with the primitive and exotic, the anxieties of female emancipation, and the sense that the world was America's oyster, an adventure waiting to happen. The headline captured the deep ambivalence that was characteristic of America during the 1920s. The country was, Lawrence Levine argues, torn between by "a belief in progress coupled with a dread of change; an urge towards the inevitable future combined with a longing for the irretrievable past; a deeply ingrained belief in America's unfolding destiny and a haunting conviction that the nation was in a state of decline."[2] The 1920s seem modern, the real beginning of the twentieth century. However, the decade witnessed not just prosperity, technological advances, and the loosening of social mores, but also countermovements that sought to re-inscribe a nostalgic vision of an older, more communitarian, more homogenous America. These latter forces found their most famous expression in Prohibition, but they also manifested themselves in a rise in anti-immigration sentiment, the re-establishment and growth of the Ku Klux Klan, and a surge of religious fundamentalism. This ambivalence between the new modernism and the yearning for an imagined tradition permeated all strata of society. Intellectuals deplored both the machine age and the provincialism of small-town America. Politicians and businessmen "boosted" American industry while decrying the immigration that provided their workforce. Middle Americans mourned the loss of community while spending their leisure time in automobiles and movie theatres.

No figure better epitomized both the exhilarating new freedoms and the destabilizing of established mores than the flapper. Frederick Allen Lewis, in *Only Yesterday,* dedicated most of his chapter on "The Revolution in Manners and Morals" to changes in women's behavior at home; in the workplace; in fashion, leisure, and sexual behavior.[3] While the 1920s may be remembered for Prohibition, gangland wars, and jazz, the decade is personified in the figure of the flapper, with her aura of modernity, cynicism, and amorality.

It is instructive to contrast the flapper with another female figure, one that, unlike the Gibson Girl, condensed the meanings of unwelcome social change in America. The flapper's foremother could be identified as the New Woman, a figure that had been characterized as a social threat for at least two generations before the 1920s. The first generation studied in the all-female colleges during the last decades of the nineteenth century until they emerged as a network of the first female professionals. They staffed women's colleges and settlement houses, working as social workers, writers, industrial reformers, feminists, journalists, and suffragettes. Independent women came under suspicion with the advent of sexology. Havelock Ellis, Ellen Key, and Sigmund Freud asserted that women's heterosexual drives were both natural and necessary for healthy physical and emotional life. These sexologists and psychoanalysts pathologized romantic and sexual relations between women and questioned the mental and emotional health of independent single women. The second generation, such as those who flocked to Greenwich Village in the early twentieth century, claimed heterosexual fulfillment as a right and a necessity but had to recognize that female sexual and romantic relationships were increasingly stigmatized. By the 1920s they had gained some respectability, but at a cost. Although more women attended college, more also married, fewer went on to graduate school, and fewer worked after marriage.[4] Articles in the popular press profiling professional New Women emphasized their femininity as well as their achievements and often used domestic images to characterize their work.

The threatening female figure of the twenties, therefore, was no longer the earnest social reformer, or the ideologically driven sexual experimenter, or the new professional, but was the "flapper." Although the flapper was sometimes subsumed under the rubric New Woman—and both terms connote women operating beyond the norms of traditional femininity—there were significant differences between the two. The flapper was a more widespread phenomenon than the upper-middle-class, educated, (mostly) white professional New Woman or the even smaller number of radical New Women who inhabited "bohemia" in Greenwich Village. Most significantly, the flapper was a product of mass culture, not of higher education. Her clothing was factory-produced in standard sizes. Movie stars provided her models. She aimed neither for self-development nor for the betterment of the world, but for freedom and heterosexual fun, with marriage as the end goal. An unprecedented range of venues offered the requisite fun, including movie houses, automobiles, speakeasies, and jazz clubs. Now, as Allen put it, "not only the drinks were mixed, but the company as well."[5] For those individuals nostalgic for more demure images of femininity, the flapper represented a dangerous assault on traditional norms.

The figure of the young woman as symbol of the 1920s was not a post-hoc construction; it was created in the decade itself. The first best seller of the decade, Sinclair Lewis's *Main Street,* began with a glimpse of its heroine Carole Milford (soon to be Kennicott), in a state of "suspended freedom" on the banks of the Mississippi. "Credulous, plastic, young; drinking in the air as she longed to drink in life," such a "rebellious girl," Lewis declared, "is the spirit of that bewildered empire called the American Middlewest." Despite Lewis's success, Carole Milford Kennicott, with her improving aspirations, did not become the spirit of the decade. Instead, the young women who propel Amory Blaine to self-knowledge in F. Scott Fitzgerald's *This Side of Paradise* captured the popular imagination. Myra St. Claire, Isabelle, Rosalind Connage, Eleanor Savage, and Jill, who were "casually . . . accustomed to be kissed," at "three-o'clock, after dance suppers in impossible cafes, talking of every side of life with an air half of earnestness, half of mockery, yet with a furtive excitement that . . . stood for a real moral letdown." They were the girls who led Amory to describe "the cities between New York and Chicago as one vast juvenile intrigue." But more than anything, these girls were cynical about love. Rosalind rejected true love "cooped up in a little flat" with the reliable Dawson, who will be good to their children and give her "sunshine and pretty things and cheerfulness."[6] Idealistic, married Carole Kennicott was anachronistic within months of being launched. The action had moved to the young, frivolous, smart, and fast.

When Franz Boas set Margaret Mead's "problem" for her first field trip as "the psychological attitude of the individual [adolescent girl] under the pressure of the general pattern of culture," he was asking her to reflect on what was widely perceived to be a social problem—the rebelliousness and loose morals of young women.[7] Boas was too much of an anthropologist to be particularly concerned with the moral outrage, but he was interested in the extent to which adolescent "sullenness," "sudden outbursts," "desire for independence," "excessive bashfulness," "crushes among girls," and "romantic love" could be induced by cultural constraints. Boas did not pressure Mead to resolve these questions in favor of cultural influences. The letter suggests genuine curiosity about the question rather than a prejudgment of the outcome.

Equipped with a "problem" and having located a "people," Mead arrived in Samoa on the last day of August 1925. She spent the first weeks in Apia learning the language and getting to know both the local expatriates and the Samoans. She soon determined to relocate to Tau in the Manu'a group in order to work with adolescent girls. Mead was clearly in her element in Samoa, although she complained bitterly about the heat and its effect on her efficiency. Once in Tau, young people surrounded her from dawn until well after dusk. Her letters home reveal that she thoroughly enjoyed their company but felt sometimes daunted by the intellectual challenges with which their lives presented her. Although she devised a number of ways of recording "data" on the adolescents and on village life, Mead struggled to integrate her vivid perceptions of the way of life she was observing with the scientific tools she had been trained to use.

Already aware of the demands of her readership, Mead tried to solve the problem of presenting her material so that it would be credible to scholars. She wrote Boas a long letter, remarkably prescient in the light of subsequent debates, which agonized over how to produce a scientifically transparent report from her study of less than sixty children. As she explained the problem, "Ideally no reader should have to trust my word for anything, except of course in as much as he trusted my honesty and averagely intelligent observation. I ought to be able to marshall [sic] an array of facts from which another would be able to draw independant [sic] conclusions. And I don't see how in the world I can do that." She elaborated for three pages, giving examples of various, obviously inadequate statistical analyses that she could perform on the data. Two weeks later, as the crisis of confidence deepened, Mead wrote again, throwing herself on Boas's mercy. "I have no idea whether I am doing the right thing or not, or how valuable my results will be. It all weighs rather heavily on my mind. Is it worth the expenditure of so much money? Will you be dreadfully disappointed in me?"[8] Displaying a pattern of girlish behavior in relation to older men, Mead used a ploy she would later adopt in her dealings with William Ogburn, A. Radcliffe Brown, and Bronislaw Malinowski.

Boas sent the desired reassuring letter by return mail, "I am very decidedly of the opinion that a statistical treatment of such an intricate behavior as the one you are studying, will not have very much meaning." He suggested an approach, "I rather imagine that you might like to give a somewhat summarized description of the behavior of the whole group" which could then "set off the individual against the background." Recommending the very course that Mead intended to follow, Boas's solution satisfied both a scientific and a popular readership. When Mead received his letter, she replied, indicating that her work was almost complete and that she was leaving to join her husband in Europe.[9]

Mead returned to New York in the autumn of 1926 and took up her new post as Assistant Curator of Ethnology at the American Museum of Natural History. She settled down with characteristic speed to write up her fieldwork, using Boas's suggested framework. Having written her report for the Social Science Research Council, she quickly determined that she would use the book to proselytize for anthropology. She sought the advice of George Dorsey, an anthropologist who had found popular writing so profitable he had quit his academic job to pursue it full-time. In December she wrote to E. S. C. Handy, the ethnographer of Samoa at the Bishop Museum, to whom she had promised a more scholarly ethnographic text on Samoa: "I've only been back a couple of months and what with getting settled in civilization and starting work here, I've not had time to touch the ethnology at all. The adolescent girl however flourisheth and merely awaits Dr. Boas' return from Europe to be sent questing for a publisher. The need for mixing propaganda for the ethnological method and human interest somewhat obscures what ethnology is scattered through its pages."[10] Having chosen to write in an accessible style that would extend her readership beyond other anthropologists, Mead already apparently understood that she might have to justify her approach to other anthropologists.

Not until March 1928 was Mead able to write to her mother: "William Morrow has taken my book. It will come out in the fall, I hope with an English edition at the same time. It's like having the first baby in a new hospital, the whole staff takes a personal interest. It's to have eight pages of illustrations." This turned out to be a fortuitous and productive partnership for both the new author and the new publisher. Having seen the original manuscript, Morrow must have known he had an extraordinary author in Mead. Her writing style was both lively and evocative. There is no doubt that she enjoyed others' writing and was able to communicate a rare, if perhaps not always accurate, sense of place and peoples.[11] Morrow enhanced the likelihood that the book would attract a popular readership by suggesting that she use material developed for an evening course at the New York League of Girls' Clubs as the basis for an introduction and a final chapter drawing out the implications of her findings for American society.

As Mead was completing her manuscript, she was preparing for her field trip to the Admiralty Islands with Reo Fortune, who would become her second husband. The press picked up on her new trip and reported it to the public with another set of sensational headlines: "American Girl to Study Cannibals"; "Will Shows Fear of Cannibals: Dr. Mead Inserts Clause in Case She Meets Death in South Sea Study." Stories about the expedition to Manus appeared in papers and magazines across the continent, with some articles syndicated over and over again. Her notoriety led to requests for magazine articles, including a proposal from H. L. Mencken for a series of pieces for *Smart Set.* Paradoxically cautioning her about "arranging for magazine articles" because she needed to "take into account the possible attitude of your fellow-scholars whose opinions would count in connection with your future career as a scholar and scientist," Morrow advised her to publish in wider circulation magazines such as *Cosmopolitan.*[12] Despite her awareness of the possible risks, Mead wrote a number of magazine articles for publication during her absence. She departed for the South Pacific, leaving Benedict to cut, collate, find photographs, deal with editors, and disburse payments.

Seeking to maximize its sales, Morrow announced the publication of *Coming of Age* with endorsements from Dorsey, Malinowski, Ellis, and John B. Watson. Ellis called it "fascinating"; Dorsey said it was "extraordinarily brilliant and unique." The combination of a fashionable social problem, sex, exotic location, and girl adventurer proved irresistible to the American press. The response was sensational in both senses of the word. Attention-grabbling headlines announced the book: "Samoa Is the Place for Women—Economically Independent, Don't Have to Cook and Go Home to Dad When Husbands Get Tiresome" and "Where Neuroses Cease from Troubling and Complexes Are at Rest." It even inspired a cartoon in *Esquire* in which a young woman wearing only a hat and bracelet explains to her hostess, "But I came of age in Samoa."[13] The popular press was, in general, uncritical, enthusiastic, nonjudgmental, and either earnest or humorous. Mead was praised for her scholarship and her courage. Her conclusion—that adolescent crises were cultural rather than biological—was widely and accurately reported and generally accepted as having important implications for modern families. Mead returned from the Admiralties in late 1929 to a backlog of requests for interviews, articles, and speaking engagements. To continue that favorable media response, Mead signed on with agencies for placing articles, public speaking, and radio engagements.

As though to suggest that Morrow's fears had been realized, Mead's anthropological colleagues expressed more reservations about her work. Alfred Kroeber admired *Coming of Age in Samoa,* but the powerful Robert Lowie, editor of the *American Anthropologist,* would not accept it as an ethnographic account because of the links she drew between Samoan and American life and her glossing over colonial influences. Edward Sapir, her estranged former lover, published an indirect but vicious attack in which he dismissed her work and Malinowski's as the "smart and trivial analysis of sex by intellectuals who have more curiosity than intuition." Not content with attacking her work, he reviled feminists and lesbians, whom he accused of being both frigid and insatiably ambitious. Sapir completed his attempted demolition of Mead's work and her reputation by accusing "emancipated women" of being little better than prostitutes.[14]

The elite magazines treated her anthropology of contemporary America more skeptically than the popular press. Mencken, for example, described *Coming of Age in Samoa* as "a sweet story, but Miss Mead finds it somewhat difficult to apply its lessons to American life." In his view, "Miss Mead's book would have been better if she had avoided discussing the woes of American high school girls and confined herself to an objective account of life in Samoa." Complaining about the tendency of social scientists like Mead to homogenize American youth, Mencken referred to the Scopes trial as an example that might merit the same scholarly investigation that Mead had accorded to Samoa. As his review concluded, "We know far more about the daily life of the Pueblo Indians than we know about the life of Mississippi Baptists. Whenever, by some accident, light is let into the subject there is gasping surprise, and even horror. This happened, typically, when a gang of slick city jakes descended upon the primitive mountain village of Dayton, Tenn. at the time of the Scopes trial, and found it full of Aurignacian men clad in dressy mail-order suits, with Bibles under their arms."[15] Perhaps responding to such criticism, Robert and Helen Lynd would soon publish *Middletown,* a book that has stood the scholarly test of time better than

Mead's but lacked the popular appeal that catapulted *Coming of Age* into the popular imagination.

Whatever the response—prurient, critical, wistful, or comic—*Coming of Age in Samoa* became nationally and then internationally renowned. To Morrow's delight, it became a best seller. Everyone may not have approved, but everyone had an opinion, which created further interest. In 1932 Morrow sold the rights to Blue Ribbon Books, which produced cheap copies of best sellers that could be sold for a dollar each. By 1933 *Coming of Age in Samoa* would be nominated as one of the hundred best books by women "in this century of progress." Of the seven social science books selected, it is the only one still widely recognized and still the subject of lively and ongoing debate.[16]

No doubt *Coming of Age in Samoa*'s attractiveness can be partially attributed to Mead's evocative writing style that Evans-Pritchard characterized as "the rustling of the wind in the palm trees" form of ethnography. The accessibility of the writing drew comments from academic and press reviewers and was certainly a factor in the book's success. Couched in novelistic language is an account of the daily life of Samoan girls as they "change from babies to baby tenders, learn to make the oven and weave fine mats, forsake the life of the gang to become more active members of the household, defer marriage through as many years of casual love-making as possible, finally marry and settle down to rearing children who will repeat the same cycle." Certainly Mead's description of the freedom and fluidity of sexual relations enhanced its appeal when she wrote about a culture familiar "with sex," which recognized the "need of a technique to deal with sex as an art, have produced a scheme of personal relations in which there are no neurotic pictures, no frigidity, no impotence, except as a temporary result of illness, and the capacity for intercourse only once a night is counted as senility."[17] Arriving in the midst of intense popular discussion about sexual techniques and the importance of sexual pleasure to relationships, *Coming of Age in Samoa* delivered a message that many wanted to hear.

According to Mead, Samoans had not only a crisis-free adolescence but also one that took place in a way of life characterized by leisure, simplicity, and relative freedom from care for adults as well. In an unconscious paraphrase of Marx's famous maxim, Mead's Samoans followed "a morning of work and afternoon of sleep with an evening of song and dance." By the time this article came out, its readers would have known what followed the song and dance—"the rhythmic beat of the surf," the "soft perfume of frangipani blossoms, [and] low-voiced protestations of love." The appeal of this vision of paradise seems self-evident. To understand its particular meanings in America in the 1920s, it is useful to step away from Samoa and sex and consider the ontology of the modern self.

If the most potent symbol of America in the 1920s was the young girl, that feminization of the nation can be linked to the more pervasive feminization of the self. As many writers have noted, the modern self is conceived in spatial terms. The "true" self is that which is inside, hidden, mysterious. The very interiority of the self constitutes a challenge to rational modernity, which means it is subject to technologies that purport to uncover or reveal it, most importantly to its possessor. The thematic of a selfhood created in childhood, lost to consciousness but able to be found, healed, or explicated, had pervaded both popular understandings and professional theorization in the decades leading up to 1928. This was a relatively recent phenomenon. In psychology, case histories only began to include life histories in the 1890s, and it was only after Freud's visit to America in 1909 that case histories began to attend to childhood and sexual life.[18] As a psychology major who had transferred to anthropology at Benedict's invitation, Mead may have been especially able to fashion an interpretation of Samoan culture based around life histories that particularly engaged her readers in the 1920s and throughout the self-conscious twentieth-century United States.

As Carolyn Steedman has argued, over the period from 1780 to 1930 "the figure of the child" came to personify the self over a range of literary forms and public policy debates. This points to a further dimension of the self conceived not only spatially but temporally as well. The "interiorized self" is "created by the laying down and accretion of our own childhood experiences, our own histories." Steedman's exegesis of versions of the child acrobat Mignon in Goethe's *Wilhelm Meister* shows how an aesthetic that eroticized the "littleness" of the child's body and the "mystery" of the child's past echoed, not just in literary forms, but in public debate about stage children and young street hawkers. This aesthetic increasingly mapped onto concerns about, and models of, the self. The child-figure is not just any child but a specific constellation of attributes: almost always female, but often ambiguously embodied, little, although she may be any age from toddler to twenty-five, and with a mysterious history of damage by an adult, a history that, if not explicitly sexual, certainly has pedophilic undercurrents. It is this specific figure that condenses the meanings of modern selfhood with its attributes of interiority, historicity, sexuality, mystery, and damage.[19]

Steedman's account of Mignon resonates uncannily with Mead's Samoan girl. The Samoan girl is constituted in terms of surface and depth, history and time, wholeness and damage. She thus slotted into tropes of selfhood already in wide circulation. Those tropes included not just the stereotypic vision of exotic maidens, but also one that was more subtly embedded in popular culture and more consciously theorized in the versions of psychoanalysis propounded in both the popular and academic media. According to Steedman, the Mignon "girl-figure" personified a self understood to be already damaged. "If we repress [our children], heaven knows what complexes their turbulent minds may develop," one reviewer wrote about *Coming of Age in Samoa.* Both professional and popular versions of psychoanalysis in the 1920s proclaimed that damage was unavoidable. When Mead wrote that the larger Samoan family "seems to ensure the child against the development of the crippling attitudes which have been labeled Oedipus complexes, Electra complexes, and so on," she was able to do so without challenge from either the psychological professionals or the popular press.[20]

Another part of the book's appeal reflected the unconscious colonial assumptions about primitive peoples that had often been expressed toward Indians, African Americans, or other "lesser" peoples. In considering how the relationship between the "Samoan girl" and the American "self" operated in the understanding of Mead's reviewers and readers, one of the first things to be noted is that the Samoan girl stood in a simple relation to "Samoa" itself in *Coming of Age in Samoa.* Mead continually reiterated the continuity between the Samoan individual and the culture or society as a whole. This is one of the principal messages of the study. Where there is no conflict between individual and social values, there is no adolescent rebellion. Samoa was what Mead called a homogeneous society. What was said of the Samoan girl was often what Mead said about Samoa. As a result, Samoa and its adult members were juvenilized, if not directly by Mead, surely by the popular press, aided by Mead's descriptions of them. "The Polynesian is not a true adult," wrote an anonymous reviewer in the *Philadelphia Record.* "He is rather an overgrown child, gay, happy, and thoughtless of the morrow." Likewise, a review in the *Brooklyn Eagle* stated that the book "takes us far into the South Sea Isles and bids our fancy rove, free and unrestrained among the child people of the Samoan group."[21] Samoa and the Samoan girl were interchangeable as discursive objects in a colonial and racial discourse that portrayed colonized groups as children or adolescents, as compared with "mature" Europeans or Anglo-Americans.

In contradistinction to the modern self constituted by personal history, the Samoan girl, and indeed Samoa itself, appeared in *Coming of Age in Samoa* and in the popular commentaries on Mead's work as being both history itself and without history. Both Mead and reviewers invoked the well-documented association of the "primitive" with the past in subtle and not-so-subtle ways. The widespread use of the terms "primitive," "simple," and "natural," as distinguished from "progress," "complex," and "civilized," located Samoa as archaic: "It is not entirely primitive, nor yet is it debauched by the most pernicious forces of Western civilization." But while Samoa was "in the past," it was also located in the present, creating the paradox of a timeless past. In an interview, for example, Mead reportedly claimed that "[l]ife is peaceful, orderly, and gracious in Samoa. It seems timeless. Nothing has ever changed. Nobody is troubled. No voice is ever raised. It would be a fine place to die." Even where there appeared to be change, she argued that it was only apparent and not real. In one of the few academic articles she wrote on Samoa, she made an impossibly confused argument. Because change is easy in Samoa, it hardly happened at all. "The ever-yielding, ever accommodating social structure has remained much the same, generation after generation, while [those] with original minds and social ambitions slid, sated with too easy victories, into undistinguished grooves." Her refusal to acknowledge change, the impact of the missionaries, or the U.S. Navy presence was a major feature of some contemporary and much more recent criticism.[22]

Suspended in a timeless past, Samoa was said to be a place without that source of modern ills: repression. "The Freudian diseases, of course, cannot occur in Samoa," wrote a reviewer in the *St. Louis Globe Democrat.* Another reviewer, more humorously, claimed that Samoa represented "Erotics without neurotics, Oedipus is not Rex, and laughter goes hand in hand with desire." Mead herself wrote that Samoan children "grew up painlessly and almost unselfconsciously." Like its children who were undamaged, Samoa itself was undamaged. This place, which is history but which has no history, is also a place in which the very repression that modern psychology identified as fundamental to modern mental illness, deviance, and just plain unhappiness cannot occur. Samoa became a place where the self was not damaged and where, notably, there was little consciousness of self. Samoan culture was shallow, concerned with surface and form. This lack of depth was set in place in early childhood "when the child is handed carelessly from one woman's hands to another's, the lesson is learned of not caring for one person greatly, not setting high hopes on any one relationship."[23] Endowed with these attributes, Mead's Samoa and Samoans lacked a defining feature of modern selfhood—depth or interiority.

Lack of strong attachments in childhood led to a similar casualness about adult sexual relationships. Samoans concerned themselves with sexual technique, ignored romantic love, and treated jealousy contemptuously. This "shallowness" was not confined to intimate relationships but could be found in the society as a whole. As Mead explained for the benefit of her readers, "However much we may deplore such an attitude and feel that important personalities and great art are not born in so shallow a society, we must recognize that here is a strong factor in the painless development from childhood to womanhood." Shallowness, according to Mead, might have its advantages: "So high up on our list of explanations we must place the lack of deep feeling which the Samoans have conventionalized until it is the very framework for all their attitudes toward life."[24] Imputing complexity to the culture of her readers, Mead provided yet another reason for *Coming of Age in Samoa* to be read by its intended audience.

The purported absence of pain, repression, and strong emotions evoked the most divergent reactions between the popular and the elite press. In many of the newspaper reviews, "civilization" explained modern ills, and the "blessings" of civilization became mixed indeed. As reviewers interpreted *Coming of Age in Samoa,* Samoans "know nothing of industrialism, bureaucratic abuses, modern warfare and the abolition of leisure." The absence of such pain and stress brought forth a strong moralistic reaction among some commentators in the "quality" press. Aldous Huxley, for example, wrote: "Margaret Mead's picture of such a life makes one wonder whether, after all, contentment may not be bought at too high a price. No great civilization has ever taken a cheerful view of life." As Huxley added, "Not to pursue happiness, but to bear unhappiness with fortitude, has been the lesson of all practical philosophers."[25]

Another reviewer, using an accentuated version of colonialist discourse, explicitly contrasted the savagery of the Samoans with the putative civilization of his readers, with whom he identified by using the first-person plural. According to the *Psychoanalytic Review,* "Savages have solved their problems apparently with very much more sense than we have and in a way to produce much less conflict" and were "to all intents and purposes without neuroses and psychoses." He nevertheless consoled his neurotic readers: "The fact remains that they are savages and that it is very possible that it is because of these easy-going solutions that they remain so. Civilization is not a process that comes upon us passively. It is the result of active, energetic attack upon problems and situations." Speaking a language that Mead would also use in writing about other peoples, including the Balinese, the reviewer explained, "One cannot make war against any aspect of reality without taking the risks that go with such a course of conduct, and the results when success is not the outcome are in many instances neuroses and psychoses."[26] Obviously providing work for psychoanalysts, he reassured his readers that mental and emotional distress was a mark of their striving for higher goals than the Samoans.

While no single piece by Mead or other writers dealt with all of these motifs of time, depth, and consciousness, they recurred in the various texts in the 1920s. This is, of course, not a new picture of "the native." It constituted a constellation of traits that had repeatedly surfaced through a variety of discursive practices. As Freda Kirchwey commented in her review of *Coming of Age in Samoa* in the *Nation,* "You will probably be astonished to discover how like a South Sea island that South Sea island can be."[27] What emerged was a girl figure that was both the adolescent girl and Samoa itself. Unlike Mignon, who personified the post-Oedipal self, this girl figure exemplified the prior self—before repression, before the personal history, which was also the personal mystery, had accrued. The Samoan girl became the undamaged, whole, sexually free, unselfconscious self without interiority. As a nostalgic incarnation, she represented lost bliss and unfettered sexuality. In her more moralistic incarnation, she posed the need for taming and disciplining sexuality in order to achieve modern selfhood and adapt to contemporary civilization. Alternating between these modes, Mead created a text that spoke to moralists and advocates of sexual liberalism alike.

What Mead suggested as a civilized substitute or compensation for the loss of unfettered sexuality was rationality—or what she termed "education for choice." In a complex society, the stresses of adolescence could be minimized by giving the young the intellectual tools to make intelligent choices among an array of values unavailable in the simple society of Samoa. Indeed, this was to have been the principal message of the book for its American readers, who looked to it for a contemporary moral. For her scholarly readers, she affirmed that the practice of science and rational, critical self-reflection distinguished them from the people of all so-called "primitive societies." Making full use of the first person plural as though to join with readers, Mead wrote, "We have one great superiority" over those whom she described as "primitive peoples." Justifying the work of cultural critique, she described the Samoans as unable to see their customs as anything other than "given, ordained, immutable." Moving "unselfconsciously within the pattern of their homogeneous, self-contained societies" was not an unalloyed benefit. To feel troubled or discontented was a mark of superiority. "We, caught almost as completely in a far more complex pattern, have acquired the ability to think about it."[28] Presenting the primitive as body and the civilized as mind, Mead provided reassuring dichotomies that flattered her readers.

Coming of Age in Samoa evoked a complex set of reactions in late 1920s America. It was interpreted, unsurprisingly, in different ways by different kinds of media. The popular press read it with nostalgia for a simple and unfettered sexual economy, while elite magazines reiterated the more "civilized" values of emotional depth, forbearance in the face of adversity, and a go-ahead, rational, problem-solving approach to life. These distinctive readings neatly encapsulated the ambivalence between nostalgia and progress that was particularly characteristic of 1920s America. In a sentence that could have been written with reference to *Coming of Age in Samoa*, Levine wrote that Americans' fascination for filmic representation of "native" and black cultures enabled them to simultaneously "feel superior to those who lacked the benefits of modern technology and to envy them for their sense of community, their lack of inhibitions, their closer contact with their environment and with themselves."[29]

Coming of Age in Samoa was launched at the height of this cultural oscillation between the values of the golden past and the promise of the golden future. Its problem, the flapper with all her associations of freedom, modernity, and rejection of the constraints of traditional femininity, represented lost gender, generational and sexual certainty. Its solution, the application of science and progressive education to the problems of social complexity and change, offered a reassurance that progress could resolve these conflicts. Mediating between the bucolic past and future progress, Mead's *Coming of Age in Samoa* supported both traditionalist yearnings and utopian hopes.

As there was a slide between "the Samoan girl" and Samoa in *Coming of Age* and its reception, there was also a slide between the author and the Samoan girls she studied. The most recurrent theme in contemporary descriptions of Mead is her youth. Repeatedly described as "exceedingly young," a "girl anthropologist" or "girl scientist," and as "the youngest anthropologist," Mead herself played on her youth and childlike appearance. She emphasized in the book, in interviews, and in feature articles the importance of her smallness and slightness to her acceptance within the girls' groups in Samoa, much as she also played the dutiful daughter to Boas's paternal mentor. The identification went beyond Mead's youthfulness. A number of reviewers wrote that the success of her research was due to the fact that she had "identified herself so completely with Samoan youth." More critical reviewers saw this identification as a drawback: "Mead forgets too often she is an anthropologist and gets her own personality involved with her materials," wrote one reviewer in a comment that would prefigure one of Freeman's criticisms of Mead's research methods.[30]

Numerous pictures of Mead in Samoan garb, which appeared in newspapers and feature magazine articles, reinforced her popular identification with the Samoan girl. These articles told how Samoan families and/or chiefs had adopted her or made her a princess; how she had learned to dance and lived "cheek by jowl" with "natives," sleeping on the floor of grass huts. "Little" Margaret, looking very much like a flapper with her bobbed hair and low-slung dresses, was also the "girl scientist," a "prodigy," going intrepidly where no white "girl" had gone before.

Circumscribing the array of statements and images about Mead and her work is the language of science. Categorization, comparison, analysis, critique, objectivity, and rigor are words and ideas that locate Mead both directly and by implication on the side of civilization. Juxtaposing descriptions of her sympathetic identification with Samoan girls and commentary on her scientific objectivity, the portrayal of Mead, like the content of the book, played to the paradoxical desires of a culture that longed for the return of the old-fashioned feminine and celebrated the flapper as a harbinger of the new.

The popularization of psychoanalysis that had taken place in America over the previous two decades contributed to the success of *Coming of Age in Samoa* in the late 1920s. Freud had downplayed the darker side of his theory in his American lectures, emphasizing instead the "ameliorative, pragmatic qualities of health and good functioning." In popular parlance, "repression" lost its association with the unconscious and the Oedipal crisis and became any force that restricted the individual. "Repression" came to signify something unhealthy, which could be eradicated; and the psyche became entrenched as a field of enterprise for the growing industry of physicians, therapists, psychologists, and advice givers—developing into what Lacan termed "the psychology of free enterprise." American know-how could "fix" the self.

Mead's work, of course, fed directly into this set of beliefs and practices. If Americans saw themselves as conflicted between lost *communitas* and modern progress, Mead assured them that through careful consideration and the application of the results of science, they could secure them both. "The girl scientist" and her book embodied the American dream.

11 / Erasures and Inclusions

Whether by conscious design or intuitive arrangement, Benedict and Mead evolved an intellectual division of labor during their twenty-five-year intellectual collaboration that assigned to Benedict primary responsibility for addressing issues of "race," while Mead researched and wrote about "male and female." However, Benedict helped to shape Mead's work on gender, including *Coming of Age in Samoa, Sex and Temperament in Three Primitive Societies,* and the work in preparation just before Benedict's death, *Male and Female.* In addition, Benedict usually left the subject of sexual repression to Mead, although in *Patterns of Culture* she linked tolerance for sexual deviance to compassion for those individuals who found their own cultures "uncongenial." Mead's *Sex and Temperament in Three Primitive Societies* echoed Benedict's warnings against "rigidity in the classification of the sexes." Although Mead was convinced that biology played a major role in shaping human behavior, she accepted Benedict's and Boas's warnings about the dangers of introducing "politically loaded discussions of inborn differences" into her analysis of gender. Aiming at "a moral which would be applicable to modern American society," Mead eschewed feminism while creating a key text that, in its emphasis on the cultural construction of gender, would become important to second-wave feminism.[1]

In the years between 1922 and 1948 only rarely did Mead address directly issues that involved definitions of "race," although in her 1932 study of the Omaha Indians in Nebraska, *The Changing Culture of an Indian Tribe,* she described the society as "broken" and "a cultural shipwreck." She traced the problem to "a long history of mistakes in American policy toward Indians"—implicitly evoking a criticism of white racism. In a review of the book, Benedict agreed with Mead's characterization of the unfortunate federal "handling of a subject people" and urged a change in policy.[2]

In 1939, as it became apparent that Boas, in his eighties, would never write the popular work on race that he had long intended to do, Benedict took over the project. A year later she published *Race: Science and Politics,* published in England as *Race and Racism.* Her use of the term "racism" is credited by the *Oxford English Dictionary* as only the third published use of the word in English. Writing during a time of national controversy, Benedict distinguished between race as a matter for scientific study and racism as "an unproved assumption of the biological and perpetual superiority of one human group over another." In 1946, again responding to what seemed a national need, this time in the shadow cast by Hiroshima, Benedict published *The Chrysanthemum and the Sword,* her study of Japanese culture. Soon translated into Japanese, Benedict's book would come to occupy a space in the popular Japanese imagination comparable to the position that *Coming of Age in Samoa* assumed in the American imaginary.

In 1949, as the Cold War hardened into ideological combat, Mead published *Male and Female.* While appearing to yield to the anti-feminism of the Cold War era and the ardent defense of the status quo, Mead repeated her earlier insights into sex and temperament. Her efforts to incorporate feminist assertions of the need to "feel as a whole human being" while rejecting feminist activism in the book would enrage feminist readers like Betty Friedan, who accused her of contributing to the construction of the "feminine mystique."[3]

The following essays scrutinize the writings of Mead and Benedict with regard to whiteness, individualism, and liberal feminism. Focusing on Mead's *Coming of Age in Samoa* and *Sex and Temperament in Three Primitive Societies,* Louise Newman argues that Mead continued to perpetuate some aspects of the ideologies and values that linked her to missionaries, pioneering anthropologists, and female travelers. Christopher Shannon's critique of Benedict's *The Chrysanthemum and the Sword* indicts what he describes as liberalism's repressive tolerance. Jean Walton examines *A Rap on Race,* Mead's dialogue with Black writer James Baldwin, to analyze Mead's production of herself as a "white woman." Walton also investigates the connections between sexual closeting and racial difference in Mead's creation of a public persona as mother of the world. Read together, these essays provide insights into the ways Benedict and Mead sought to erase racial difference while bringing gender and sexuality into public debate.

4 Coming of Age, but Not in Samoa

Reflections on Margaret Mead's Legacy for Western Liberal Feminism

Louise M. Newman

> In the case of anti-colonial critique, it is the similarity of past and present that defamiliarizes the here and now and subverts the sense of historical progress.
>
> Nicholas Thomas, *Colonialism's Culture*

One of the most famous and popular works ever published by an American anthropologist, *Coming of Age in Samoa* first appeared in 1928 when its author, Margaret Mead, was twenty-seven years old. By 1935, when she published *Sex and Temperament,* Mead had gained a national reputation as an expert on "primitive cultures" and was recognized by the public, if not by her colleagues, as one of the leading anthropologists of her day. Prolific, outspoken, charismatic, unconventional, provocative, controversial, and brilliant, Mead achieved widespread public renown that was remarkable for a woman who constructed herself as a scientist and intellectual. She recognized instantly that her audience extended far beyond the elite worlds of the university and museum, and she cultivated her public by publishing hundreds of articles on domestic issues and international politics in such venues as the *American Anthropologist, Natural History, Redbook, Vogue, Good Housekeeping, Seventeen,* and the *New York Times Magazine,* to name just a few. From the time that *Coming of Age in Samoa* appeared until her death fifty years later, journalists sought Mead for her opinions on marriage, homemaking, child rearing, feminism, civil rights, and race relations.[1]

Among the general public old enough to remember her, Mead is probably best known for the role she played in the 1930s in prompting Westerners to question their sense of cultural superiority, using so-called primitive societies to critique patriarchal gender relations in the United States. Mead was not alone in this endeavor, writing at a time when other artists, professionals, and elites drew from such cultures to reinvigorate Western arts—literature, music, dance, visual arts, photography, and film. Historians of anthropology remember Mead as one of Franz Boas's many students who helped bring about a paradigm shift in anthropology from evolution to cultural relativism by challenging biological explanations of cultural differences and refuting the explicit racism in eugenics and mainstream physical anthropology.[2]

This essay, however, situates Mead in a different intellectual context. In addition to seeing her as someone who helped foster cultural relativism within anthropology of the 1930s, it will place Mead within a history of feminism and, more specifically, within a tradition of white feminist thought on racial questions. Repositioning Mead in this way requires seeing Mead as an integral part of a Victorian tradition that combined notions of white or "civilized" women's sexual restraint and black or "primitive" men's bestiality to reinforce the dominant cultural taboo against miscegenation. Although historians of anthropology usually understand Mead as challenging the racism implicit in such constructions, nonetheless such dualisms informed her work. In other words, this essay reconsiders the nature of Mead's antiracism, highlighting the continuities between Victorian and modern anthropology. Such an approach to Mead's work risks eliciting criticism from scholars who view Mead solely as a cultural relativist and racial egalitarian, an opponent of Western ethnocentrism and racial bigotry. To grasp the central point—that Victorian racial politics shaped Mead's work—it is necessary to understand that oppositional movements retain residues of that which they oppose. To put it bluntly, Mead's substitution of cultural theories for biological explanations of difference did not purge her work of its Western ethnocentric and white racial biases.

Mead's work also needs to be situated within a canon of historical feminist writings.[3] In contrast to the scholarship that traces the way feminists brought their ideology to the practice of social science, this essay will examine the way feminist social scientists in the late nineteenth century used the prestige of science to legitimize their feminism. In the 1920s and 1930s, Mead's work continued an intellectual tradition that empowered "civilized" women to assert their superiority to those primitives with whom they compared themselves. Indeed, Mead's legacy has proved to be a lasting one: the discourse of Western liberal feminism, to which Mead was an important contributor, continues to position Western societies as superior to non-Western ones in terms of "women's freedom" as defined by individuality and choice.

Mead's early monographs, *Coming of Age in Samoa* and *Sex and Temperament in Three Primitive Societies*, broke with an earlier tradition called evolutionary or Victorian anthropology.[4] From at least the mid-1800s to the early twentieth century, evolutionary anthropology expressed and reinforced Anglo-American definitions of themselves as a superior race because of their supposedly unique, race-specific, biological forms of sexual difference. Mead's work helped overthrow this central tenet of evolutionary racism. Yet while Mead challenged Anglo-Americans' belief in their inherent biological superiority to primitive peoples, she did not contradict their belief in the cultural superiority of Western civilization. Mead invoked primitive societies to criticize gender relations in the United States, but at the same time she negatively characterized primitive societies like Samoa or Bali as lacking freedom and circumscribing individual choice. For Mead, primitive societies provided Americans with conceptual alternatives to reflect on, but she never advocated that the United States remake itself in the image of the primitive. Instead, as in the case of Manus, she praised Pacific cultures for emulating the United States. Mead's work on Samoa and New Guinea incorporated nineteenth-century ideas in which race, sexual difference, and sexuality were linked, enabling her to transform, without transcending, the racist formulations of evolutionary anthropology.

Mead's work, like her persona, can be understood as the culmination of three nineteenth-century female avocations: the missionary, the explorer, and the ethnographer. These avocations helped to solidify a role for Anglo-American women as legitimate practitioners of anthropological science, a profession that grew out of their previous activities as Christian civilizers and governors of the primitive. Relating Mead to these traditions helps to explain how she could become so popular at a time when many still questioned the suitability and capability of women as scientists.

In the context of academic anthropology in the 1920s and 1930s, both *Coming of Age in Samoa* and *Sex and Temperament in Three Primitive Societies* can be described as remarkable books that helped contribute to the dissolution of evolutionary anthropology. Mead insisted that primitive peoples, rather than just being evidence of a shared and fortunately transcended past, had something valuable to teach Americans about reforming their institutions. In her autobiography, *Blackberry Winter*, Mead described her training at Columbia in the early 1920s, where she studied with Franz Boas and Ruth Benedict: "We had, of course, had lectures on evolution. . . . But we went to the field not to look for earlier forms of human life, but for forms that were different from those known to us." She further claimed that her generation of professionally trained anthropologists "did not make the mistake of thinking, as Freud [had] . . . that the primitive peoples living on re-

mote atolls, in desert places, in the depths of jungles, or in the Arctic north were equivalent to our ancestors."[5] Rejecting Victorian evolutionary anthropology's belief that all societies followed the same path of development, Mead emphasized that she had gone into the field prepared to find differences that would not be explained in terms of an evolutionary hierarchy.

As Mead understood, evolutionary anthropology located primitive societies at an earlier stage of development than civilized societies and often measured a society's relative position by "woman's status" or "condition." Evolutionary accounts held that the "progress of a race" depended on the adoption of specific sex roles that in turn would produce specific manifestations of sexual difference. Thus another indicator of a society's higher evolutionary ranking was the existence of pronounced physical, moral, and sexual differences between men and women. In the words of William I. Thomas, a social scientist at the University of Chicago in the 1890s: "The less civilized the race the less is the physical difference between the sexes." As historian Nicholas Thomas has argued, "the degradation of women was a measure for the degradation of a society and enabled it to be mapped against others in a region. Gender was thus central to the evolutionary ranking of societies." Or as historian Gail Bederman summarizes this ideology: "Savage (that is nonwhite) men and women were [taken as] almost identical, but civilized races had evolved the pronounced sexual differences celebrated in the middle-class doctrine of 'separate spheres.'"[6]

Mead challenged these evolutionary beliefs, which characterized savage or primitive races as either lacking in sexual differentiation or exhibiting uncontrollable, rampant sexuality. In *Sex and Temperament in Three Primitive Societies* she argued that primitive societies differed substantially from one another in how they understood sexual differences and sexual drives and in the ways they structured gender relations. As she put it, "the personalities of the two sexes are socially produced."[7] Mead concluded that sexual differences varied so substantially from one society to another that they must be understood as culturally, not biologically, determined.

Together with other anthropologists, including Boas and Benedict, Mead helped consolidate a new paradigm in anthropology, which scholars often refer to as cultural relativism. Anthropologists developing this new paradigm understood cultures as developing along different, incommensurable lines, and they no longer held up Western practices as either the inevitable future for all cultures or as the only morally legitimate forms of cultural arrangements. Despite her commitment to this new paradigm, Mead was neither a moral relativist nor did she attempt to write value-free ethnographies. Her understanding of cultural relativism did not prevent her from making moral distinctions among various cultural practices. A term like *cultural comparativism*[8] may more accurately describe Mead's perspective, for it retains the idea that Mead did not study other cultures to argue that all social arrangements were equally valid but to determine which arrangements represented better ways of living. Such a perspective meant that Mead and others who worked within this paradigm resisted normative judgments of the sort that automatically called for primitives to adopt civilized gender roles, but they did not suspend all judgments. Mead wanted to expand Americans' repertoire of conceivable alternatives so that they might envision new ways of reforming their social institutions.

Positing a distinction between "social constructs," which was Mead's term for culturally specific beliefs about sexual differences, and "biological facts," as she referred to those universal aspects of sex difference manifest in all known societies, Mead assisted anthropology and feminism in creating new distinctions between culture and biology. Evolutionary anthropology had preferred the term "civilization" to culture and used "civilization" to refer to all social practices passed from one generation to another, partly through learning and partly through heredity. As anthropologists helped to replace the term "civilization" with "culture," culture became fully distinguishable from biology. In both anthropological and popular discourses, the older term, civilization, eventually lost any connection to a notion of heredity. Social scientists understood culture as passing from one generation to the next by social processes, not through biological inheritance.

Mead used "culture" to account for the differences among people and "biology" to account for the similarities. The fact that all women could lactate and bear children she deemed an attribute of biology or "sex" as indicated in the title, *Sex and Temperament in Three Primitive Societies.* The fact that some women in specific societies she studied were passive or gentle was an attribute of culture. Mead's particular contribution to the new paradigm was to show how varied different societies' views of sexual differences could be and thus to shift these differences from the category of biology to the category of culture. According to Mead, Europeans were not the only ones to consider (civilized) men and women fundamentally different from one another; other peoples considered their men and women fundamentally different as well. Indeed, sexual differences varied so much from culture to culture that it was no longer possible to account for sexual difference by appealing to biological notions of innate maleness and femaleness. Mead separated the idea of the racial superiority of Anglo-American whiteness from sexual difference identified with genteel Anglo-American gender relations, which had been linked concepts in Victorian evolutionary schema, and emptied them of their usual content.

In sum, where Victorian evolutionists believed that civilization was a racial trait, inherited by advanced white races, Mead assisted in differentiating what was culturally transmitted through teaching and learning—including many racial and sexual differences—from what was genetically or biologically transmitted (innate sexual traits). Along with other social scientists of the 1920s and 1930s, Mead argued that sex and race were not significant variables of biological transmission. As Mead wrote, "One by one, aspects of behavior which we have been accustomed to consider invariable complements of our humanity were found to be merely a result of civilization, present in the inhabitants of one country, absent in another country, and this without a change of race."[9]

Such theoretical innovations had profound consequences for the development of feminist analysis, for they permitted new critiques of Western patriarchy. Previously, evolutionary feminists such as Charlotte Perkins Gilman, Mary Roberts Coolidge, and Elsie Clews Parsons had invoked the primitive to argue for the elimination of the primitive traces that remained in the United States' patriarchal civilization. For them, evolutionary or social progress for the United States meant increasing its cultural distance from existing primitive groups.[10] Mead's revaluation of primitive societies made possible a new strategy for Anglo-American feminists. The primitive could now be invoked as an alternative to be emulated, rather than as a vestige to be eliminated.

Mead challenged an assumption shared by many experts at the time that adolescence was an inherently and inevitably stressful stage of biological maturation about which nothing could be done. Mead explicitly situated her study in opposition to works like G. Stanley Hall's *Adolescence,* which ascribed the restlessness and rebellion of young people to inescapable maturation processes. As Mead succinctly summed up her doubts about a biological-developmental explanation of young people's behavior: "Were these difficulties due to being adolescent or to being adolescent in America?"[11] Answering the question she set for herself, Mead attributed the pain of adolescent American women (implicitly understood to be white, middle-class, and heterosexual) to the changing social mores of the 1920s.

Mead identified another factor as the youthful lack of sexual experience and knowledge. Women no longer felt compelled to adhere to traditional forms of heterosexual marriage but were also not encouraged to choose from among a much broader range of possible marital arrangements. Once married, the American girl was less likely than her Samoan counterpart to experience a satisfying sexual life because American society had such a limited notion of what constituted normal sexual behaviors. Samoan society, by contrast, had a wider range of acceptable sexual practices, which prevented sexual problems of both an individual and social nature—including guilt, frigidity, marital unhappiness, and prostitution. Samoan girls acquired the knowledge and experiences that enabled their adolescent discovery of sex to be pleasurable rather than a painful experience of sexual repression, as it was for American girls. As Mead explained, the acceptance of a "wider range as 'normal' provides a cultural atmosphere in which frigidity and psychic impotence do not occur." These alternatives included premarital sex, open marriages, which would include extramarital sexual relations, trial marriages that could end voluntarily after a period without divorce proceedings, companionate marriages, marriages without children, marriages combined with a career, and traditional marriages, that is, those between a bread-winning husband and a wife who stayed home to take care of the children. She told her American readers that "the acceptance of such an attitude" would help to solve "many marital impasses" and might empty "our park benches and our houses of prostitution."[12]

Despite believing that Samoan society enabled its girls to enjoy sex without shame or guilt, Mead found troubling the seeming contradiction that Samoan society demanded more "conformity" and allowed less "individuality" among its women. Samoa could serve as a model in one sense—demonstrating that young women could enjoy sex—but not in another sense, since sexual enjoyment for Samoan women entailed restrictions on women's individuality and freedom. In Mead's view, the difference in sexual enjoyment could be boiled down "to the difference between a simple, homogenous primitive civilization," where there was "but one recognized pattern of behavior," in which all Samoan girls had no choice but to engage, and "a motley, diverse, heterogeneous modern civilization," in which various types of sexual behaviors for girls were possible.[13] In other words, Samoa served Mead, paradoxically, as a means of pointing out what was wrong in American gender relations but, at the same time, as an example of a more extreme gender oppression than that which existed in the United States.

Although Mead rejected certain assumptions from evolutionary anthropology, she retained its tendency to use gender to encode and assess cultural progress. This practice of measuring the status of a society by the degradation of its women has had a long history in Western imperial and anthropological thought and, in particular, in the way Westerners understood Pacific societies, including Samoa.[14] Although Mead rejected the crude judgments of nineteenth-century anthropologists that debased primitive women, nonetheless implicit in her work was the belief that Samoa was a flawed society because it restricted the freedom of its women. In making this claim, Mead helped foster a liberal feminist critique of American society, which attacked patriarchy for placing restrictions on women's expression of sexuality and conceptualized a free society as permitting women's choice in how they lived their sexual lives.

Mead never advocated that the United States model itself after Samoa. The United States could not make its culture simpler or less diverse, and Mead would have viewed any attempt to do so as authoritarian and repressive, involving an elimination of social options and resulting in less freedom for women. She also believed that young women in the United States had more freedom than did young women in Samoa, and she would not have been willing to trade freedom for happiness. However, Mead believed that individual freedom and personal happiness need not be at odds, particularly if one defined these ideals in terms of being free to choose what best suited one's innate temperament. The American girl's pain and suffering, Mead argued, resulted not from having too many choices, but from being unable to live out, without social stigma or economic repercussions, the option(s) that best suited her nature or temperament. A subtle observer of her own society, Mead realized that some of the available alternatives were often not livable possibilities, and she realized too that class, race, ethnicity, and religion, among other factors, prescribed the choices of particular groups and individuals.[15] What the United States must do, Mead believed, was to make the sexual alternatives that were conceptually available real options for all women.

In 1935, with the publication of *Sex and Temperament in Three Primitive Societies,* Mead moved from a perspective that emphasized sexual conformity within a given primitive society in which all men adhered to the same ideal of masculinity and all women adhered to the same ideal of femininity to a perspective that emphasized differences among primitive societies. Her analyses of the Arapesh, Mundugumor, and Tchambuli stressed that each had a different understanding of what constituted natural sexual differences. Both the Arapesh and Mundugumor believed that men and women shared a similar temperament, but the Arapesh assumed that both sexes were gentle and unassertive, while the Mundugumor understood both men and women to be violent, competitive, aggressively sexed, jealous, and quick to avenge insult. In contrast, the Tchambuli, like Americans, believed in innate or natural sexual differences between men and women but had "a genuine reversal of the sex attitudes of our own culture." The Tchambuli woman was "the dominant, impersonal, managing partner," while the man was "the less responsible and the emotionally dependent person." Arguing that such variability eliminated "any basis for regarding such aspects of behavior as sex-linked," Mead concluded that "human nature is almost unbelievably malleable," responding to cultural conditions rather than biological determinants.[16]

Mead drew an explicit lesson for the United States from her study of these societies, arguing that they showed that many so-called sexual traits of American men and women were arbitrary and not an inevitable outgrowth of biological difference. In other words, it was possible to change how men and women behaved and to eliminate many forms of apparent sexual differences. In the conclusion to *Sex and Temperament in Three Primitive Societies,* she wrote that American society could "take the course that has become especially associated with the plans of most radical groups: admit that men and women are capable of being molded to a single pattern as easily as to a diverse one, and cease to make any distinction in the approved personality of both sexes. Girls can be trained exactly as boys are trained, taught the same code, the same forms of expressions, the same occupations." Once Americans realized that temperament was not sex-linked, Mead queried her readers, "is it not reasonable to abandon the kind of artificial standardizations of sex-differences that have been so long characteristic of European society and admit that they are social fictions for which we have no longer any use?"[17] Mead used her research to argue to American readers that there were other ways to structure gender relations than the ways that most middle-class white Americans felt to be natural, inevitable, and good.

Thus Mead mounted an explicit challenge to American beliefs that men and women had sex-linked differences in temperament that were impossible to change. "Standardized personality differences between the sexes" became "cultural creations, to which each generation, male and female, is trained to conform." Mead was not quite the social constructionist in the fluid way contemporary scholars use the term, because she believed that innate personality differences, which she referred to as "temperament," powerfully shaped individual personalities. Nonetheless, Mead's research findings in *Sex and Temperament in Three Primitive Societies* represented a significant break from the earlier evolutionary belief that primitive peoples were blatantly sexual beings, unable to exercise any restraint over sexual impulses, an idea still vestigially present in *Coming of Age in Samoa. Sex and Temperament,* which Mead would later describe as her "most misunderstood book," unsettled readers in a way that *Coming of Age* had not.[18] Some readers had difficulty grasping the distinctions that Mead was trying to make between sex and temperament, in which *sex* meant sex-associated differences, some of which were biologically transmitted through heredity and some of which were not, and *temperament* designated innate individual endowments, which were not sex linked but which were nonetheless biologically transmitted.

Some readers mistakenly thought that Mead was denying the existence of any innate biological sex differences, when all she hoped to show was that most sexual differences thought to be biologically transmissible were not. Other readers found this point obvious, even trite. The sociologist Hortense Powdermaker called for additional work assessing the significance of those sex differences that were innate and universal. In 1949 Mead answered Powdermaker's challenge by publishing *Male and Female,* only to find that readers used this book to discredit the previous arguments she had made in *Sex and Temperament in Three Primitive Societies.* Responding to this debate, Mead wrote a new preface to the 1950 edition of *Sex and Temperament,* taking the opportunity to explain: "In our present day and culture . . . there is a tendency to say: 'She can't have it both ways. If she shows that different cultures can mold men and women in ways which are the opposite to our ideas of innate sex differences, then she can't also claim that there are sex differences." Answering her critics, Mead wrote, "Fortunately for mankind, we not only can have it both ways, but many more than both ways."[19] Trying to include biology and culture, Mead found it difficult to convince her readers that male and female roles could be shaped by both.

Sex and Temperament in Three Primitive Societies clearly represented a major break from *Coming of Age in Samoa* by introducing variability into "primitive" sexuality, but in other ways it shared a basic premise with the earlier book. In both works, Mead's overarching purpose was to prove that alternatives to American gender relations existed elsewhere and thus could serve as examples to Americans seeking to reform U.S. institutions and practices. For Mead, despite her stated belief in the incommensurability of different cultures, the point of intercultural comparisons was to broaden Americans' vision of possible alternatives in gender relations. She used this knowledge to argue that Americans could and should reform their culture and themselves, establishing a fundamental principle that still operates within Western feminist anthropology today.

Mead never relinquished her belief that intercultural comparisons could be put to the use of social reform. Nor did she challenge Americans' belief in the cultural superiority of Western civilization. Instead, she proposed a new set of criteria on which to base that judgment. Mead believed that the United States was superior to the societies she studied, not because of the existence of sexual differences, the means by which it judged its "progress" from primitive societies, but because only it had the potential—due to a presumably greater complexity and sophistication—to maintain a larger range of gendered behaviors from among which individuals could choose. One of Mead's associates, Jeannette Mirsky, embraced this very point in her review of *Sex and Temperament in Three Primitive Societies:* "The author concludes that by assigning definite and different traits to the sexes or by setting a single pattern for men and women, we get misfits, persons of either sex who cannot fit into their defined roles. Her plea is for a variety of roles open to both men and women so that everyone will have institutionalized backing to express his temperament and talents." Mead had been insisting on women's right to choose since 1928, for she had ended *Coming of Age in Samoa* with the following injunction. "Samoa knows but one way of life and teaches it to her children. Will we, who have the knowledge of many ways, leave our children free to choose among them?"[20]

These were controversial arguments in the 1920s and 1930s, but Mead's work became and remained popular, in part because she skillfully maneuvered within a discourse that continued to differentiate the civilized from the primitive on the basis of white women's moral purity and nonwhite people's unrestrained sexuality. From the eighteenth century on, these racialized sexual differences provided the basis on which the Anglo-Americans understood themselves as constituting a superior race and a civilized nation. Mead brilliantly redeployed these constructions of primitive sexuality to prompt her readers to reconsider their own understandings of racial differences and gender relations. She thereby used science to intervene in feminist theory, which had, to this point, accepted the dominant view that civilized societies were more advanced than primitive ones because of white women's sexual purity and moral superiority.

However, Mead's credibility as a "scientific" authority would not have been possible had a renegotiation of the social relations between white women and so-called primitive peoples not already taken place. Evolutionary theorists at the turn of the century, like the anthropologist Otis Mason and sociologist Lester Ward, had begun to reassess woman's role in preserving and passing on civilization and race traits. As Ward wrote in 1888, "Woman is the race and the race can be raised up only as she is raised up." According to Ward, "True science teaches that the elevation of women is the only sure road to the evolution of man." As expressed in this discourse, Anglo-American women became central to the civilizing process, in part because of their purported special attributes as moral guardians and teachers, in part because they held up white civilization as a model of gender relations to be imposed on primitives for emulation, and in part because scientists like Ward imbued them with new biological functions as the transmitters of civilization and racial traits. Anglo-American women had responsibility for transmitting civilization, not just to their own children, but also to other peoples who, in their supposed simplicity and naïveté, were assumed to resemble children. This was a "progressive" view in the 1880s and 1890s because of its assumption that primitives could be civilized.[21]

Despite this view, Anglo-Americans continued to perceive primitives as dangerous and wild, not in control of their sexual feelings and always lusting after white women. In the late nineteenth century, prominent Anglo-American women reshaped this discourse through their work as missionaries, ethnographers, and explorers. Clarissa Chapman Armstrong, the wife of Richard Armstrong, worked as a missionary and educator in Hawaii for several decades. In the early 1880s Armstrong published an account of her experiences in which she stressed her personal vulnerability in the face of primitive sexuality. Alice Fletcher, a special agent for the Department of the Interior in the 1880s, became recognized as an expert ethnographer on Indian cultures in the 1890s and portrayed herself as a maternal protector who could help save American Indians from being overcome by "evolutionary processes." May French-Sheldon, a leader of her own safari to East Africa in 1892, publicized this "adventure" into the 1920s, emphasizing the personal sense of empowerment she felt among Africans. Her adventures helped to propagate a feminist fantasy that white women could live with more dignity and freedom among so-called "savages" than they could by staying at home among the "civilized." These women and others negotiated the tensions that arose from their culture's taboo against miscegenation, eventually overturning cultural beliefs that insisted that white women needed protection from primitives. They helped bring about a new role for white women as protectors of the primitive, a role that Mead later used in building her reputation as a scientist. By the time Mead embarked for Samoa in 1925, the idea that white women had a valid cultural role in protecting and liberating the primitive was firmly etched into her public's consciousness.

Anglo-American women found missionary work appealing because it gave them unprecedented authority to speak publicly and to assume an explicitly political role. Missionary work also terrified Anglo-American women, frightened by primitives' dark skin color and nakedness, which they interpreted as signs of unrestrained and wanton sexuality. This fear, experienced as sexual vulnerability, was exacerbated in the post-Reconstruction era by the highly publicized lynchings of black men, which whites often justified on the grounds that the black male victims had attacked or molested white women. Missionary reports sent back to the United States told stories of white women fending off unwanted sexual advances of "savages," continuing longstanding themes of Indian captivity narratives. Because of the cultural taboo against miscegenation and the imperative to remain sexually chaste, white women denied any feelings of sexual attraction they may have felt toward these men through the projection of an exaggerated and aggressive sexuality onto the primitive.

A vivid illustration of how such repression and denial operated may be ascertained from a close reading of Clarissa Armstrong's "Sketches of Mission Life," which contained the story of her relation to Papatutai, a "savage" her husband designated as her protector when he was away from home. This story was published in the *Southern Workman,* the school newspaper of Hampton Institute, which her son, Samuel Chapman Armstrong, had founded as a vocational training school in Virginia to educate newly freed slaves. Despite the strong feelings of repulsion she purports to feel toward Papatutai, Armstrong nonetheless decides to sketch his portrait so that she can send a picture of him home "for friends to see what sort of neighbors [she] had." For this purpose, she had "Papatutai stand, spear in his hand" as if he "were about to thrust it into a victim"—and at her request or insistence, Papatutai dons a war costume. According to Armstrong, "His appearance as he thus stood . . . was revolting beyond expression."[22] This savagery was quite literally Armstrong's creation. She positioned Papatutai in what she thought to be a suitable and typical pose, with little clothing, a fierce expression, and a weapon about to be launched. Emphasizing his imposing height, his "erectness," Armstrong's selection of details reveals how great an interest she took in Papatutai's physical person, an interest she never could have registered as directly were she describing a white man.

These sketches reveal Armstrong's anxious anticipation of being left alone with Papatutai, an event that finally took place without incident. Although the anticipated rape never occurs, it nonetheless served as the backdrop to the account, adding drama and tension to what otherwise would be an uneventful narrative. A subtext strains against the bounds of Victorian sexual propriety. By imagining that it was Papatutai who desired her, and not the other way around, Armstrong secured her identity as a sexually chaste Christian woman. She believed that her proper ladylike comportment, her Christian faith, and her civilized womanhood provided a shield through which Papatutai's "spear" cannot penetrate.

As an Indian reformer and ethnographer, the unmarried Alice Fletcher found opportunities to experience herself as a powerful political and intellectual leader, helping her to overcome her frustration with the patriarchal aspects of white culture. Fletcher became active in the Indian reform movement after having spent more than a decade in the woman's movement. In the 1880s, Fletcher helped formulate the rationale for severalty or allotment policies, which broke with the federal government's previous policy of denying Indians citizenship on the grounds that they were inherently unfit to be assimilated into American society. Instead, the Dawes Indian Act of 1887, for which Fletcher claimed credit, was based on the premise that savages could be civilized, making U.S. citizenship rights and individual titles to land conditional on the abandonment of tribal customs. The Dawes Act also specified that tribal lands not allotted to individuals, the so-called "surplus" lands, would be opened to white settlement, thus forcing the transference of a large portion of Indian territory to the United States. Having earlier allotted the Omaha lands, Fletcher served as a special agent for the Department of the Interior, administering this legislation among the Winnebago of Nebraska and the Nez Perce in Idaho, where she was accompanied by E. Jane Gay.[23] Under Fletcher's authority, Indians found themselves strongly encouraged, when not forcibly required, to conform to white middle-class gender relations and to establish monogamous patriarchal families.

The seeming paradox between Fletcher's evident desire to free middle-class white women from patriarchal family structures and her active support for the introduction of patriarchal family structures into Indian societies can be explained by her belief in evolution. Convinced that no primitive society could advance to higher levels without going through the intermediary stages, Fletcher thought patriarchal family structures would propel Indian societies upward along the evolutionary hierarchy toward civilization. She felt a sense of urgency about this process, fearing that if Indians were not civilized quickly, they would die out as "evolutionary processes" overtook them. Fletcher argued that Indian cultures should be compelled to adopt the supposedly more advanced forms of civilized gender relations. In practice, this meant monogamous sexual relations, male support of women and children through farming, the domestication of Indian women in Western-style homes, the learning of English and the conversion to Christianity, Western education for children, and the adoption of Western styles of dress and appearance. Fletcher personally assumed the role of the economic and sexual protector of American Indian women, whom she argued were being abused and exploited by Indian men and ignored by white male politicians and reformers.

In the 1890s and early 1900s, Fletcher earnestly strove in her ethnographic work to document the existence of the cultural forms that she had earlier helped to curtail. In part, Fletcher could safely shift her views because of a growing sense of security among white elites that Indians could no longer offer violent resistance to the U.S. government. These ethnographic writings coincided with Fletcher's increasingly intimate relationship with Francis La Flesche, a member of the Omaha tribe, with whom she eventually shared her home and life as her "adopted son." Perhaps this relationship showed Fletcher another side to assimilationist policies, which she now acknowledged as helping to bring about the disintegration of traditional Indian familial structure and a lowering of the status of Indian women. Her later ethnographic work may have been motivated in part by a desire to atone for her previous governmental activities, but it also represented a new way to command authority at a time when assimilation was becoming less viable. Although saddened by the consequences she had helped bring about, she never doubted the necessity of assimilationist "solutions." Until her death in 1923, Fletcher believed that patriarchy was a crucial and inevitable step in the evolutionary advancement of all societies.

Born a decade later than Fletcher, May French shared a similar worldview but found a different way to make the primitive serve her feminist purposes. She was born in Beaver, Pennsylvania, to a wealthy and prominent family whose wealth came from sugar, cotton, and tobacco plantations. In the 1860s the family took an extended trip to Europe to round out the education of the two girls and also perhaps to lessen the traumatic impact of the Civil War. By the mid 1880s, May French-Sheldon lived in London with her new husband, Eli Lemon Sheldon, a prominent banker and publisher. In January 1891 the forty-three-year-old French-Sheldon began to plan an expedition to East Africa, writing to the explorer Henry Morton Stanley for advice. Despite having secured letters of introduction from Stanley, French-Sheldon met with a great deal of resistance from British colonial authorities. Financed by her husband, she eventually set forth on a historic expedition that lasted three months, from March to May 1891. It would be the first of her journeys to Africa. After her husband's death in 1892, French-Sheldon would live with a female companion who also served for many years as her personal assistant.[24]

Although more than one hundred African men accompanied May French-Sheldon on her safari, most press accounts repeatedly stressed that she traveled "alone," meaning without white male escorts. This extraordinary departure from convention initially subjected her to public ridicule and suspicion in regard to her motives for undertaking such an audacious act. Commentators insinuated that she was naive for ignoring the sexual risks of such a venture, or worse, secretly desirous of illicit contact with primitives. Perhaps to counter this stigma, French-Sheldon wrote a series of descriptions of her trip, calling attention to her skill at disciplining African men. She constructed herself as the leader and protector of her porters, although on several occasions she conceded that the porters had saved her life. She also stressed the many and ingenious stratagems she devised to elude the sexual advances of the sultans who "courted" her. Her most popular lectures from the 1920s, "Thrilling Experiences in Savage Africa," "Camp Life with Natives in the Jungle," and "Thrilling Adventures of a Lone White Woman in Savage Africa," emphasized these same themes.

Obituaries following her death in 1936 marked the significance of her traveling "alone" among "head-hunters and cannibals." American newspapers represented French-Sheldon as an emancipated woman who also worked for the emancipation of African natives. The *New York Times* called her "a pioneer among women explorers in Africa" and "one of the few of either sex who in her generation returned with kind words for the natives." "For many years she argued the cause of the blacks and [the] decrease in the cruelty with which natives were handled was in some part attributed to her championship." Despite veiled references to the danger represented by the primitive, the hint of miscegenation, and the subtextual rendering of the primitive as rapist, these obituaries suggested that encounters with the primitive could symbolize a realm of freedom and power, personal independence, and control over others that was not available to women in Britain or the United States. By the 1920s May French-Sheldon would be heralded as "one of the most remarkable of women," a model for feminists and other unconventional women of the 1920s, women like Margaret Mead.[25]

Mead clearly understood herself as representing a break from all three of these nineteenth-century traditions. By the time of her writings in the 1920s and 1930s, the virtual abolition of what had once been deemed primitive in American Indian cultures made possible a nostalgia for primitivism that was not possible when Indian submission was still in question. Although her upbringing was steeped in Protestant evangelism, Mead was not a missionary and did not want to convert others to Christianity. Nor did she conceive of her own scientific expeditions as a way to demonstrate white women's independence and courage to a skeptical world, although the mainstream and feminist press reported on her achievements under headlines that recalled those that publicized French-Sheldon's exploits, with headlines like "'Going Native' for Science" and leads that began: "Here's the only white woman to live alone among cannibals." Indeed, the *New York Times* would place her among the group of "women explorers" who could "more than hold her own," suggesting that Mead shared some aspects of French-Sheldon's persona.[26]

Determined to establish herself as a professional scientist, Mead went to extreme lengths to differentiate her own scientific practices from those of amateur ethnographers like Alice Fletcher. Opposed to what she called "culture-wrecking," Mead characterized the earlier ethnography as poor science, distorted by assimilationist goals. According to Mead, the "pure" anthropologist "does not want to improve them, convert them, govern them, trade with them, recruit them or heal them."[27] Constructing herself as an objective scientist carefully trained in modern scientific methods, Mead presented her own practice as politically detached, morally neutral, theoretically valid, and empirically sound. Despite her disavowals, however, Mead depended on nineteenth-century traditions to maintain her status as a scientific expert in the eyes of her Western audience. In part, anthropology attracted Mead because it empowered her to act as a cultural mediator between the civilized and the primitive. Rather than Fletcher's goal of assimilation, she wanted to protect primitive cultures from "contamination" by the West.

Despite Mead's avowals that the pure scientist must remain uninvolved, as a citizen sensitive to the injustices of Western imperialism, she often found a neutral stance impossible. Like Fletcher, she took advantage of imperial power relations when it served her purposes to do so. In 1932 she explained how her second husband, Reo Fortune, coerced unwilling men to help them carry their belongings by "unearth[ing] their darkest secrets, which they wished kept from the government, and then order[ing] them to come and carry." Audiences in Manus and Port Moresby, Papua New Guinea, shouted Mead down in 1953 because she refused to acknowledge that she had made her fortune by "telling their stories around the world." Her biographer, Jane Howard, defended Mead by observing that she would have been the first to encourage New Guineans to write down their own stories. Indeed, during the 1940s, Mead had urged Western anthropologists "to study with members of other cultures," cautioning them that it was "imperative to phrase every statement about a culture so that those statements are acceptable to the members of the culture itself."[28]

Like Armstrong and French-Sheldon before her, Mead played to Western fears that a single white woman among primitives was always at some physical risk. This aspect of Mead's self-presentation can be most clearly discerned in her radio conversations with James Baldwin, transcribed and published in 1971 under the title *A Rap on Race.* In these exchanges, Mead recollected an incident from her fieldwork in New Guinea in which she felt she had to retrieve a book of matches that strange men from another village had stolen from her. Recalling that she was all "alone in a village where there wasn't a single white person within two days' walk," Mead recounted: "I had to get that box of matches back. If I didn't, I would have been as good as dead. White people who let a thief go used to be killed; they had shown themselves as weak." Thus she "stormed up to the end of the village" in a "fine exercise of sheer white supremacy, nothing else," and demanded the matches be returned. She received them. "Then we were all safe." She added, "If I had made one misstep I'd have been dead, and then the administration would have sent in a punitive expedition and they would have been dead." Assuming responsibility for everyone's safety, Mead took on the Anglo-American female role of protector that Fletcher had also assumed, claiming power over the New Guinea men whose fate she believed lay in her hands—not her fate in theirs.[29]

Mead's sense of having greater power, knowledge, and skill than the primitive within the primitive's own world linked Mead to French-Sheldon and Fletcher. Attributing agency and authority to herself, she simultaneously denied the primitive corresponding agency, knowledge, and power. It did not occur to Mead that the village men might have understood the risk of retribution by the colonial government. She immediately added in her conversation with Baldwin that she imagined that this was what it must have been like for white women in the antebellum American South. "This is the burden, in a sense, that in this country the black man and the white woman carried in plantation days. If a white woman made a mistake, or didn't remember who she was every single second, everyone would suffer."[30] No doubt made more aware of the complexities of race in the political turmoil of the 1960s, Mead may have understood her complicity in maintaining the racist boundaries required by the strictures of white supremacy in a way that clearly Armstrong, French-Sheldon, and Fletcher did not.

Whatever Mead's understanding of racism, her work owed its cultural power to the persistence of nineteenth-century constructions that continue to be of great relevance to contemporary feminists. For example, Mead vested responsibility in women as mothers to abolish racial discrimination and oppression. Downplaying economic and social structures that perpetuated racial oppression and conflating all forms of racism with individual prejudice, Mead argued that prejudice served primarily as an "educational device" that the "average mother [uses] to bring up her children." Believing that all forms of racial oppression could be overcome through an alteration in child-rearing practices, Mead recommended that mothers not make negative references to other groups as they raised their children. Mead insisted that the "big differences that exist in any society" were "all originally worked out in the home," despite acknowledging that some of her audience might think that she was avoiding the "major issues."[31]

Mead's claim of being a neutral observer marked a point of considerable instability on which she seesawed throughout her career. The objectivity she claimed as a "pure anthropologist" would sometimes disappear in her critiques of patriarchal relations in Western societies. Mead's understanding of the relationship between primitive and Western societies was equally problematic. As she wrote in "Human Differences and World Order," "Practices that are repugnant to our ethical system, often [are] also to the natives who practice them. . . . It is very interesting to see the way these practices which are most repugnant to humans disappear quickly when primitive people are given a chance at something else."[32] Mead wanted to protect primitive societies from the contaminating influences of Western colonialism and modernization, yet she also expected primitive societies to alter practices that fell short of the standards of Western feminism. She did not see any contradiction in this perspective.

Much recent debate surrounding Mead's work centers on the question of whether she produced accurate ethnographies.[33] The debate over Mead's competency as a scientist, as well as the relative neglect that Mead has suffered at the hands of feminist historians, has foreclosed other types of assessments. In *Coming of Age in Samoa* and *Sex and Temperament in Three Primitive Societies,* Mead contributed important ideas about the "unnaturalness" of sex differences in the United States and argued—before the conceptual language was available—for the cultural construction of gender and sexuality. Nonetheless, Mead's legacy also clearly demonstrates the pitfalls of an unexamined liberal framework, one that naively assesses women's freedom by the seeming number of choices they have.

Although feminist anthropologists today may acknowledge anthropology's implication in the history of Western imperialism in ways that Mead could not, we should not be too quick to deny our historical connection to the liberal, feminist, and ethnographic traditions that Mead so centrally embodied.[34] Mead grappled with critical theoretical and methodological questions that are of ongoing concern. If the objective stance of the modern "pure anthropologist" is no longer available, what stance may the postmodern anthropologist assume? This critical and unresolved question is all the more pressing because the problems of gender and racial oppression that confronted Mead continue. Mead's reflections on these problems, flawed as they were, continue to shape scholarly and popular feminist discourse to this day.

5 "A World Made Safe for Differences"

Ruth Benedict's *The Chrysanthemum and the Sword*

Christopher Shannon

> We are willing to help people who believe the way we do, to continue to live the way they want to live.
>
> Dean Acheson

Liberals have always prided themselves on their acceptance of diversity. At no time was this value more central to American liberalism than during the Cold War era.[1] The racial and totalitarian ideologies that seemed to precipitate World War II brought forth a postwar vision of what Ruth Benedict called "a world made safe for differences," in which cultural freedom would serve as the basis for world peace.[2] Throughout the Cold War era, liberal policymakers, no less than their critics, decried the gap between the theory and the practice of tolerance, but few critically analyzed the idea of cultural tolerance itself. Historical treatments of tolerance have suffered from the same deficiency.

In this essay, I would like to shift the historical debate over tolerance from the social to the intellectual level through a close reading of one of the earliest articulations of the postwar doctrine of tolerance, Ruth Benedict's *The Chrysanthemum and the Sword.* A study of Japanese culture intended to bring intercultural understanding to bear on the reconstruction of Japan, *The Chrysanthemum and the Sword* offers a critique of conventional cultural imperialism that fosters a subtler imperialism of "culture" as a general social-scientific mode of perceiving all particular cultures. Benedict's book demands that the Japanese become, not Americans, but in effect, anthropologists; it demands that the Japanese learn to view their culture with a certain scientific detachment and see their received values as relative and therefore open to revision in the service of consciously chosen ends. Ultimately, the imperial vision of Benedict's "world made safe for differences" lies not in any covert imposition of American values on the Japanese but in the overt and uncompromising call for the subordination of all cultures to the demands of individual choice.

A critical re-examination of Benedict's conception of cultural tolerance appears particularly pressing in light of the current vogue of multiculturalism in left/liberal scholarly discourse. The rise of multiculturalism has brought a renewed interest in Benedict's work as well as a rehabilitation of the broader Cold War rhetoric of "pluralism" and "tolerance" so deeply indebted to that work.[3] The dominance of multiculturalism has forced contemporary social thought back to the stale liberal dichotomy of tolerance/intolerance—a dichotomy established as central to American democracy in Benedict's work, yet one that has proven itself incapable of allowing for relations among people that transcend individual choice. The persistent equation of tolerance with some self-evidently neutral conception of "freedom" has served merely to naturalize the social relations of modern Western individualism. It is this fundamental social value commitment that links Cold War intellectuals to contemporary multiculturalists.

The unreflective bias in favor of tolerance has prevented recent scholarship on Benedict from critically examining the meaning of tolerance in her work. Scholars in our multicultural times write as if the events of the last forty years had not cast some suspicion on the virtues of the tolerant or "open" society. One need not accept Herbert Marcuse's "repressive tolerance" thesis entirely to concede the possible coexistence of domination and tolerance; the new left revolt against the liberal welfare/warfare state directed itself not only to the overt military domination that culminated in the Vietnam War but also to a subtler cultural domination that, in Marcuse's words, "encourages non-conformity and letting-go in ways which leave the real engines of repression in the society entirely intact." Marcuse and other new left critics place much of the responsibility for the "tolerant" side of this repressive regime on social therapists who preached self-fulfillment via "adjustment" of the individual to dominant cultural norms. This critique presented social engineering as the representative mode of domination in "a society of total administration" that pacifies political dissent through a therapeutic regimen of bureaucratic social control.[4] New left politics shaped a substantial body of historical writing that saw in the development of large-scale bureaucratic institutions, not the triumph of liberal reason, but the rise of a sinister "corporate liberalism" that extended capitalist domination from economic life to cultural life in general.[5]

The work of Richard Handler and Virginia Yans-McLaughlin places Benedict firmly in the context of the rise of social engineering but fails to deal adequately with the issues of power raised by new left scholarship. Handler credits Benedict's colleague Margaret Mead with realizing that "social engineering was but a step from . . . fascistic social control" and praises her distinction between a "beneficial social science" and "an antidemocratic and manipulative one"; he then judges Benedict's and Mead's anthropological social engineering as "beneficial" by virtue of its commitment to a "society that would make use of diverse human abilities without branding any as deviant." Yans-McLaughlin addresses this problem as well and likewise places Benedict in the "beneficial" camp by virtue of her insistence on the supreme worth and moral responsibility of the individual.[6]

Handler and Yans-McLaughlin recognize that social engineering raises certain questions of power, yet they tend to see power in terms of simple coercion or intolerance. This failure to address the problem of power at a more complex level stems in part from the biographical orientation of their work; both Handler and Yans-McLaughlin seek primarily to understand Benedict's place in the established context of the debate on social engineering rather than to unpack such provocative phrases as "socially engineered tolerance." Handler ultimately sees social engineering as having provided an opportunity for Benedict to discover "her convincing sense of selfhood . . . in the role of the technical expert, the scientific creator who puts her individual talents at the service of the collectivity." Similarly, Yans-McLaughlin sees in Benedict's propaganda work for the Office of War Information a "pragmatic choice" of "patriotism over passivity," a "passionate effort to maintain democracy as the only way of life that made scientific inquiry possible."[7] This scholarship provides a sympathetic account of Benedict as an intellectual struggling with the great issues of her time but fails to deal adequately with the cultural significance of social engineering.

The weakness in Handler's and Yans-McLaughlin's work lies, not in the failure to recognize the possible political abuses of anthropology, but in the failure to address the politics of anthropology as a structure of thought apart from its use or abuse by institutions of power. In playing these scholars off Marcuse, I do not mean to suggest that we return to the critical language of repressive tolerance. In retrospect, that critique seems as naïve as the official tolerance it directed itself against. Even a cursory reading of "Repressive Tolerance" reveals Marcuse to be deeply indebted to the classical liberal definition of freedom as autonomy. He makes bold assertions about the possibility of distinguishing true from false tolerance that today seem about as convincing as Mead's insistence on the distinction between beneficial and manipulative social engineering.[8] Repression/liberation, like intolerance/tolerance, assumes a single, universal standard of freedom as self-determination and autonomy from nonconsensual social relations. I do not doubt that Benedict valued this freedom and sought to extend it to as many people as possible; however, as an anthropologist, she worked primarily on non-Western cultures not organized around this understanding of freedom. For all of its "tolerance" of diversity, Benedict's anthropology set the acceptance of freedom-as-autonomy as the price of inclusion in a world made safe for differences and excluded all who would not submit to this single standard. Through a close reading of *The Chrysanthemum and the Sword,* I hope to examine the social relations of Benedict's anthropology in such a way as to repoliticize words such as *tolerance* and *difference.* I will argue that we understand these words best, not as "repressive" in a psychological understanding of power, but as more broadly coercive as representative of a particular social/intellectual structure.

The ideology of tolerance that informs *The Chrysanthemum and the Sword* grew out of the understanding of cultural relativism that rose to prominence in anthropological circles during the first half of the twentieth century. As formulated by Franz Boas, Benedict's teacher at Columbia and the leading anthropologist of the early twentieth century, cultural relativism set itself against the reigning orthodoxies of evolutionary anthropology and scientific racism. Boas rejected the evolutionary framework of Victorian anthropology, which viewed the whole "way of life" of human culture as a single, developing entity to be divided into various stages, ranging from the savage to the civilized. He argued that culture must be seen as a plurality of integrated wholes, each organized according to a distinct and unique pattern of values. Boas argued for the relativity of cultures in relation to one another, but he nonetheless insisted on the absolutism of "culture" as a way of understanding social organization. In this, he explicitly rejected racial explanations for the differences among peoples. Culture, not race, determined behavior; and culture was not innate or biological but was a pattern of values learned in daily life. The learned quality of culture gave it a flexibility that notions of race seemed to lack, and Boas hoped that an awareness of the diversity and malleability of cultures would inspire a general tolerance for cultural differences.[9]

A student of Boas, Ruth Benedict embraced not only his anthropological theories but also his commitment to liberal reform. No single work did more to popularize Boas's humanist agenda than Benedict's *Patterns of Culture,* published in 1934. Benedict's book presented an account of three different "primitive" cultures as a way of reflecting on the relativity of all cultures, particularly American culture. Benedict insisted that values differ not only among cultures but also within cultures over time. In light of the Great Depression, she urged Americans to reconsider the permanence of some of their own cultural values, such as the association of economic competition and limited government with moral virtue. In support of the New Deal, Benedict argued that traditional American values of liberty and equality could be preserved only if Americans gave up their equally traditional hostility to cooperative planning and government regulation of the economy.

In the late 1930s the national focus shifted from economic depression to war. As American involvement in World War II appeared inevitable, Benedict attempted to translate her understanding of culture into practical programs that would aid the war effort. The New Deal flirtation with "planning" had created a hospitable environment for experts and academics in government, and Benedict was one of the hundreds of social scientists who offered their services to Washington for the coming war. Along with Mead, Benedict formed the Committee for National Morale in 1939 to examine ways to apply psychology and anthropology to the problem of building morale during the war. In 1941 she joined the war effort in a more official capacity as a member of the Committee on Food Habits, a joint venture of the Department of Agriculture and the National Research Council designed to improve the food habits and nutrition of an American population mobilizing for total war.[10] War, like culture, came to be seen as a whole way of life.

For all of her attention to American culture, Benedict's most significant wartime work would come in the more conventionally anthropological role of interpreting non-Western cultures for Westerners. In June 1943, Benedict replaced her friend (and Japan expert) Geoffrey Gorer as head analyst at the Overseas Intelligence division of the Office of War Information (OWI). Benedict's job was to prepare cultural profiles of various countries as requested by either an operational division of the OWI or the Bureau of Overseas Intelligence. Restricted from conventional fieldwork by the war, Benedict prepared her reports by examining available studies of the country and interviewing first- and second-generation immigrants living in America. Through 1943 and early 1944, Benedict prepared brief reports on Thailand, Rumania, and the Netherlands; each report contained suggestions for "psychological warfare" based on the pattern of culture particular to each country.

In June 1944, the psychologist Alexander Leighton, a friend of Benedict's from New York and head of the Foreign Morale Division of the OWI, assigned Benedict to conduct a study of Japan, despite her lack of any particular expertise in Japanese language or culture. After the war, Benedict revised and expanded her OWI report and published it as *The Chrysanthemum and the Sword* (1946).[11] A kind of cultural guidebook for American officials overseeing the reconstruction of Japan, Benedict's book argues for a greater understanding and tolerance of all cultures as the key to preserving world peace.

Without casting suspicion on the sincerity of Benedict's plea for tolerance, the circumstances of the writing of *Chrysanthemum and the Sword* do bring into special relief issues of power that have plagued anthropology since its birth as a distinct intellectual discipline. Benedict's contribution to the psychological warfare efforts of the OWI recalls nothing if not the role of anthropology in helping European colonial administrators subdue indigenous peoples during the nineteenth century. Proponents of a "humanistic science" like Benedict and Mead were well aware of anthropology's imperial past; however, they believed that a self-critical anthropology directed toward the true "dignity of man" could serve as the basis for a truly democratic form of social engineering that would foster mutual respect among all the nations of the world.[12] This ideal of a self-critical anthropology decentered the West in relation to the East only to center the detached perspective of anthropology itself in relation to both West and East. Within Benedict's Boasian framework of cultural relativism, the old anthropological opposition of civilization to savagery gives way to a more abstract dichotomy of the anthropological and the non-anthropological, with anthropology as a neutral, placeless principle of rationality distinct from all particular cultures.

In Benedict's work, anthropological detachment, like "civilization" before it, serves as a universal standard by which to judge all the peoples of the world. Consequently, Benedict located her savage "other" in non-anthropological attitudes toward culture. In *The Chrysanthemum and the Sword,* Benedict identified many of the traits conventionally associated with the savage—the lack of self-consciousness, the neurosis, the immaturity—with the modern West itself, particularly with America. Lacking a proper understanding of the anthropological notion of culture, Americans "still have the vaguest and most biased notions, not only of what makes Japan a nation of Japanese, but of what makes the United States a nation of Americans." Such ignorance fosters fear and insecurity. Americans become "so defensive about their own way of life that it appears to them to be by definition the sole solution in the world," and they "demand that other nations adopt their own particular solutions." This "neurotic" attempt to repress others leads Americans to repress themselves. By demanding cultural uniformity, Americans "cut themselves off from a pleasant and enriching experience" and deny themselves "the added love of their own culture which comes from a knowledge of other ways of life."[13] This childish, neurotic fear of difference threatened to leave America culturally isolated and stagnant, as America's inability to appreciate other cultures ultimately issued in a self-destructive inability to appreciate its own.

Even as Benedict rejected the single standard of American culture, she constructed a single standard of a certain attitude toward all cultures:

> It sometimes seems as if the tender-minded could not base a doctrine of goodwill upon anything less than a world of peoples each of which is a print from the same negative. But to demand such uniformity as a condition of respecting another nation is as neurotic as to demand it of one's wife or one's children. The tough-minded are content that differences should exist. They respect differences. Their goal is a world made safe for differences, where the United States may be American to the hilt without threatening the peace of the world, and France may be France, and Japan may be Japan on the same conditions. To forbid the ripening of any of these attitudes toward life by outside interference seems wanton to any student who is not himself convinced that differences need be a Damocles' sword hanging over the world. Nor need he fear that by taking such a position he is helping to freeze the world into the status quo. Encouraging cultural differences would not mean a static world.[14]

Benedict's concept of "ripening" implied not only a temporal narrative of maturation but also a spatial narrative of detachment. The peace of the world depended on replacing the neurotic "insider" perspective of nationalism with the rational "outsider" perspective of anthropology. According to Benedict, all nations, including America, existed within particular cultures; they took their cultural values "for granted" as if those values were "the god-given arrangement of the landscape," and thus they cannot properly evaluate either their own culture or another nation's culture. In contrast, the discipline of anthropology operates outside of all particular cultures and thus may examine them objectively.

Benedict conceptualized this neutral, scientific detachment through metaphors of vision. Culture is the "lenses through which any nation looks at life." As the "oculist" of culture, the anthropologist is "able to write out the formula for any lenses we bring him." Benedict even predicted that "someday, no doubt we shall recognize that it is the job of the social scientist to do this for the nations of the contemporary world."[15] Benedict clearly offered *The Chrysanthemum and the Sword* as a model for this kind of anthropological mediation. Her choice of metaphor is more appropriate than she may have intended. Oculists not only describe formulas, they also *prescribe* them. In writing prescriptions, an oculist analyzes the deviations of individual vision so as to correct that vision. The understanding of and sympathy for the variety of deviations in vision in no way prevents the oculist from correcting vision with proper prescriptions. Benedict followed something of the same procedure in *The Chrysanthemum and the Sword*. The description of difference serves as prelude to the prescription of uniformity.

To be fair, most of the book remained true to the ideal of detached description that Benedict wished to evoke with her metaphor of the oculist. Roughly half of the chapters devote themselves simply to mapping out some particular aspect of "the great network of mutual indebtedness" that Benedict presented as the pattern of Japanese culture. Separate chapters dealt with such topics as the debt owed to parents, the debt owed to one's own reputation, and the conflicts between equally valid obligations. The intricacy of the networks Benedict describes threatened to play into contemporary stereotypes concerning the rigidity of Japanese culture, yet Benedict insisted throughout that submission to these obligations stems from the cultural value of common loyalty, not arbitrary authority. Benedict saw in the obligation network of Japanese culture a flexibility absent from both the authoritarianism of America's other major wartime enemy, Nazi Germany, and the caste hierarchies of other non-Western societies, such as that of India. To counter the image of the Japanese as machinelike automatons, Benedict included a chapter entitled "The Circle of Human Feelings," informing her readers that "the Japanese do not condemn self-gratification. They are not Puritans." She concluded her study with an account of the intimate relations between parents and children as seen through Japanese child-rearing practices.[16]

Still, the concern to refute stereotypes did not lead Benedict to gloss over the substantive differences between Japan and America. She began her account of Japanese culture proper with a chapter entitled "The Japanese in the War," and her initial analysis reveals a culture completely at odds with American values. According to Benedict, the "very premises which Japan used to justify her war were the opposite of America's." Japan fought because it saw "anarchy in the world as long as every nation had absolute sovereignty" and felt "it was necessary for her to fight to establish a hierarchy—under Japan, of course"; however, America fought because Japan "had sinned against an international code of 'live and let live' or at least of 'open doors' for free enterprise." From this initial opposition of hierarchy to equality flows a series of sharp contrasts. The Japanese rely on "a way of life that is planned and charted beforehand," while Americans expect "a constantly challenging world." Unlike Americans, the Japanese value fixed social stations over social mobility, the spiritual over the material, honor over profit, and the external sanctions of shame over the internal sanctions of guilt.[17] Through these oppositions, Benedict creates a picture of the Japanese as a people bound together by a complicated network of ordered, hierarchical, personal obligations that contrasts sharply with the easygoing individualism of American culture.

This contrast between America and Japan is as predictable as it is extreme. Oppositions such as hierarchy versus equality and shame versus guilt have structured America's conception of itself in relation not only to non-Western cultures but to Europe as well.[18] The differences that Benedict observed were variations on the differences between the Old World and the New offered by commentators at least as far back as Crèvecoeur. Before these oppositions served as a cultural narrative contrasting East and West, they served as a historical narrative of the passage from the medieval to the modern. Indeed, much of the hostile criticism of Japan on the part of Western observers can be seen as an extension of the self-critique that drove the modern West to throw off its own primitive, childish, neurotic feudal past. Commentators ranging from the filmmaker Frank Capra to General Douglas MacArthur spoke of the negative aspects of Japanese culture as residues of "feudalistic forms of oppression," and the anthropologist Geoffrey Gorer specifically compared Japanese attachment to the emperor with the attachment of medieval Catholics to the pope.[19] In general, the degree of hostility expressed in wartime accounts of Japanese culture depended on the relative weight given to its premodern dimensions, as opposed to its more palatable modern aspects, such as its enthusiastic embrace of Western science and technology. This hostility often slid into racism when premodern traits came to be seen as rooted in the biological makeup of the Japanese people, but the demonizing of these traits themselves predates and transcends racial categories.

Benedict avoided the racist excesses of other American commentators by taking a historical approach to the seeming inconsistencies of Japanese culture. Through her use of history, Benedict replaces the demonized "other" with a relatively domesticated "self." That is, Benedict presented the Japanese as essentially historical beings who share with Americans a basic, fundamental human character trait: the ability to adapt to change. In her sympathetic account of Japanese modernization under the Meiji Reform of the late nineteenth century, Benedict interpreted Japan's acceptance of modern science and technology, not as a capitulation to Western values, but as an adaptation of enduring Japanese cultural patterns to changing circumstances. She praised the "energetic and resourceful statesmen who ran the Meiji government" for rejecting "all ideas of ending hierarchy in Japan." Meiji reformers, she said, "did not unseat hierarchical habits" but simply "gave them a new locus" in economic, scientific, and technological development.[20] Benedict's account ultimately endorsed neither hierarchy nor technology so much as the general cultural process of hybridizing often contradictory values. Japan's participation in this process served as evidence of its basic humanity against its racist detractors.

In Benedict's account, even Japan's flaws appeared as common human flaws. Despite the admirable cultural "ripening" of the Meiji period, "Japan's nemesis came when she tried to export her formula." In launching a campaign of aggressive foreign expansion, Japan "did not recognize that the system of Japanese morality which had fitted them to 'accept their proper station' was something they could not count on elsewhere" (96). Thus, the Japanese have demonstrated both the common human capacity to adapt to change within their own culture and an all-too-common human unwillingness to accept the differences among cultures. The Japanese made the mistake of committing themselves to specific values rather than to the general process of organizing values into specific patterns, and this is the very mistake that Americans were in danger of making as they assumed a leadership role in world affairs. Thus, both in virtue and in vice, Japan served as a mirror for America: a model of what Americans were as cultural beings and an example of what Americans might become if they refused to accept their status as cultural beings.

Held to the same standard of cultural consciousness, Japan and America nonetheless stand in different proximity to this standard. America's potential intolerance on the international scene contrasted with the easygoing, live-and-let-live attitude within America itself. America's cultural ripening simply required Americans to extend their tolerance of individual differences within America to the individual nations of the world. Japan's proven intolerance, however, stemmed from its cultural values and its inability to accept individual differences within its own culture. According to Benedict, the reverence for hierarchical social ties that sustained the Japanese through the transformations of the Meiji period carried within itself the seeds of imperial aggression:

> Social pressures in Japan, no matter how voluntarily embraced, ask too much of the individual. They require him to conceal his emotions, to give up his desires, and to stand as the exposed representative of a family, an organization, or a nation. The Japanese have shown that they can take all the self-discipline such a course requires. But the weight upon them is extremely heavy. They have to repress too much for their own good. Fearing to venture upon a life which is less costly to their psyches, they have been led by militarists upon a course where the costs pile up interminably. Having paid so high a price they become self-righteous and have been contemptuous of people with a less demanding ethic. (315)

Benedict's analysis of Japanese intolerance paralleled the psychological profile of American intolerance with which she opened her study. For both Japan and America, intolerance stems from a repression rooted in group allegiances and the fear of new experiences; however, while Benedict traced America's intolerance to a kind of free-floating cultural chauvinism, she traced Japan's to a concrete cultural pattern geared toward repressing the individual. Japanese and Americans will have to give up their chauvinistic ties to their respective nations, but the Japanese will also have to give up the irrational ties to "family" and "organization" that have proven incompatible with world peace.

The achievement of an anthropological detachment from one's culture thus ushers in an affirmation of the very American value of individual freedom. For Benedict, the U.S. victory in World War II will be complete only if the United States can ensure this individual liberation for the people of Japan. As Commodore Perry's gunboats opened Japan to modern science and technology, so the administrators of U.S. postwar occupation must open Japanese culture to the modern Western value of individual freedom. The "new goals" that Benedict suggested for this cultural reorientation include "accepting the authority of elected persons and ignoring 'proper station' as it is set up in their hierarchical system" and adopting "the free and easy human contacts to which we are accustomed in the United States, the imperative demand to be independent, the passion each individual has to choose this own mate, his own job, the house he will live in and the obligations he will assume." That a middle-class U.S. intellectual takes these goals for granted should come as no surprise, but it is not clear what would be left distinct to Japanese culture after the achievement of these "new goals." The liberation of the individual—and thus the peace of the world—would seem to require the full-scale acceptance of modern U.S. social relations on the part of the Japanese.

With this vision of cultural ripening, the trajectory of Benedict's encounter with difference becomes clear. For Benedict, Japan's values were incommensurable with America's, yet the Japanese shared in the common human practice of organizing values into coherent patterns. Respect for the Japanese must be rooted, not in respect for their values per se, but in respect for their ability to organize and reorganize values into coherent patterns. Having abstracted a general cultural process from particular values, Benedict figured cultural freedom not so much as the freedom of different values to flourish but as the freedom of different peoples to exercise active and creative roles in the shaping and reshaping of their culture. Thus Benedict insisted that "the United States cannot . . . create by fiat a free, democratic Japan." Americans may set the goal of democratic freedom, but the Japanese "cannot be legislated into accepting" that goal (314).

Against those who thought the Japanese incapable of becoming democratic, Benedict pointed to encouraging "changes in this direction" that the Japanese themselves had made since the war. "Their public men have said since VJ-Day that Japan must encourage its men and women to live their own lives and to trust their own consciences. They do not say so, of course, but any Japanese understands that they are questioning the role of 'shame' (haji) in Japan, and that they hope for a new growth of freedom among their countrymen" (315). For Benedict, the Japanese had proven themselves capable of taking their culture into their own hands and of uplifting themselves to the internal authority of conscience required of a responsible democratic citizenry. The Japanese had earned a certain autonomy from the United States by exercising a certain autonomy from themselves—that is, from their culture. Outside interference by the United States would only inhibit the growth of this democratic autonomy.

The very sincerity of Benedict's demand for Japanese self-determination suggests the need for a fundamental rethinking of the power relations of anthropology as a mode of intercultural mediation. The imperial vision of *The Chrysanthemum and the Sword* lay not in making Japan a cultural colony but in making Japan a cultural equal. Benedict asked of Japan simply what she asked of the United States: that it open itself to the "pleasant and enriching experience" of another culture (in this case, the United States itself) and that it come to "know the added love of [its] own culture which comes from a knowledge of other ways of life" (16). For Benedict, cultural autonomy requires a certain detachment from culture. Japan and the United States must both learn to detach themselves enough from their own cultures to appreciate different cultures, yet they must somehow still remain engaged enough with their own cultures to appreciate them as well.

Benedict herself served as a model of this detached-yet-engaged autonomy for Americans. True to her commitment to self-determination, she offered indigenous Japanese models for such autonomy. Benedict's anthropological account of Japan found its Japanese equivalent in the autobiographical accounts of two Japanese women confronting American culture. Like Benedict's anthropology, these autobiographies figure the contrast between Japan and America as a conflict between tradition and modernity. In *My Narrow Isle*, Sumie Seo Mishima tells of the conflict between her desire to study in America and what Benedict describes as "her conservative family's unwillingness to accept the *on* [the incurred obligation] of an American fellowship" (225–26). Mishima ends up going to Wellesley despite her parents' objections, yet she feels out of place.

The teachers and the girls, she says, were wonderfully kind, but that made it all the more difficult, she felt. "My pride in perfect manneredness, a universal characteristic of the Japanese, was bitterly wounded. I was angry at myself for not knowing how to behave properly here and also at the surroundings which seemed to mock my past training. Except for this vague but deep-rooted feeling of anger there was no emotion left in me." She felt herself "a being fallen from some other planet with senses and feelings that have no use in this environment where I was completely blind, socially speaking." It was two or three years before she relaxed and began to accept the kindness offered her. Americans, she decided, lived with what she calls "refined familiarity." But "familiarity had been killed in me as sauciness when I was three" (226). A microcosm of Benedict's argument as a whole, Mishima's story begins with the contrast between tradition and modernity, only to end with the passage from tradition to modernity. On the personal as well as the geopolitical level, peace depends on the acceptance of the easy-going relations of American individualism.

Mishima functioned as a synecdoche for Japan. Benedict placed Mishima's story at the end of a chapter on Japanese social obligations entitled "The Dilemma of Virtue" and insisted that it embodies "in its most acute form the Japanese dilemma of virtue." Such a reading of Mishima's story, however, works against Benedict's own interpretation of Japanese culture. Throughout the chapter, Benedict described the Japanese dilemma of virtue in terms of the conflict between competing and equally valid obligations within a single culture, while she presented Mishima's experiencing a qualitatively different conflict between a way of life in which formalities and obligations are important and one in which they are not. In making Mishima a representative figure, Benedict translated this conflict between cultures into an inevitable historical progression within Japanese culture: "Once Japanese have accepted, to however small a degree, the less codified rules that govern behavior in the United States they find it difficult to imagine their being able to experience again the restrictions of their old life in Japan" (227). Benedict conceded that some Japanese experience this change as loss; however, as gain or loss, she failed to show how this experience offers any "added love" for Japanese culture itself. A culture would already have to place a premium on new and enriching experiences in order to benefit from an encounter with American culture, and clearly Japanese culture does not. By the single (and very American) standard of individual autonomy, Japanese culture came off as something second-rate.

Obviously Mishima's experience is not representative in any empirical sense. Most Japanese do not go to Wellesley and do not write autobiographies. What Mishima does represent is the experience of marginality with respect to one's own culture that Benedict deems necessary to living in a world made safe for differences. In such a world, marginality must be central; it must be the representative experience of individuals within cultures. These individuals may not actually become anthropologists or write autobiographies, but ideally they would achieve a certain kind of anthropological and autobiographical consciousness, an ability to detach themselves enough from their own culture to examine their relation to that culture in some kind of objective manner. Within this universal form of marginality, difference lies in the specific content of marginality, the variety of loci for marginality. The individuals of the world must all become marginal figures, but they will always be marginal to different cultures in different places at different times.

Benedict closed her account of Japanese culture by offering an example of how difference might persist despite the uniformity of marginality. Once again she turned to the story of a Japanese woman confronting Western freedom—in this case, Etsu Inagaki Sugimoto's account of her experiences in a mission school in Tokyo, *A Daughter of the Samurai.* Sugimoto's epiphanic encounter with Western freedom came when her teachers gave each girl at the school a plot of ground on which to plant anything they wanted. Benedict quoted Sugimoto: "This plant-as-you-please garden gave me a wholly new feeling of personal right. . . . The very fact that such happiness could exist in the human heart was a surprise to me. . . . I, with no violation of tradition, no stain on the family name, no shock to parent, teacher or townspeople, no harm to anything, was free to act" (294).

After this experience, Sugimoto could not look at her own culture in the same way again. She came to realize that although at her home "there was one part of the garden that was supposed to be wild . . . someone was always busy trimming the pines or cutting the hedge." According to Benedict, Sugimoto's new awareness of the "simulated wildness" and "simulated freedom of will" of Japanese culture initiated a "transition to a greater psychic freedom" through which she discovered the "pure joy in being natural." Like Mishima's experiences at Wellesley, Sugimoto's detachment from tradition, family, teacher, and townspeople can serve as a model of "natural" social relations for postwar Japan. A "self-respecting Japan" will be a Japan that respects the self, or more precisely, the individual. To be a responsible member of the family of nations, Japan must "set up a way of life which does not demand the old requirements of individual restraint" and must move toward a "dispensation which honors individual freedom" (294–96, 150).

Just when Benedict seems to have completely dismissed Japanese culture, she pulls back. Insisting on respect for the individual within Japanese culture, Benedict also insists on the individuality of Japanese culture within the global family of nations. Adopting a universal formal relation between the individual and culture, Japan must nonetheless cultivate a particular cultural content consistent with its own cultural traditions. Benedict insists not only that Japan must work toward greater self-respect and greater psychic freedom but also that Japan "will have to rebuild her self-respect on her own basis, not ours," and "purify it in her own way." Benedict even suggests that "certain old traditional virtues . . . can help to keep [Japan] on an even keel" as it purifies its culture (150, 295–96). Condemned as constraint, culture may be recovered as resource.

Benedict took Sugimoto's story as an occasion for reflecting on the possibility of purifying two central symbols of Japanese culture: the chrysanthemum and the sword. Both symbols embodied the contradictions of Japanese culture, and these contradictions needed to be resolved in order for Japanese culture to ripen. Chrysanthemums, for all of their beauty, "are grown in pots and arranged . . . with each perfect petal separately disposed by the grower's hand and often held in place in a tiny invisible wire rack inserted in the living flower." Purification demanded that the Japanese "put aside the wire rack" and accept that "chrysanthemums can be beautiful without wire racks and such drastic pruning." Conversely, the cult of the sword, which fostered international aggression, contained within it "a simile of ideal and self-responsible man." The sword stood not only for the warrior code but also for a more general code of personal conduct, and the Japanese "have an abiding strength in their concern with keeping an inner sword free from the rust which always threatens it." Viewed in this nonaggressive sense, the sword is "a symbol [the Japanese] can keep in a freer and more peaceful world" (295–96). Purified of their repressive connotations, these symbols can serve as a particular language through which the Japanese may participate in a single, universal conception of freedom to be shared by the nations of the world.

The fate of the chrysanthemum and the sword revealed the necessity that serves as the basis for Benedict's vision of cultural freedom in the postwar world. The idea of a world made safe for differences allowed for the retention of indigenous cultural symbols, provided they be abstracted from the often restrictive normative contexts that originally gave them meaning and then be instrumentalized in the service of the greater psychic freedom of the individual. Naturalized as a neutral universal principle of order, Benedict's prescription for cultural freedom nonetheless demanded that the Japanese learn to internalize both aesthetic and ascetic authority in a manner suspiciously like that of the Cold War liberal elite emerging in postwar America. The chrysanthemum and the sword stood for the "soft" and the "hard" side of the personal ethic that would come to define establishment liberal intellectuals during the Cold War: on the one hand, an aesthetic of cultural cosmopolitanism open to the beauty of all cultures while being bound to none; on the other, an ascetic of "responsibility" able to transcend the easy answers of both the left and the right and adopt a "realistic" attitude toward questions of foreign and domestic policy.[21]

An attempt to democratize this ethic, *The Chrysanthemum and the Sword* argued that the peace of the world depends, not on a liberal intellectual elite controlling world events, but on the peoples of the world controlling themselves in accord with the values of that elite. The achievement of this self-control depended on the peoples of the world being able to achieve a certain anthropological detachment from their own culture and to see their culture as a means to the end of individual growth. Benedict's conception of a world made safe for differences ultimately reduced to a vision of a world made safe for the personal ethic of the cosmopolitan, responsible, liberal intellectual.

6 White Maternity, Rape Dreams, and the Sexual Exile in *A Rap on Race*

Jean Walton

Recent re-evaluations of Mead's relationship to U.S. feminism, of her contributions to anthropological accounts of gendered subjectivity, and of her implication in U.S. and international race politics have suggested that Mead is a prime candidate for an investigation of white women's fantasies of racial difference. Louise M. Newman has argued for the need to "view Mead as an integral part of a Victorian tradition that combined notions of white or 'civilized' women's sexual restraint and black or 'primitive' men's bestiality to reinforce the dominant cultural taboo against miscegenation." She undertook a reconsideration of "the nature of Mead's antiracism, highlighting the continuities between Victorian and modern anthropology." In order to understand how Mead's work was "implicated in and shaped by [the] Victorian race politics" to which it was ostensibly opposed, Newman insisted that "oppositional movements retain residues of that which they oppose." To put it most simply, Mead's substitution of cultural theories for biological explanations of difference did not purge feminist theory of its Western ethnocentric and white racist biases.[1]

Micaela di Leonardo, an anthropologist, offered trenchant comments about the same issue: "If we consider . . . the long history of Mead's changing anthropological politics, we see that feminist or antifeminist, liberal or conservative, what she most consistently sold was the Primitive as commodity—and herself as authoritative anthropological interpreter of the uses of exotic merchandise."[2] Like Newman, di Leonardo has eloquently traced the residues of racism in Mead's work, giving a nuanced picture of her role in the ongoing tradition of the commodification of the "dusky maiden" in American culture. Di Leonardo documented how Mead's work contributed to anthropology's refusal to interrogate how ethnographic research is both enabled by and reproduces imperialist relations between post-industrialized nations and the "third-world" locales they have helped to "underdevelop."

Despite these insights, however, neither Newman nor di Leonardo explored the relationship between Mead's sexuality and her production of herself as a "white" woman. Hilary Lapsley's study of Mead's professional and personal partnership with Benedict explores the issues of gender, sexuality, and female relationships but fails to examine race.[3] By foregrounding Mead's self-presentation in her writings, we can discover her understandings of herself as a white woman among "natives," as an anthropologist enmeshed in a complex kinship system with these natives as well as with other anthropologists, and as the mother who could take the "world" in her arms—but, paradoxically, also as a modern sexual subject. Mead's heterosexual, maternal public persona is structurally motivated by, and built around, her private practice as a lesbian, including her lifelong intimate relationship with Benedict. The sexual closet she inhabited throughout her life was necessarily inflected by her implication in the racializing structures of anthropology.

The relationship between sexuality and race in Mead's life may be elaborated in part by an interpretive reading of her sense of inhabiting the dual kinship system of those racially marked as other on the one hand, and those professionally marked as same (i.e., anthropologists) on the other.[4] But it may also be explored by turning to a document that is particularly rich with implications for understanding the relationship between the sexual closet and racial difference in Mead's fantasy life: her *A Rap on Race* with James Baldwin, a discussion whose ostensible focus was race and racism, but whose subtext, given its two interlocutors, could be nothing other than the repercussions of sexual closeting. Newman and di Leonardo have already mined this extraordinary text for its evidence of Mead's implicit racism. *A Rap on Race* allows a reflection on how Mead's accounts of the field, her ethnographic observations and conclusions marshaled for the purpose of intervening in U.S. public attitudes and practices around gender and sexuality, are informed by her unexamined racial fantasies. The same text also enables its readers to explore how the maintenance of her whiteness is implicated in the maintenance of her closetedness, again through a public persona as mother of the world. Finally, by noting how Mead rhetorically figured her body in relation to its environment in *A Rap on Race,* it becomes possible to gather much about the whiteness of her bodily ego, or in other words, of how she thinks of her body in terms of its boundaries, its strengths, its vulnerabilities, its difference from other bodies, its solidity or permeability, whether it is self-possessed or possessed by others, and how it figured in her fantasies of racial difference.

In August of 1970, Mead and Baldwin met for a two-day recorded discussion on "race and society," the transcript of which was published as *A Rap on Race.* The blurb on the back of the book claimed that Mead contributed her "knowledge of racism as practiced in remote societies around the world," while Baldwin brought "his personal experience with the legacy of black American history." Here "Baldwin's creativity and fire" conversed with "Mead's scholarship and reason." The blurb already reinforced a certain racist fiction: one that attributes rationality to whiteness and emotionalism to blackness and that sees the white speaker as capable of knowledge and the black speaker as capable only of experience. Mead spoke in this dialogue as much from "experience"—the experience of "whiteness"—as Baldwin, but her experience of being in a "raced" body is obscured by the racial dichotomizing of the blurb. The interest in this dialogue is not, in fact, Mead's "knowledge of racism as practiced in remote societies" but her "experience" of this racism as the anthropologist who is implicated but not fully aware of how it inflects the knowledge she produced.

Race circulates in Mead's reflections as something about which one is by turns conscious or unconscious, that makes a white person "self-conscious." A number of the anecdotes that she shared with Baldwin will illustrate this trope of "consciousness." When among the Samoans on her first field trip, Mead remarked, "I slept on their beds and went fishing with them and dancing with them. I was much smaller than they were, so I could work with adolescent girls and act as if I was a fourteen-year-old when really I was twenty-three years old. But then I remember I was surprised when I went back just before I left, to the village that I had stayed at in the beginning, and of course by this time I spoke the language fluently. One of the girls said to me, 'We watched you when you came. We watched you and we saw what you did. I offered to lend you a comb and you took it. We watched.' But I hadn't been conscious at all."[5] As usual, Mead depicts herself as having completely assimilated to Samoan culture. This fantasy of affiliation and fusion with the Samoans is only disrupted when one of them told her later that she was being "watched." Her impression that she had blended in, and thus was the origin rather than the recipient of the gaze, was at this moment contradicted. She realized now that she had not been "conscious at all." There is an ambiguity about the object of that unconsciousness. Is it that she had not been conscious that her whiteness was perceived as such by the Samoans? Or is it that she had not been conscious of the very existence of racial and cultural difference, so absorbed had she been in her impression of coalescence with the Samoans?

Earlier, Mead told a similar anecdote as a way of demonstrating that she shared a certain experience with Baldwin. Baldwin had just remarked that at the present moment "black people no longer care what white people think. I no longer care, to tell you the truth, whether white people can hear me or not. It doesn't make any difference at all." "I can really understand this," Mead responded. "When I'm in New Guinea, when I'm living in a village completely related to everyone in the village, they're the people that matter, and nobody else matters. I'm not particularly conscious any longer that I'm white, that I'm an anthropologist from a long way off, because there are too many other things to be conscious of." She illustrated this with what seems to be a favorite anecdote, since she reproduced it more than once in her published writings. She described being surrounded by New Guineans in a house from which she can see a boat arriving on the river:

> All around me would be lots of New Guinea people: mothers with their babies in their arms—whole groups of people. And white men would get off that boat. I could see that they were white before I knew who they were, because they wear different kinds of clothes and they are much bigger and so forth. The unknown white men would get off the boat and they looked to me like paper dolls. They didn't look like real people at all. And as they came closer, and I could see who they were and either recognize them or not recognize them, they became individuals. And then the people around me turned into paper dolls. You see, the break was so great; the break was so great that you couldn't look at both worlds at once. So one of them became unreal, two-dimensional, flat.[6]

If Mead knew, like Baldwin, what it is like not to "care what white people think," it is by virtue of a complete identification she believed she had made with her anthropological subjects. In making this identification, in other words, she imagined she was able to cast off the "cares" or "thoughts" of white people. Mead thus enjoyed an "unconsciousness" with regard to her racial whiteness most notably when she was "in a village completely related to everyone." The "relatedness" is doubly marked when she picked out of the crowd of "lots of New Guinea people" the "mothers with their babies in their arms." This particular way of emphasizing "relations" with a stress on the maternal is a frequent rhetorical trope in Mead's writings. For the moment, it is important to note that when alone with her village people, there was, for Mead, no racial difference. And yet, when other whites arrive on a boat, thus making Mead aware of her own whiteness, racial difference was so predominant as to be the definitional boundary between two "worlds," only one of which can be "real" at a time.

A third anecdote illuminates this question of Mead's consciousness or unconsciousness with regard to race: the role of a racist history in the United States as it does or does not impinge on Mead's sense of her whiteness. When working among the Arapesh, Mead recalled, "there was just one of the teenage boys I knew very well who looked a little bit more like an American Negro. Just a little more, just a touch more, and I always had a slight awareness, you know. I was holding their babies in my arms, and nursing them when they were sick, and they were carrying me around, and everything. Yet with this one boy I would notice in myself just a touch of self-consciousness. That's all it was, but it was self-consciousness."[7] Once again, Mead seemed to feel "completely related" to the people in the village, insofar as she was "holding their babies" in her arms, while in turn they are "carrying" her around, and we may presume that she was "unconscious" of her whiteness. As usual, the unconsciousness of race is bound up with maternal images of carrying and being carried, or the sense of a maternal affiliation with the Arapesh. Racial difference does not make itself evident except in the form of one individual who, Mead imagined, resembled "an American Negro"—and who thus reminded her that she was an American white. Apparently, it is only when reminded of the U.S. context that she can acknowledge asymmetrical political arrangements that produce racist dynamics: the United States is the space of racism; the tribal locale is outside this space. By projecting all racism into the "container" of the United States, was Mead able to disavow the racist implications of the anthropologist's participation in a colonialist relation to the objects of the ethnographic gaze? If this is the case, then anthropology would be the means by which Mead may be white among blacks and yet not think of herself as occupying racist structures, not be "conscious" of race, not be conscious of her whiteness.

The question of consciousness shifts in another pair of anecdotes in which white supremacy figures as something that Mead must deploy mainly for the benefit of nonwhites. Here Mead explained what she meant by a "consciousness that there are different kinds of people, and you've got to remember it." When left solely in charge of two hundred men in New Guinea, Mead recalled that she "had to give them orders based on absolutely nothing but white supremacy." This led to a story that figured as a defining scenario in Mead's sense of her racial relation to the objects of her study. Some strange men came to the village and wanted to sell her wormy beans. She refused to buy them, believing that to allow herself to be swindled would put her in danger. When the men left, she noticed that a box of matches was missing. She was convinced that if she failed to retrieve the matches, she "would have been as good as dead," since "white people who let a thief go used to be killed; they had shown themselves as weak." She thus set out to get back what had been stolen from her. "So I stormed up to the end of the village. This was a fine exercise of sheer white supremacy, nothing else. I didn't have another thing. I didn't have a gun." Mead found the men, demanded her matches, and one of them returned them to her. "Then we were all safe," she concluded. "Now if I had made one misstep I'd have been dead, and then the administration would have sent in a punitive expedition and they would have been dead."[8] Congratulating herself on her resourcefulness, Mead turned herself into the heroine of the episode who saved a village from destruction.

It is important to note that the "white supremacy" Mead referred to in her anecdote was something she experienced as external to herself. It is something the circumstances imposed on her, or something that she adopted as a purely "conscious" role. She played the role of the superior white in order to save the lives of the black men who had threatened her: "I have never been in the position of believing that I had any rights because I was white. I have been in the position of acting out white supremacy in New Guinea to save everybody's lives, because I was in a situation where it was necessary. But I never felt one moment of white supremacy in New Guinea, and I simply do not have the feeling which is one component in this country."[9] Her consciousness of whiteness, we are to understand, differed from the way in which her compatriots in the United States experienced whiteness. She "acted out" white supremacy, while most other white Americans presumably "felt" it. As a conscious performance, white supremacy was not understood by Mead to be something constitutive of her subjectivity, not something that racially defined her.

She was different because she believed herself to have been, as she put it, "lifted out" of racism. Indeed, the image she used to convey what she believed to be her immunity to an unconscious embeddedness in whiteness and in racism is worth examining for its evocation, once again, of maternal relations as they pertain to race: "As a child I was lifted out of the situation that grips this country and is destroying it. I was lifted out of it as a child. I was never permitted to grow up in it, and neither was my own child ever permitted to grow up in it."[10] Mead thus saw herself as a link in a chain of maternal influences, each mother in the chain assuring that her child is "lifted out" of racism, that is, out of a certain consciousness of race that leads one to "feel" rather than merely "act out" white supremacy.

That Mead privileged her mother as the agency by which she was "lifted out" is suggested by a striking story Mead tells to illustrate how her racial unconscious differed from that of most other whites in the United States. "I learned about race when I was a child," Mead began, early in her dialogue with Baldwin, stressing that she had "completely Northern ancestry" and that her grandfather "fought in the Civil War on the Northern side." Her father bought a house in Pennsylvania that "had been a station on the underground railroad. This history we regarded as very good, romantic, good Northern behavior."

Then Mead began to tell the story that her mother had passed on about a local African American couple whose "very black wife" had a "half-white son." As Mead told Baldwin, "what I was told by my mother—who believed in telling children the truth and telling it correctly so they wouldn't get it wrong—was that she'd been raped by a white man. You see, I had the reverse picture that most Americans have, because most white women picture a rapist as a black man." When Baldwin told her to continue, she explained, "So whenever I dreamt of rape, I dreamt of this black woman being raped by a white man. And our mother insisted on our calling her Mrs.—this is 1912. And whenever she turned up in a dream, I knew exactly what I was dreaming about. This is a straight reversal of ordinary American experience."[11]

That Mead considered this recurrent rape dream as the "straight reversal of ordinary American experience" is worth noting. There would seem to be a presumption here that the "ordinary" experience of Americans is a white one. Even as she tried to distance herself from a white supremacist position, the inability to conceive of "ordinary American experience" as anything but white betrayed her tacit complicity with racist structures.[12]

Perhaps even more importantly, it must be stressed that this "reverse picture" that Mead carried with her and replayed in her dreams was a legacy of her mother. Her mother's insistence on a certain "truth" about the black woman "lifts" Mead "out" of racism in two ways: at the conscious level by alerting the young Mead to a history of white on black racial violence, and at the unconscious level by installing in Mead's psyche the racialized fantasy that inflected her own sexual and racial constitution. That Mead thought of herself as differing from most whites at the level of the unconscious is indicated by her observation that the "reverse picture" of rape bequeathed her by her mother is something that recurred at the level of her dream life. Whenever she dreamt of rape, she said "I dreamt of this black woman being raped by a white man." Moreover, she insisted that "whenever she turned up in a dream, I knew exactly what I was dreaming about."

But Mead did not specify what the dream signified, as though it should be obvious to her interlocutor. Perhaps Mead meant to imply that the dream reassured her (and is proof) that she has an accurate, and therefore nonracist, picture of race relations in the United States? That unlike white women who fantasize that, contrary to historical truth, they might be raped by black men, she had dreams that simply represented reality as it is, a state of affairs in which white men are the brutal aggressors and black women (and by extension all blacks in the United States) their hapless victims? If this is the case, it would seem that, in spite of her knowledge of Freud, she has forgotten or set aside a psychoanalytic sense of how dreams function.

Indeed, her evocation of the notion of "consciousness" throughout her musings on race, whiteness, and white supremacy is markedly divorced from any psychoanalytic concept of the relation between conscious and unconscious processes.

If her dream of rape is treated, for example, as a wish fulfillment, we would want to ask how it might be functioning in Mead's unconscious, regardless of how she thought it indicated her exemption from racism. What do we make of the recurrence in Mead's dream life of this "very fat, very black wife" of a former slave, being brutally raped by a white butcher? And why is it, once again, that, reversed or not, it is a fantasy of rape that functions as the paradigmatic scenario marking a white woman's racial identity? One way to approach this recurrent dream is to consider it as not so different after all from the "ordinary [white] American experience." While it is reversed, it nevertheless indicates something about how racial difference is imbricated with sexual difference in the constitution of a white woman's subjectivity at a certain historical moment.

It will be instructive at this point to compare Mead's dream with that of a patient of the British psychoanalyst Joan Riviere, whose 1929 essay "Womanliness as a Masquerade" received serious attention in the 1980s and 1990s from feminists who sought in it a more flexible account of femininity than that offered in more orthodox Freudian accounts. According to Riviere, women like Mead who engage in public displays of competence in a professional arena reserved for men may follow that display with flirtatious behavior toward men they perceive as hostile to their proficiency. By seducing "father figures," these women hope to ward off retaliation for "stealing" the penis that is rightfully a man's. In this context, "womanliness" is a compensatory—not essential—behavior; the masquerade offers self-protection in a patriarchal social sphere. Yet the analysand that most preoccupied Riviere had fantasies of being attacked by a "negro" whom she would seduce and then hand over to the "authorities." "This phantasy . . . had been very common in her childhood and youth, which had been spent in the Southern States of America; if a negro came to attack her, she planned to defend herself by making him kiss her and make love to her (ultimately so that she could then hand him over to justice)."[13] The true "father figures" in this imagined scenario set in the "Southern States of America" would not be the attacking "negro" but rather the white male authorities representing "justice." To propitiate the (white) fathers, the white woman fantasized that she could substitute the black male body for her own. This suggests that the "masquerade" involves a degree of identification and desire across imagined racially defined differences—indeed, a trafficking in the eroticized black male body.

If Riviere's fantasy is indicative of what Mead called the "ordinary [white] American experience" (because "most white women picture a rapist as a black man"), then is Mead's fantasy atypical because she figured the rapist as a white man? Or is it that Mead's dreams belong within and function to elaborate the same cultural fantasy shared by Riviere's patient? It will help to consider, first of all, the discursive context in which each dream is narrated to an interlocutor or produced as evidence of something. In the case of Riviere's patient, the dream is told as part of an exploration the analyst is making of her patient's sexual subjectivity. The racial component of the dream is of no interest to the analyst, and the dreamer is an explicit protagonist in her dream, the one who engages in "masquerade" as a defense mechanism against retaliation for having "stolen" the phallus. The black man in her fantasy functions alternately as the sacrificial figure who will take the blame for unauthorized possession of the phallus, or as the illicit sexual partner whose blackness is the means by which a prohibited lesbian desire is indirectly indexed.

In the case of Mead, however, while an exploration of this dream could no doubt tell much about her sexual subjectivity, she offered it as evidence of a decidedly non-racist subjectivity. Mead, in fact, curtailed analytic exploration of the dream by announcing that she already knew "exactly what I was dreaming about." Moreover, Mead was familiar with the deployment of such dreams in theoretical discussions of racism. While it is unlikely that she knew Riviere's text, she may quite well have intended for her presentation of her dream to resonate with Fanon's discussion of white women's rape fantasies in *Black Skin, White Masks,* since this is a text that early in her dialogue with Baldwin she mentioned having read. Presumably, she knew that, as a white woman, to dream of being raped by a black man is to express one's masochism by making the "Negro" the "predestined depositary of . . . aggression." She has no doubt learned from Fanon that, as she puts it, "when a woman lives the fantasy of rape by a Negro, it is in some way the fulfillment of a private dream, of an inner wish. Accomplishing the phenomenon of turning against self, it is the woman who rapes herself."[14] No doubt intent on differentiating herself from the negrophobic, masochistic white women of Fanon's text, Mead produced a rape fantasy that would seem to exempt her both from perversity and from racism at the same time.

Yet if Mead really did, as she implied, regularly have this dream, what subject position did she occupy in the dream? Given that dreams are always overdetermined, it is not unreasonable to suppose that Mead identified, by virtue of her gender, with the black woman who is being raped, and by virtue of her whiteness, with the white male aggressor. While this is quite likely, there is a third position implied by the dream that is worth exploring in more detail and that helps to understand where white femininity is located vis-à-vis this ritually repeated scenario of racial violence.

Referring back to the black woman much later in her dialogue with Baldwin, Mead stressed having been brought up with "tremendous concern for every person who was poor or different" in her community and emphasized that her happiness was due, not to her being white, but to "the fact that my mother did insist that I call the black woman who worked for us Mrs." This is the second time that Mead has stressed her mother's insistence on the designation "Mrs." The Mead household, then, included not only Mead's parents and siblings but also a black woman whose sexual life was vividly dramatized for Mead by her mother: she was the victim of sexual violation by the white butcher, which no doubt helped to define the "difference" that called for "concern"—a concern that is to be expressed in a very significant way. She is to be called "Mrs.," that is, given a title that conveys respect, of course (this is what Mead is stressing), but that conveys this respect by marking her married status, by situating her in a kinship relation to her black husband that is on a par with Mead's own parents' marriage. If Baldwin's suffering, Mead says, "was a function of the fact that a caste position was forced on you," she herself has learned from her mother "a denial . . . or a refusal of a caste position," and that refusal is expressed through the performative speech act of calling the black woman employed in their white household "Mrs." While this woman is bound up with the kinship structures of white people in troubling ways (as the victim of rape; as the "poor" or "different" employee of Mead's family), to call her "Mrs." promises to locate her in a black kinship system of her own, giving her a certain autonomy from the whites who would otherwise exploit her. "The reason I've been happy is not because of color," Mead claims.[15] Yet it is precisely her whiteness that provided her the luxury of being able to inherit and pass on the proper ethical relation to those who are "poor or different" from her white family.

Thus, the interracial rape scenario that dwelt in Mead's unconscious may be thought of as culminating in Mead's intervention in it. The first part of the account consists in Mead's envisioning the "very fat, very black" woman being raped by the "brutal character" of the "white butcher." The only descriptors Mead used to characterize the woman serve to define her as a figure of excess. This excess contrasted markedly with the apparent pride Mead takes in her own small size as a young woman: the smallness that allowed her to enter imperceptibly, as she saw it, into other cultures, to be carried by her native informants, and to pose as an adolescent.[16] Having witnessed the violation of this very black, very fat woman by the white man whose profession is to slaughter and dismember animals for consumption, Mead then made an appearance in the scenario as the white rescuer who, through her magical interpellation of the black woman, is able to "lift" her "out" of her racial and sexual victimage. Calling the black woman who worked for them "Mrs." is the logical conclusion of the dream, where Mead, via her mother, at once rescued the black woman from the white butcher but also re-subordinated her to her black husband (as his "Mrs."). If the black woman suffered at the hands of the brutal white man, she is restored through the agency of the concerned white woman to her proper position in a black kinship system.

It is possible that Mead also fantasized having rescued the woman from the excessiveness that characterizes her body. In "pronouncing" the black woman a wife, Mead at once mitigated her racial and bodily excessiveness and saved her from what seems almost to be a cannibalistic fate, the fate of being the victim of the butcher. The fantasy of consuming and being consumed is thus followed by a fantasy of salvation via a heteronormative speech performance. I would add that this scenario is complemented by Mead's oft-retold account of her own white body as it is endangered by, then becomes the savior of, black male bodies that pose a threat to it: an account that is not, after all, "the reverse picture that most [white] Americans have" but its duplicate.

In these accounts, while it appears that Mead's very life rather than her sexual integrity was threatened by the black men, the imagery Mead used implied a sexual undertone. The story of the matches is the most obvious example of this. Like Riviere's patient, in this account Mead was left alone in a village, when a group of black men came, not to attack her, but to sell her wormy beans. When she refused, they took her matches, those symbols of the sexual conflagration that has marked other white women's fantasies of interracial encounters (as in Riviere and Fanon) but that Mead is at pains to keep out of her own account. At this point, Mead had to retrieve the matches or, as she believed, be killed for her weakness. Given her status as white in the village, she was obliged to appear as though she had the phallus, retrieving the matches precisely insofar as they signify the phallus that had been wrongfully stolen from her. The relation between the black men and the white woman is mediated or even determined by the implied presence of a third party, the true wielders of the phallus, the white male authorities who hover in the wings, ready to be deployed when necessary. Rather than handing over her black thieves to the authorities, Mead imagined that she rescued them from the "punitive expedition" that the "administration" would have sent in. This she does by exercising the white supremacy that she shares with this white male administration, allowing her to appear as though she already has the phallus in order to retrieve the matches that prove it.

The foregrounding of Mead's body as the site of racial difference results in the obscuring of her body as the site of sexual difference. At no point did Mead explicitly depict herself as either object or subject of sexual desire in her encounter with the black men who came when she was alone in the village. Elsewhere in *A Rap on Race,* Mead once again cast herself in a maternal role that portrayed her, like the black woman who worked for her family (who had a "half-white son"), as the nurturer of brown offspring: "When I'd lived in New Guinea too long or for a long time, I came back here and I didn't like babies being so huge and pale. I didn't mind other people, but I didn't like such white babies because I'd been holding all those thin little brown babies in my arms and they were so beautiful and our babies looked too fat and too big. They looked like whales and I didn't quite like it."[17] If a cross-racial desire is experienced in the field, it is not expressed in Mead's accounts of her dealings with black men but rather through images like this, where Mead is so enchanted by the "thin little brown babies" she holds in her arms that she comes to find white Western babies too "huge," "pale," "fat," and "big." One wonders if this repugnance extended even to her own daughter, who would, after all, have been one of the first babies she held after her many sojourns among the brown babies who populated her field work. In any case, "whale-like" or not, her daughter is a crucial element in the mimicry of whiteness-as-heteronormativity that legitimized Mead as the white maternal holder of all babies, white, brown, or black.

It is precisely this image, that of her body as maternal embracer, that Mead used to assert what she believed to be a crucial difference between herself and Baldwin. If Baldwin has suffered, she says, it is because of his race. Her own happiness, on the other hand, is not due to her race but rather to that proper ethical position her mother has taught her vis-à-vis those who suffer as a result of their race. She experiences "felicity," but not as a result of how she has been raciated. The image she uses is a familiar one and leads to a discussion of the question of exile. "You see," Mead said, "I could go anywhere in the world. I can take any people in my arms." "*You* can!" Baldwin interposed. "I have." Baldwin interrupted: "We are both exiles," he said. Mead resisted this depiction, tenaciously holding fast to the fantasy that she can "take any people in my arms," that she held the world in her arms. But Baldwin insisted again that "*there is an area where we both were exiled.* You said you weren't, but you are because of what you know" (Baldwin's emphasis). Mead's zealous protestations to the contrary are worth recording here:

> Mead: I am not an exile. I am absolutely not an exile. I live here and I live in Samoa and I live in New Guinea. I live everywhere on this planet that I have ever been, and I am no exile.
>
> Baldwin: What you mean is that you refuse to accept the condition of being an exile.[18]

Mead's vehement refusal to entertain Baldwin's observation suggests that he came too close to revealing the very point of identification between them that would, indeed, threaten Mead with exile. While she was white and he was black, while she was a woman and he a man, they shared a queer relation to the heteronormative status quo in the United States and elsewhere. One wonders if Baldwin, suspecting or knowing of Mead's lesbian lovers, was encouraging her in this passage to avow her kinship with him, thus making visible that version of her body that might, indeed, make an exile of her as well. This is a kinship that Mead refused, as though it is the one affiliation that would threaten to undo the powerful illusion she had fostered of inhabiting the position of mother to the world in a meticulously articulated series of heterosexual kinship relations.

Mead's sense that she had absolute freedom of movement "on this planet," that she could "go anywhere" in the world and "take any people" in her arms, that she thus was the white maternal agency whereby the black world, in particular, is saved, seems predicated, in her mind, on the maintenance of her official status as heterosexual. Her adamancy that there is absolutely not "an area where we both were exiled" registers the inviolability of the closet occupied by that area. If Mead were to remain the legitimate producer of knowledge about babies everywhere, she must also appear in the world as a "Mrs." Her refusal of the condition of exile depended on the exiling of her lesbian body from public discourse.

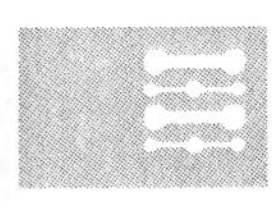

III / Imperial Visions

Portrayed in a recent book as selling the "primitive as commodity," Mead has been criticized for her uncritical portrayal of the naval administration of American Samoa and her erasure of colonial influences in her research on Samoa, New Guinea, and Bali. Commentators from the societies she studied have joined other critics. Although Benedict's shorter career and her more limited research into Pueblo and California Indian cultures gave her greater immunity from such criticisms, Micaela di Leonardo has argued that Benedict displayed the tendency to ignore the historical context shaped by "power-laden interactions" with American and other colonizers.[1]

Benedict and Mead also stand accused of engaging in "social engineering" and of advocating the exercise of power by disinterested "experts" like themselves. They volunteered to serve in federal agencies during World War II, applying their professional expertise to the study of enemy and allied cultures. Their undertaking patriotic service invited questions about whether they served the imperial ambitions of the United States. Anthropological critics accused Benedict, Mead, and other advocates of the "culture and personality" school of ignoring the effects of power and economic forces on the people they studied.[2]

The contributions that follow will address the issues of whether Mead or Benedict colluded with imperialist endeavors and engaged in wrongful social engineering. Gerald Sullivan's contribution examines the research Mead engaged in with her third husband, Gregory Bateson, in Bali. Contrasting Mead's depiction of the Balinese as "feys" as against her self-designation as a "cultural planner," Sullivan suggests that Mead accepted social engineering as an attribute of a progressive culture. Nanako Fukui explores the origins of Benedict's *The Chrysanthemum and the Sword,* identifying it as a sympathetic account that would foster postwar reconciliation and reflect Benedict's own delight in cultural discovery. C. Douglas Lummis seeks the reasons for its longevity despite its failure to understand the complexity of Japanese culture and society. Margaret Caffrey views Mead's *New Lives for Old* as advocating a progressive form of Westernization as the only alternative to more regressive forms of colonialism. While Lummis argues that *The Chrysanthemum and the Sword* reflects an imperial vision, the other contributors portray Benedict and Mead as trying to produce more positive outcomes to the trauma of war, decolonization, and postwar reconstruction by seeking to educate American readers "that the society we live in is not necessarily the only possible model."[3]

7 Of Feys and Culture Planners

Margaret Mead and Purposive Activity as a Value

Gerald Sullivan

Although most interpretations of Mead's writings focus on the published works such as *Coming of Age in Samoa*, her letters from the field and other unpublished materials should be read as carefully as her books. In letters written in New Guinea in 1933 and a field trip to Bali that began in 1936, Mead developed her theory about the interaction between temperament and character (which together yield personality) and their relation to culture. This theory, which included a diagram identifying four major temperamental categories that became known as "the squares," grew out of intense discussions between Mead and Gregory Bateson, who became her third husband and her companion on the trip to Bali. Reo Fortune, her estranged second husband, angrily denounced the theory, and Benedict and Franz Boas warned her never to publish it. Thus it never appeared in its fully developed form in any of her published writings, including those most directly shaped by the theory: *Sex and Temperament in Three Primitive Societies* and *Balinese Character: A Photographic Analysis.* Ultimately, however, some of its conceptual underpinning would influence Benedict's description of Japan in *The Chrysanthemum and the Sword* as an "iron-clad" culture. One can gain insights into Mead's and Benedict's intellectual development as well as particular insights into Mead's values by analyzing her comparison of what she called the "fey" culture in Bali and the traits of Western "culture planners."

Mead arrived in Bali in early March 1936. On October 1 of that year, almost seven months later, she wrote to her friend, English writer Geoffrey Gorer, "I continue to think that European civilization may be going somewhere, whereas Balinese [civilization] distinctly is not and never has been. They demonstrate very prettily under what conditions a fey culture can survive, by means of an iron-clad social system, and I am interested in documenting that, but I don't think God meant feys as culture planners." European civilization, she wrote, "may be going somewhere" because, by implication, it was inhabited by culture planners who might adapt their social system in order to respond to changing social conditions. Planning is an essential part of the capacity to go somewhere. In contrast, for Mead, Balinese civilization is not—indeed, never has been—"going somewhere"; rather, it survives because the fey culture of the Balinese is imbedded in an iron-clad social system, one that, again by implication, is not and probably will not be altered in order to adjust to changing social conditions. Feys, according to what Mead takes hyperbolically to be divine fiat, apparently cannot plan.[1]

Using a term that would not appear in her other writings, Mead apparently identified herself positively with "cultural planners." She probably defined a culture planner as someone like the members of Thorstein Veblen's "Soviet of Technicians," who assess what should be done to bring about the most efficient outcome for society at large on the basis of the best available knowledge and techniques of the sciences, including the social sciences. Because her parents studied with Veblen at the University of Chicago and her father continued to enjoy his books, Mead undoubtedly was familiar with his ideas. Mead also was friendly with Howard Scott, the leading member of Technocracy, Inc., a group founded in part by Veblen and dedicated to establishing a Soviet of Technicians in the United States. She admitted that Scott had influenced her thinking. Mead thus knew of Veblen and Scott's notion of the benefits of the rule of disinterested experts. She had yet to act in such a capacity, however, except as the author of books popular with the reading public, as a public speaker who suggested ways to improve American society, and as a member of a group of reform-minded social scientists who gathered around Lawrence Frank of the Rockefeller Foundation and the Social Science Research Council, including Gorer, John Dollard, and Helen Lynd.[2]

Unlike the term culture planner, "fey" appeared regularly in Mead's unpublished notes and papers beginning in 1933, but not in her published work. Most often, fey referred to one of the four primary psychological types of temperament and character that Mead and Gregory Bateson designated according to the cardinal directions and placed in their "square system." By referring to Balinese culture as fey, Mead implied that those psychological characteristics had become pervasive within the ethos of Balinese society. Alternatively, Mead on occasion wrote of persons of one sort of these psychological types as "going fey," that is, becoming subdued and unresponsive as a result of an encounter with a person of another psychological type. A fuller explanation of this usage, given space limitations, is beyond the scope of this essay. For Mead, the Balinese "demonstrate very prettily under what conditions a fey culture can survive, by means of an iron-clad social system, and I am interested in documenting that." She wished to engage herself and her collaborators, primarily Bateson and I Made Kaler, their Balinese secretary and aide-de-camp, in joint research intended to further prove this notion, which by October 1936 Mead had already decided was correct.

By the time she wrote Gorer, Mead's and Bateson's analysis of this set of psychological types had progressed significantly. They could write to select friends and family without having to explain their terms. More importantly, they had developed a scientific project for the study of what Mead called, solely for reasons of communicative ease, "the personality and culture problem," with significant theoretical underpinnings, combining biological, psychological, and cultural elements. They had also set forth a plan of the study in a setting with what Mead hoped would be suitable controls and had improvised a set of methods. The biological sciences of the day did not, in any case, have the capacity, theoretical or technical, to undertake the studies, in the field no less, necessary for a full exploration of the relevant matters.

My purpose is this essay is not to delineate this project in depth. Rather, I examine what made the opposition between fey-ness and culture planning important to Mead in order to better understand both Mead and her attitude toward the Balinese and toward the liberal culture she represented. My discussion will return to other portions of the letter to Gorer and also to Mead's debate with Gorer in letters concerning the happiness of the Balinese and Balinese art. Before I do that, however, I need to explain Mead's initial formulation of the difference between chromosomally defined sex and what would now be called gender.[3]

In 1933, while working in New Guinea among the people she called the Tchambuli, Mead had written a document entitled "Summary Statement of the Problem of Personality and Culture." The document listed Reo Fortune, then her husband, as one of the authors, but he later disavowed its contents. The document does not mention Bateson, although the Baining among whom he had done fieldwork are discussed. Nor does the document refer to either the Balinese or the so-called squares.[4]

In this text, Mead held that both American culture and the cultural understandings of the day assumed that men and women, qua men and women, displayed different capacities and inclinations and hence responded to other people, notably spouses and children, in different ways. These differences arose directly from purported biological, specifically chromosomal, differences between male and female. The research activities of these sciences were organized accordingly. Though she had previously agreed with this general assessment, indeed held that she "was innocent of any suspicion" that it might be wrong, Mead, beginning in 1933, decided that these assumptions were incorrect. She then proposed that "there are two types of human beings: each of these types includes males and females, probably in equal ratio. One type is characterized by the psychology, physiology, etc. of the type usually called maternal, the other paternal." Elsewhere in the document she identified her two types as Female and Male, because these two types of human beings, each of which included both male and female persons, corresponded to the expectations that American science and culture misattributed to male and female persons qua male and female persons.[5]

According to Mead, each of these two types of human beings displayed a distinct or particular temperament. Drawing implicitly on the work of William McDougall, Mead used "temperament" to refer to those psychological predispositions and emotional propensities or tendencies that arise directly out of the person's innate biological inheritance. McDougall had adopted conceptions developed by C. G. Jung, using as his prime examples of such innate propensities introversion and extroversion. Mead had known McDougall's work since she was an undergraduate psychology student at Barnard in the 1920s. Jung's book *Psychological Types,* which contained his lengthiest elaborations on introversion and extroversion, was translated into English in 1923 and discussed by various people in the circles of psychologists, sociologists, and other social scientists with whom Ruth Benedict and Margaret Mead had become increasingly active from the mid-1920s onward.[6]

For Mead, temperament, like physiology, descended from ancestors within family lines. In a handwritten document from the spring of 1933, Mead speculated on the circumstances by which multiple strains of inheritance could become sufficiently characteristic of a population that it would obtain a recognizable coherence. The population would not only have to remain endogamous over a period of time but also become well adapted to the conditions—food, diseases, the rhythms of life—prevailing locally. This adaptation would have to be maintained in the face of both external threats and internal difficulties. Among the threats facing smaller societies in particular were war, cultural intrusions, and economies too small to be sustainable. Internal difficulties could arise from recessive types, climatic change, and the cumulative effects of invention. Such a stability was, then, delicate and not easily preserved.[7]

While Mead noted European attempts to maintain such "culturally achieved balances" under the rubrics of race and caste, the important conclusion of this reasoning was that a given temperamental type could become a general characteristic of a population and influence the ethos of its society. Because the balances achieved differed, the temperaments, with their respective strengths and weaknesses, differed as well. "Fluidity of habit," Mead wrote, "is the refuge and strength of F[emale] types," while "fixed habits and a persistent and obstinate activity" were "the refuge and strength of M[ale] types."

Temperament, being innate, differed from character, or the more or less coherently organized pattern of what John Dewey called "the interpenetration of habits" developed over the course of a lifetime. Mead adapted this notion of character, derived from her study of McDougall and her familiarity with Dewey, to reflect the wide variety of child-rearing practices found in human societies. Mead distinguished between small-scale societies, such as those she had studied in Oceania, in which child rearing is a public enough activity that these practices are consistent, and larger scale, more industrialized societies, such as the United States, where child rearing is more private and idiosyncratic. In these smaller societies, as child-rearing practices differ from society to society, so character develops differently from society to society in accordance with the specificities of the pattern of psychological values that children learn as a result of the ongoing encounter between themselves and those who care for them. In these smaller societies, then, character differs systematically as culture does.[8] These ideas organize *Sex and Temperament,* written just before her marriage to Bateson and their field trip to Bali.

By 1935 Mead and Bateson's proposed project held that personality arises from the interaction of temperament and character. Temperament meant the individual's original constitution, or the innate propensities toward specific psychological tendencies inherited from his or her ancestors. Character referred to the pattern of his or her habits developed over the course of a lifetime, arising from the individual's upbringing in a specific society, the accidents of the individual's life, and what Bateson called "the norms of behaviours in the culture in which he has grown up and lives . . . even though he may react violently against it." By working in a closed "area of homogenous culture and common racial background," Mead hoped to find a society in which temperament and character joined together in a sufficiently coherent and clear pattern, itself consistent with the culture as a whole, that she and Bateson could adequately control for variations of temperament and character.[9]

In "Summary Statement of the Problem of Personality and Culture," Mead focused her attention on male and female types and their divergent qualities, notably their emotional possessiveness and responsiveness. Elsewhere Mead and Bateson speculated that the interface between this pair of temperaments defined a range of relationships broadly describable as sadomasochistic. Mead made no attempt in this document to reconcile her analysis with the Jungian categories of introversion and extroversion. In the four-fold, or square system, this pair occupied the north-south axis.[10]

The fey, as a psychological and cultural type, is neither wholly maternal nor paternal but rather one possible result of the mating of people of Male and Female temperament. Feys are "at first demanding of attention from M[ale] and F[emale] and [are] entirely uncareful" of others. Feys are unlike the other combination Mead postulated, the Turk. Among the Turks, the descendants of persons of Male temperament attain some of the Female temperament's carefulness about people, while persons of Female ancestry attain some of the Male temperament's possessiveness and caring. Mary Catherine Bateson, no doubt recalling conversations with her parents, held that in this context "'caring' indicated concern for one's own feelings," while "'careful' indicated concern for the other people's feelings." Each type would tend toward its own "perversions." Fey perversions included "narcissism, masturbation, and exhibitionism," those that demonstrate a self-involvement at odds, potentially, with the necessities of social life. Feys are narcissists, that is, introverts; while Turks are extroverts. This pair occupied the east-west axis of the square system.[11]

All four types could become stabilized through the requisite combination of breeding and upbringing, given the "repeated process of hybridization and isolation" wherein temperament and character attain an ongoing pragmatic cultural unity. Within any given culture, the interrelation of these types is fixed. In a society where one type has become dominant, each of the other types will "be in given position" to the culture. Hence, persons of a given type "will behave in one way in one culture and in an opposite way in another culture," where a different type has become dominant. Such changes in behavior will be consistent with the changes in patterns of dominance among the types. The types thus formed a variable structural set. This set of contrasting and dialectically related types allowed Mead to pursue her idea of systematic variation by comparing and contrasting the various implications for what she called "the individual's relationship to culture." According to this view, the "person" is not so much a collective representation as he or she is the product of a pragmatic encounter that gives shape to both the person and the culture involved. The body is shaped by a series of ongoing cultural processes, including the selection of a preferred body type and temperament, and is acted on by the local patterns of child rearing. The person experiences a locally specific process of embodiment. The body is never "precultural," but it always is biological. The deviant is someone whose temperament or character conflicts with the local system of value.[12]

Almost three months after her letter to Gorer of October 1, 1936, Mead and Bateson outlined their "main problems." First was the necessity to define a fey culture in terms of its developmental picture, sexual relations, social functioning, symbolism, the position of the deviant, and who was likely to be perceived as deviant.[13] Mead and Bateson made good on most of this plan, publishing their findings both separately and together in two books and a series of articles. The only part of the plan not realized was the final element.

Mead also referred to the feyness of the Balinese when she wrote to Helen Lynd on February 6, 1937: "The growing child is treated as a sexual object but in a strange way; his responsiveness is played upon, as the adult Balinese has no responsiveness (that was one of my errors about feys) to human beings at all; he is tossed and cozened and petted and teased and titillated, made enormously sexually conscious, and at the same time insulated from human feeling." Feys are thus more than "at first demanding of attention from M[ale] and F[emale] and [are] entirely uncareful" or "dangerous to the pure M[ale] and F[emale]." Feys, thus understood, are in Eugen Bleuler's terms dissociated (not responsive, or displaying inappropriate affect) and in Ernst Kretchmer's terms schizoid (both touchy or overly sensitive, and cold or distant at the same time), in short, prone to what was then called dementia praecox and is now identified as schizophrenia.[14]

Bleuler's use of the term *dissociation,* now largely forgotten by psychiatrists, differed from the contemporary usage of Morton Prince. Bleuler treated dissociation as a process that in various ways (e.g., inducing a negativism or ambivalence—a term Bleuler coined just as he coined schizophrenias, note the plural) separated persons from the world and gave their emotional responses a degree of inappropriateness, unreality, or flatness. Prince referred to the engendering of multiple personalities, separated from one another (hence dissociated), within a single person. Mead cautioned Bateson, while he was working on *Naven,* not to conflate "schizophrenia and dual personality," two distinct psychological formations inaptly joined by Bleuler's and Prince's contrasting uses of the term *dissociation.* Mead indicated that "we," meaning she and Bateson, had examined Kretschmer's work on returning from New Guinea in 1933. Bateson made use of Kretschmer's theory of the relation between bodily type and psychological tendency in *Naven.*[15]

In her letter to Lynd, Mead referred to what she would call a Balinese fantasy that penises are separable from human bodies. For Mead, this fantasy arose from a pattern of interaction between young boys—her "growing child"—and those caring for them. This interaction involved an adult or an older child pulling rapidly and briefly on the young boy's penis and then ceasing this act almost immediately and either turning away from the young boy or continuing on with whatever else the older person was engaged in. According to Mead, this sort of interaction left the young boy "enormously sexually conscious, and at the same time insulated from human feeling," because of the increasing frustration deriving from these repeated encounters with emotionally unresponsive adults. It is not that the young boy is originally unresponsive, far from it: "The growing child['s] . . . responsiveness is played upon," and therefore the young boy must be inherently responsive. Rather, it is that this sort of interaction eventually rendered the older child or adult male unresponsive. There are no indications in the field notes of an equivalent developmental process for Balinese females, despite Mead and Bateson's contention that adult women also became fey.[16]

In February 1937, Mead did not think she could "give a very adequate account of it all yet." For present purposes, however, the key must lie in Mead's confession. She had thought people of fey character responsive, but she now realized she was mistaken. This error must have been grave, for Mead wrote in the letter to Lynd, "I don't feel any warmth for the Balinese. They are fey—so fey, lovely creatures—lovely the way wild dear or birds are, but not like human beings. They never quarrel, or gossip, or fuss, they are tireless and graceless and easily pleased with life and highly critical toward art. When they sit, their attitudes are always beautiful. There is no internal strain, no conscience, no guilt, no drive of any sort to make them awkward or maladjusted, or—human." Whatever one may think of Mead's referring to human beings as "lovely creatures—lovely the way wild dear or birds are, but not like human beings," it is important to note that she presented a broadly culturalist argument. Like any interpenetrating arrangement of temperament and character, the Balinese have attained one among the many possible "culturally achieved balances." Mead's argument explicitly held that the circumstances she described derived from a culturally specific pattern of human development involving ongoing, hence self-reinforcing, patterns of child rearing, sexual relations, and social functioning, yielding a type of symbolism. Mead's view neither wholly separated culture from biology nor reduced culture to biology. For Mead there are, in any case, only "so-called Races of man," while the "hereditary physical types" Mead is concerned with in her square system occur "in all human groups." Mead agreed with and extended Boas on "the question of race." There was "no [racial or sexual] inheritance, there are only family lines bequeathing to their lineal descendants" temperament, "a certain physical type and an adaptive body chemistry," and providing an education that, given the vagaries of life, yielded an apposite character.[17]

There is no obvious theoretical inconsistency between what Mead wrote to Lynd and her subsequent statement of 1939 that "the evidence from primitive society suggested that the assumptions which any culture makes about the degree of frustration or fulfillment contained in cultural forms may be more important for human happiness than the biological drives it chooses to develop, suppress or leave undeveloped."[18] Mead's enduring contribution to anthropology consisted in the development of a nearly fully morphological approach to that discipline rather than in any of her more narrowly psychological studies. She paid particular attention to the interplay between temperaments (that with which humans are born), their character (the habits that become more or less coherent as they learn the world they live in), and their culture (the teachings and social organization of the wider world). Bali's ostensibly fey culture would be one possible manifestation of such a cultural selection.

By 1936 and 1937, Mead had developed the argument that elements from "the great arc of potential human purposes and motivations" became "selected material techniques or cultural traits" transmitted through the cumulative effects of child-rearing practices in each society. According to Mead, the forms of Balinese art have "taken on a definite relationship to Balinese character" through the "mediation of individual obsession, and not directly integrated with the social structure."[19] Mead referred to the "degree of frustration or fulfillment contained in [Balinese] cultural forms." This pattern rendered the Balinese without "internal strain" and hence neither "awkward" nor "human" but rather "fey—so fey." This process explained why adult Balinese were emotionally unresponsive, why Balinese social structure must be iron-clad, why the Balinese did not plan, and why the organization of Balinese life formed a limiting case for the humanly possible.

Balinese social structure must be iron-clad because its members developed without conscience, guilt, or other internalized monitors to restrain them. It required an absolute set of rules to prevent socially destructive encounters. Conversely, a society without iron-clad rules would survive only if its members learned and lived according to the dictates of conscience and guilt that determined the rightness or wrongness of every action prior to the person undertaking the action. For Mead, this monitor was an internalized image of the individual's parents. She believed that such a monitor might produce internal strife, but it would make the members of such a society human. The Balinese, however, did not internalize the image of the parent because their belief that ancestors returned to life among their descendants did not adequately differentiate the parent from the child. Writing to Gorer on August 20, 1936, Mead commented, "No, I don't think I shall ever like Bali as much as New Guinea. I suppose I put too high a premium on free energy, either free intelligence or free emotion, and these people have neither." A month later Mead wrote John Dollard that Gorer considered "the Balinese happy and I consider that happiness isn't worth buying at such a price. Not an ounce of free intelligence or free libido in the whole culture."[20] Freedom from conscience or guilt created a person bound by externalized conventions, the iron-clad system of rules, and hence neither free nor autonomous.

Mead did not deny the existence of Balinese intelligence and lust. In her letters she often described her informants, including her aide-de-camp I Made Kaler, who "knew five languages and had a vocabulary of some 18,000 words in English, although he had never before met a native speaker of English," as intelligent or libidinous or both. In her report to Clark Wissler, head of her division at the Museum of Natural History, Mead described several of their informants from Bayung Gedé. Nang Oera, for example, was "a warm intelligent, aberrantly emotional man." Djero Balian Seken was "shrewd, intelligent," and "lusty, exhibitionistic." Men Djeben was "gentle, charming, intelligent," while Men Singin was "over-endowed physically and sexually." According to Mead, "an impersonal scheme" contained the intelligence or desire of these people within a social order whose continued functions did not require the affective assent on the part of those persons who lived within it. The social order existed separately from Balinese individual obsessions and hence from any personal sources of what might, in other cultures, be called art. Within "their cultural grooves they run beautifully, almost miraculously—outside them they are practically paralyzed with fear." Albeit intelligent and libidinous, the Balinese could not be described as free or fully human.[21]

Mead did not provide an adequate definition of freedom, primarily offering a negative definition when she wrote that Balinese "symbolism comes not from repression" but from "the endless free imaginative elaboration of an obsessional pattern." This pattern developed because the Balinese found themselves prevented from enjoying "full genital activity." Genital activity aside, definitions like Mead's are common among liberal theorists. Hobbes and Locke do not differ substantially in this matter from John Dewey, who wrote that "freedom is found in that kind of interaction which maintains an environment in which human desire and choice count for something." Whether or not Mead had read Hobbes or Locke, she was a product of a society in which such beliefs formed a climate of opinion. Mead's relation to Dewey is clearer. Her parents had attended the University of Chicago when Dewey was on the faculty. Dewey later taught at Columbia University while Mead attended that institution. Benedict may have introduced Mead to Dewey's *Human Nature and Conduct;* Mead carried a copy of the book with her for some time while an undergraduate at Barnard.[22]

In accord with the climate of opinion associated with such liberal theorists, Mead developed a constellation of related traits that constituted ground against which her notion of freedom might be understood. Among these traits, Mead included a lack of intellectual enterprise, a paralysis of initiative, a refusal to perceive strange or intrusive facts, an unresponsiveness, and a preference for admitting ignorance. Freedom would be manifested as intellectual enterprise, initiative, responsiveness, and affect, as well as an unwillingness to admit ignorance.[23] In the Balinese case, the absence of freedom and conscience was made tenable by the iron-clad social system.

Continuing to develop her ideas, Mead wrote to the men with whom she often discussed such issues. As she informed Gorer in 1937, "The question remains, is the occasional good symbolic form which fits the gap in character structure worth all the other rubbish that fits the gap equally well?" For Mead, the Balinese patterns of child rearing gave "everyone profound obsessions and then [they] develop adequate symbolic forms for dealing with those obsessions so that they are no threat to the even ordering of society. If one could be sure that the symbolic form would be art, there might be something to be said for it, but judging from the Balinese material, I see no reason why it should not be gambling or cock fighting or football or bridge instead." She suggested that the "only difference between America and Bali" might be ascribed to the fact that "America only plays bridge and poker and gets drunk because they lack an artistic tradition, while Bali gambles and cock fights and has an artistic convention." She admitted, "By taste I am on the side of the culture that first develops the lack and then fills it in the most intricate way possible, but I don't think it can be justified in terms of social equilibrium or even of happiness. There must be some other referent, very possibly that much maligned Progress."[24] As her letter to Gorer of October 1936 suggested, Mead considered that progress required planning and expertise if civilization was to "go anywhere." In this sense, progress was congruent with her preferences for conscience, or the conditions necessary for forethought about the consequences of activity.

Mead had taken up the theme of Balinese art in her earlier correspondence with Gorer, where she had asserted that "a hundred years ago, Bali had no artists in the European sense, individuals who used a technique for purposes of expression. They had only skilled painters, clever dancers, versatile musicians, etc. introducing, unselfconsciously minor variations in their technical mastery of strictly defined cultural themes. One of the problems now is to find out who becomes the artist, in our sense of the word, as such people are now appearing, under European stimulus." This absence puzzled Mead because Balinese culture, with its ornate mythology, seemed to her fertile ground for artists. "For anyone with an obsession," which was normal among the Balinese, "the symbolic materials presented to him, already stylized, for use, are enormous. The step he has to take, to infuse an individual vision into the materials is slight, and possible very likely to a less gifted person than to one with aberrant obsessions who has to search further afield for his symbols or actually create them, like Blake." Mead explained that the Balinese had "no training in self-consciousness, no training in the use of individual vision, and it is possible that only when he does not have to go too far in his search, can he actually come through with a finished work of art."

Even contemporary Balinese artists such as "the great dancer" Mario, whose dance was "an individual emotional expression and feeling about the music," had tried to teach his students "not to express their feeling about the music, but to do exactly what he does." Under these circumstances, his students' performances were "almost always a failure, except in the case of pupils who have enough personality to break through his attempt to teach them to imitate exactly." Visionary artists would be unusual "individuals who used technique for purposes of expression." Their artistic expression would require some psychological opposite of emotional unresponsiveness and, in that sense, of being fey. The artist so understood would become a deviant among the Balinese: a person whose temperament—and, especially, character—would not find "the emotional and intellectual wellsprings of [their] society" within "their own psychological foundations."[25]

What is perhaps surprising is that Mead admitted her inclination toward "the side of the culture that first develops the lack and then fills it in the most intricate way possible." By taste, she would take the side of the Balinese, at least possibly, if "one could be sure that the symbolic form" resulting from the spiritual lack or obsession "would be art." For the Mead of this period, the United States lacked an artistic tradition, hence Americans only played bridge or poker and got drunk, which did not make America wholly admirable. On the other hand, she was unwilling to approve of a culture that could be "justified in terms of social equilibrium," as if functionalism justifies everything, or even by "happiness." For Mead, "there must be some other referent, very possibly that much maligned Progress." Thus the shadows of conscience and freedom—that is, strain, drive, and awkwardness—receive their justification through their contribution to progress, to the development and steady improvement to society, or what Mead might well recognize as competent culture planning, sure enough of its intentions that it need not admit to ignorance.[26]

Mead had been concerned with this opposition between equilibrium and progress for some time. Referring to her diagrams, which included directional notations identifying temperaments with the positions on a compass, Mead recounted a comment she made in a conversation with Bateson during their research in Bali: "The point about these people versus the Manus is that these people are interested in preserving an equilibrium. Your SW or NW is essentially interested in disequilibrium, in either getting more than he gives, or giving more than he gets, and that's the reason why progress is associated with western cultures." Mead's Manus and "western cultures" are variations on her Turks, both careful and caring. Mead's Balinese are concerned with equilibrium, not each with "getting more than he gives, or giving more than he gets." That is why they are not concerned with progress, despite being the "most tirelessly industrious people I have ever seen, without alibis or any sense that work is something to be shirked or avoided." Mead referred to her "considerable experience with native peoples as carriers. In most instances, all the squabbling is to get the lightest loads. In Bali, there is never a dearth of carriers, and the whole problem is to make the loads equal, so that everyone can be paid equally." Each must abandon any concern he or she has with "getting more than he gives, or giving more than he gets." As she interpreted their cultural orientation, "Their great emphasis is upon cooperative consumption rather than upon cooperative production." Such an attitude did not favor some getting more than they gave or giving more than they got. From Mead's perspective, it would not yield progress. It did not require cultural planners to respond to imbalance or to figure out ways to defuse problems arising when some worked harder while others received more. Rather, the Balinese divided proceeds equally among the membership of their various voluntary associations.[27]

Mead brought all these themes together in a letter on March 28, 1937, to Erich Fromm, a friend deeply interested in the causes of artistic production and the development of an autonomous personality:

> I still think in some obscure way that they are closer to beautiful animals than to human beings, almost always gentle, never good or bad, graceful and delightfully at ease when absolutely nothing threatens, in a panic if it does, infinitely skilled along lines they know, incomparably dumb along new ones. It is par excellence a cooperative society, with an enormous emphasis on consumption and excellent devices for penalizing both the rich and the industrious into greater consumption. I think that is the necessary point, that the energetic and the industrious should always be pushed into consumption, instead of into production, provided one wants a stable, non-competitive, non-capitalistic society.[28]

But if one wanted an unstable, competitive, capitalistic society, if one wanted the energetic and the industrious to be free, to have a society that could pursue progress, one would plan accordingly and cultivate conscience and freedom, along with the strain, drive, and awkwardness that ultimately contribute to progress. Guilt, like ambition, greed, and the desire to accumulate became, under such cultural conditions, positive virtues to Mead.

Bateson arrived at the heart of the matter early on in one of their discussions as they both tried to understand Balinese society by contrasting it with their own:

> When we say concentrate, we are speaking of an affective situation, of an act of will. When the Balinese concentrate it is non-affective and in fact may be affectively disjunctive and discontinuous, while cognitively continuous. When we say "keep your mind on your work," we mean "keep your emotions behind your work." The fey, being neither caring nor careful but unresponsive, keeps his or her emotions behind nothing. The culture planner, seeking progress, must learn how to encourage others to keep their emotions behind their work, that is, to be single minded, sacrificing all to obtain his or her goal, rather than ambivalent and ultimately fey.

Identifying themselves as members of a "civilization [that] may be going somewhere," Mead and Bateson agreed that such a civilization was the only way to progress.[29] As was frequent in her published works, Mead's study of another culture circled back to the society from which she had come.

Mead spent a significant portion of the 1930s in New Guinea and Bali. Still, her analysis displayed an affinity for the ideology of planning and progress associated with Veblen's Soviet of the Technicians, Deweyan pragmatism, and the New Deal. When she returned to the United States in 1939 to give birth to her daughter, Mary Catherine, she easily fit into wartime work; her concepts of culture, personality, and temperament offered a useful conceptual scheme that scholars might use to frame the research they produced to fulfill the demands on the part of the U.S. government for knowledge of allied and enemy cultures. The insights into American culture derived from the comparative study of the Balinese would underpin her wartime writings, including *And Keep Your Powder Dry,* and her mid-1950s praise for progress, *New Lives for Old.* Identifying herself as a believer in progress, Mead would dedicate her postwar energies to promoting positive change in the United States and the other cultures she studied for the next three decades.

8 The Lady of the Chrysanthemum

Ruth Benedict and the Origins of *The Chrysanthemum and the Sword*

Nanako Fukui

A newspaper article published in *Nihon Keizai Shinbun* in 1992 reveals the way Benedict and her reassessment of Japanese culture in *The Chrysanthemum and the Sword* continue to reverberate in Japanese debates about culture and national identity. Written by the head priest of the Tendai Buddhist sect, the article described a visit to Hieizan Temple in 1946 by a delegation from the U.S. occupying forces. According to Yamada Etai, "the visit caused a bit of commotion, for that was a period in which there was nothing with which to regale guests." The delegation included four or five military personnel, including a woman, who impressed the priest as being exceptionally tall, "well-mannered and exceedingly civil." Later the priest would hear "that Ruth Benedict, who had written *Chrysanthemum and the Sword,* had been part of the delegation."[1] Although Benedict had in fact never visited Japan, the apocryphal story illustrates the way that both Benedict and her book continued to fascinate Japanese readers five decades after the purported visit.

Providing his own way of explaining the origins of Benedict's engagement with Japanese culture, Etai described the visit to his temple. Taking the visitors on a tour, he led them to the "inner sanctum, where the images of Yakushi Nyoria and Dengyo Daishi" sat in the glow of the "light of eternity." Listening in a state of bafflement to their discussion, he asked the translator what they were saying. According to the translator, the delegation from the U.S. forces wanted to "investigate why the Japanese, with such a small, resource-poor country had been so strong in war, and been able to fight so unceasingly." According to the visitors, the visit to Mount Hiei had helped them "solve the riddle."[2]

Seeing the "immortal light in the inner sanctum that had been burning unextinguished for twelve hundred years," they recognized the connection to Shinto and Buddhism as "firmly rooted in Japanese hearts" and giving "out a strong bright light." The visitors concluded that "the power to carry out the things they believed to be right was not unconnected to the Japanese sense of religion." Claiming that he had received a "small shock from this unexpected reply," the priest told his readers that "these people from another culture" had taught him to "think in a new way of the legitimacy of religion, the preciousness of the sacred light, and of gratitude."[3] Connecting this revelation to Ruth Benedict, whom he believed to have been a member of the delegation, the priest not only emphasized the importance of his religious rituals to Japanese culture but also pointed to the way that the Japanese could learn about themselves by reading or hearing the voices of outsiders like Benedict.

Despite the error made by Etai, his faulty memory shows how Benedict and the image of the chrysanthemum had acquired a meaning independent of their origins in her analysis of Japanese culture. Translated into Japanese after being officially authorized by the occupation administration in 1948, *The Chrysanthemum and the Sword* entered Japanese discussions about their culture and identity at a particularly vulnerable moment in postwar reconstruction. Since its first publication in Japanese, *Chrysanthemum and the Sword* has sold more than 2.3 million copies and continues to be a best-seller. As the family crest of the Emperor, the flower and the word *chrysanthemum* still evoke an emotional response among Japanese people. Even today, whenever the word appears in articles, particularly when it refers to Benedict or her book, ultra-rightwing groups react quickly. For them the chrysanthemum represents the Emperor, while Benedict represents a tool of U.S. imperialism and the occupying forces. Benedict's book, through its artful choice of the chrysanthemum, continues to evoke both the positive, if inaccurate, response represented by Yamada Etai's recollections and the hostile reaction of Japanese nationalists.

Before its publication and its subsequent transformation into a potent cultural symbol, the origins of *The Chrysanthemum and the Sword* can be traced back to the 1920s, when Ruth Benedict, Margaret Mead, and others were embarking on their common search for the way to interpret and understand cultures. Although Benedict never became a specialist in Japanese culture, her presence at Columbia University in the 1920s may have acquainted her with one of the sources from which she would develop the dual and polarized symbols that her title emphasized. On the campus at the same time as Benedict, Etsuko Sugimoto lectured on Japanese cultural history and language as the first Japanese language teacher at the university. Benedict may also have read Sugimoto's best-selling book, *A Daughter of the Samurai,* which had appeared in 1925 after publication in the magazine

Asia between December 1923 and December 1924.[4] Continuing to wear the kimono until she returned to Japan in 1928, Sugimoto imparted a distinct impression of the Japanese that her writings reinforced. Her book portrayed its subject as bent under the weight of strict convention, fixing in the American mind the self-abnegating and self-disciplined images of both the Japanese samurai and Japanese women. It was a suitable book for Benedict to draw on when she looked for references to chrysanthemums and the sword.

The experiences of the 1930s and the attack on Pearl Harbor encouraged Benedict to consider more carefully the culture that had played such a significant role in leading the United States into World War II. From June 1943, Benedict worked in the Division of Cultural Analysis and Research of the Bureau of Intelligence, a part of the Office of War Information (OWI), where she initially undertook "a job on European cultures" after receiving a formal appointment as head of the Basic Cultural Analysis Section for the Overseas Branch of the OWI. By September she had begun to research Thailand and Romania as well as the use of psychiatric material to explore the cultural characteristics of the members of societies involved in the war.[5]

Developing a research process that would later become known as the study of cultures at a distance, Benedict evolved a methodology that would enable her to carry out cultural analysis without leaving the United States. Meeting other anthropologists, social scientists, and psychologists working in Washington for the government, she contributed to the emergence of a research field that would eventually become known as "culture and personality" or "contemporary cultural research," using documents, diaries, newspapers, descriptions of rituals and ceremonies, published research, journalistic accounts, interviews with informants knowledgeable about the culture being studied, and critical analysis of films, novels, and other literature as a substitute for the participant observation and fieldwork methods that anthropologists had used during peacetime.[6]

In June 1944 Benedict received instructions to commence research into "the Japanese character." She collected a variety of materials, including Japanese books translated into English, books on Japan written by Westerners, and movies produced in Japan. She interviewed Japanese raised in Japan but living in the United States and American-born Japanese to discover the elements of their culture. In September she joined the newly created Foreign Morale Analysis Division under the leadership of Alexander Leighton. Drawing together anthropologists, psychologists, psychiatrists, sociologists, political scientists, and scholars of Japanese culture and language, the division began to study the morale of the Japanese and develop appropriate policies for the war and its aftermath. After five months' intensive work, Benedict had developed a fragmentary awareness of the major constituent elements of "the Japanese character" but had yet to find the pattern or "key" to the culture. Then, in December 1944, she attended a conference on Japanese character structure sponsored by the Institute of Pacific Relations in New York.[7]

Originating in a conference on the problems of the Pacific peoples held in Honolulu in July 1925, the Institute of Pacific Relations operated as a center for Asian studies, headquartered in New York. In addition to organizing conferences, the Institute published books and journals about Asian issues. Its 1944 conference attracted prominent scholars, including sociologist Talcott Parsons and anthropologists Douglas Haring, Benedict, Mead, and Geoffrey Gorer. Parsons gave two lectures, "Culture Patterns of Japan" and "General Outline of Japanese Social Structure," to an audience including government officials concerned with present and future plans with regard to Japan; journalists like Helen Mears, the author of *Years of the Wild Boar: An American Woman in Japan;* experts on Japanese culture and language; and social scientists without any direct knowledge or experience of the Japanese but accustomed to taking a systematic approach to the problems of personality and culture. The organizers and the participants believed they could apply their methods to Japanese character as a contribution to the war effort and to postwar planning.[8]

Although the conference would not provide Benedict with the solution to her problem, the discussion pointed to the need for educating the public about Japan. The focus of much of the discussion involved a comparison between Japanese character structure and the behavior of the adolescent in Western society as the participants sought to explain the cultural origins of such character traits as the kamikaze. A psychiatrist commented on one Japanese soldier's attitude to death as recorded in his diary: "It is natural for a man to fear dying and also for a man living in a culture which exalts virility to deny fear. What is striking to us is *how* this individual adapts himself, especially the adolescent aspect of his adaptation. It is constant fluctuation. Is it essentially immaturity, or is it something in the Japanese which would be immature in us? We know of individuals who react *as if* they had no fear. In psychoanalytic practice we see people who seek death because of a psychotic 'drive'" (emphases added). The psychiatrist asked whether the Japanese have "achieved an attitude which would be psychotic—a fantasy which they can achieve by dying, more importantly than in life?" Parsons responded by commenting, "What security and stability a Japanese had is dependent on fitting into a system of culturally defined patterns of group life. This is analogous to the conformism of our adolescent patterns. If the system of patterns were to be broken up, the behavior result would be chaotic." The claim that the Japanese character structure emphasized the need for conformity and taught its members to sacrifice themselves for the benefit of the group was frequently endorsed.[9]

Continuing with the emphasis on the paradoxical nature of the Japanese attitudes toward death, the psychiatrist argued that "the great strength of the Japanese—their conformity to the point of suicide"' could also be considered "their weak point." Once again, the emphasis on the "adolescent" aspect of Japanese behavior cropped up in discussion. Alexander Leighton, who had studied Japanese reactions in relocation camps, argued that "American-born Japanese are perhaps not the best point of reference. They are like other kinds of Americans rather than like Japanese, with one or two important differences. In the first place they are more sensitive to group pressure than most other types of Americans. Secondly, they conform to our pattern in a more intense and perfectionist manner than other groups." Frank Tannenbaum, a historian married to an anthropologist friend of Mead and Benedict, referred to "the typical gang psychology in Japan where you have security in the group and complete individual insecurity outside the group. You stick with the group when it is successful and you go to pieces when it does. It's an all or nothing at all proposition." John Maki, who would publish *Japanese Militarism: Its Cause and Cure* in 1945, concurred: "The in-group, out-group idea is very important in Japan." According to Maki, politeness occurred only in the context of an "in-group" relationship. In the context of an impersonal or "out-group situation," however, the Japanese "treat strangers, or all those not classified in the hierarchy very badly." A defeated enemy would be "treated with contempt." Seeking to explain Japanese behavior in the war, the conference then turned to how the members of such a culture would respond to defeat.

Musing on these insights, the participants sought to deal with the crucial problem of how Americans could change the Japanese patterns of behavior. The discussion proceeded to the problem of potential Japanese reactions to defeat and of how the occupation might be conducted. Douglas Haring posed the question, "Can we use the older-brother, younger brother relationship of the Confucian family as an approach in dealing with the Japanese? Such an approach could regard the present war as a quarrel within the family and justify our being tough. At the same time it would give them the security implicit in such a relationship." Earnest Kris, a political scientist, pointed out the difficulty in applying that model by focusing on the way war propaganda had dehumanized the Japanese. "If American propaganda is going to assume the elder-brother role, it isn't feasible to continue the sub-human image of the Japanese in our home-front propaganda. Japanese are becoming more sub-human in American public opinion and this makes the pattern of superior but friendly relationship very difficult." Mead added another complication as she joined the discussion. "There is no relationship in America like the elder-brother relationship in Japan. How are we going to fit ourselves into this role when we don't understand it as the Japanese do?" Mead's and Benedict's close associate Gorer added, "We can't expect the Japanese to think that we are elder brothers unless we think of them as younger brothers." Another participant agreed with Mead and Gorer about the difficulties of applying a cultural concept that was not "an American idea." Such misgivings would convince an interested participant like Benedict of the need to educate Americans about the cultural assumptions of the society that they would try to reconstruct.

Listening attentively, Benedict rarely spoke during the conference. The discussion did not provide the "key" to understanding Japanese culture for which she was searching. Certainly her earlier scholarship suggested her reluctance to accept the prevailing assumption that Japanese culture was somehow immature or inferior. Perhaps she expressed frustration in one of her few contributions to the discussion: "To have peace, you don't have to have everyone love each other." She may have objected to the psychological emphasis on the Japanese as neurotic or obsessive, despite her earlier characterizations in *Patterns of Culture* that had identified the culture of the Dobu as paranoid. There is no record of her response to the comments by Douglas Haring that the "chapter on neurotics in Karen Horney's *The Neurotic Personality of Our Time*" offered a "perfect description of the Japanese," according to a Japanese expert with whom he had discussed this issue. Benedict would have been concerned that such conclusions might have influenced men like General MacArthur, who popularized such insights with his notorious comment that the Japanese had, in American terms, an average mental age of twelve. Rather than providing a solution to Benedict's dilemma, the conference simply restated the problem about educating Americans about Japanese culture that she would try to solve in *The Chrysanthemum and the Sword.*

Shortly after the conference, on January 1, 1945, Benedict became head of the Foreign Morale Analysis Division of the OWI. That same month she met Robert Seido Hashima, a former inmate of a war relocation camp employed at the Basic Analysis Section and asked him to comment on the meaning of a haiku translated into English. As Benedict carefully assembled and read English translations of Zen Buddhism, discussions of tea ceremonies, flower arrangements, economics, newspapers captured during the war, essays, and diaries written by Japanese soldiers, Hashima helped her in her struggle to find the key element that could capture the essence of Japanese culture.

Eventually Benedict found the answer, much as she had done earlier in

Patterns of Culture, by looking outside the scholarly sources. Rather than Nietzsche, whose insights she had applied to the cultures in her earlier book, she found the answer in the famous Japanese novelist Natsume Soseki, the author of *Botchan.* With the help of Hashima, Benedict seized on the Japanese concept of *on* as it was revealed in an episode in the novel. The incident involved a teacher called Porcupine, who treated Botchan "on his arrival to a glass of ice water, paying one and a half *sen.*" Later Botchan heard "a rumor that Porcupine had said a slighting thing about him." Disturbed at owing a debt of honor to Porcupine, that is, an *on,* Botchan anguished over the slight on his own honor. "One *sen* or half a *sen,* I shall not die in peace if I wear this *on.*" As he explained the meaning of *on,* the "fact that I receive somebody's *on* without protesting is an act of good-will, taking him at his par value as a decent fellow. Instead of insisting on paying for my own ice water, I took the *on* and expressed gratitude."[10] The next day Botchan relieved his dilemma by paying back the money in order to wipe out the *on.* In that novel, Benedict discovered what she believed to be the key to Japanese culture.

Benedict's epiphany resulted in Report 25, entitled *Japanese Behavior Patterns.* She built the report around a discussion of two major Japanese cultural concepts: the Japanese system of obligation, which includes *on, gimu,* and *giri;* and Japanese self-discipline, which comprises *syuyo* (mental training) and *muga,* which Benedict explained as "ecstasy with no sense of I am doing it" or "effortlessness." She explained that *muga* was the goal of those "who practice mysticism" so that "the devotee becomes 'one with the Universe' by ecstatic contemplation of Ideal which is 'not of this world.'" Explaining *muga* as a means to resolve the "culturally induced conflict within the self," Benedict saw it as eliminating those internal conflicts. Linking the concepts of *syuyo* and *muga* together, a "revealing light" could be thrown on Japanese psychological responses, including the emphasis on masochism or self-sacrifice.[11]

The extraordinary acuity of Benedict's penetration with regard to her subject is particularly exemplified by the following passage about the concept of *makoto.* As she explained, "*Makoto* means more than the English 'sincerity' and it is as important to recognize this added meaning as to recognize what is left out." Obviously drawing on information from Robert Hashima and others imprisoned in relocation camps, she wrote,

> In these same relocation camps the Japanese rank and file also made a stock accusation of 'having no *makoto*' against the Japanese group whom they had chosen to administer the camp under the self-government scheme. This clique thus put in power were for some time the pro-Japanese elders, but the camp popula-

> tion accused them of currying favor with the American administrators, of acting for their selfish interests, and not in the interest of the whole. To say that they had 'no *makoto*' covered all of this, for according to Japanese code a person who is self-seeking cannot be *makoto,* no matter how 'genuine' his profit-seeking is.

Her own earlier published strictures against cultures that admired "unbridled and arrogant egoists" may have made her particularly sympathetic to the Japanese concept of *makoto.*[12]

The final chapter to *Japanese Behavior Patterns,* "Walking the Tightrope," provided further development of the concept of *makoto.* Providing a summary of her insights into Japanese culture, Benedict wrote that "the Japanese code is formal in comparison with most cultures not only in the sense that it is regularized by precise rules of obligation and of etiquette but in the sense that it is specific to culturally defined situations and that good and evil depend in so great an extent on context. A people who live by such a situational code acquire a range of habits, each precise and accepted, which seem contradictory one to the other to persons brought up under a more absolute ethic." From that analysis, she drew a hopeful message about the way the Japanese might deal with defeat: "When situations change, however, they can throw themselves into the new with great facility." Another passage would later appear in modified form in *The Chrysanthemum and the Sword.* Trying to provide information about Japanese culture, she wrote, "The Japanese insert tiny wire racks into chrysanthemums to hold each petal in meticulous place. Chrysanthemums show just how important the notion of self-respect is to the Japanese. And without self-respect, the Japanese is ignored and even jeered by society." Using the flower to symbolize a somewhat different aspect of the Japanese "character," Benedict was already exploring the issues that would be developed at greater length in her postwar writing.[13]

As a result of her research into the Japanese character, Benedict concluded that it was not impossible that the Japanese might change, but at the same time she pointed out the dangers of American mistakes in dealing with the Japanese. Should the United States mishandle the task of guiding Japan through and out of defeat, the cultural imperative to wreak revenge in the most selflessly patriotic manner could prove a considerable problem. The future treatment of the Emperor was a focus of her concern. Benedict warned that the occupying forces should employ every means to avoid giving either the appearance of contempt for the Japanese people or statements that could be construed as libeling them. The title of the final chapter of *Japanese Behavior Patterns* conveyed a double message. The people who might be walking a "tightrope" could be the Americans, who might fall should they misunderstand the delicacy of their task in enabling the Japanese to accept defeat by treating them with appropriate respect. Even now, the tone of the argument in the final chapter can still give sensitive readers a sense of foreboding. This is Benedict at her best.

In addition to all its other valuable insights, the discussion of the significance of the Emperor demonstrated that confidence in Benedict's powers as an anthropologist was justified. Benedict compared the Emperor to the leaders of Pacific cultures and to the Polynesian pattern of division between the roles of Talking Chief and Sacred Chief. According to Benedict, the Meiji rulers had examined Western governments "with some care and they did not propose to expose the actual administrators of Japan to the criticism and danger of revolt which they saw in these nations." Instead, they developed a "shadow-administration" that mirrored "the wide-spread Pacific Islands pattern of division between the Sacred Chief and his administrator." The Sacred Chief withdrew from actual administration to become "a tribal symbol," serving "as the flag serves the United States" with "the added advantage that he was flesh and blood and could command a religious loyalty while still being a human being who, in theory, reciprocated with appreciation the worship of his followers." Based on these insights, Benedict warned her official readers: "Veneration of the Imperial House is a strict religious tenet of Japan and, however much it offends nations which espouse other tenets, it commands the deep loyalty of the Japanese. Every job to be done in rehabilitation will be less difficult according to the degree to which it has the sanction of the Emperor behind it."[14] Whether or not Benedict's advice directly influenced U.S. decision makers in regard to the decision to allow the Emperor to continue to preside over the Japanese nation, she sought to encourage that course of action.

Before the war had ended, Houghton Mifflin had sent a request to Benedict to publish the results of her research on Japanese culture. Had *The Chrysanthemum and the Sword* appeared under the initial title proposed, it might have succeeded in achieving Benedict's goal, but it would most likely never have achieved the same longevity it continues to enjoy in Japan. The first title suggested by the publisher of Benedict's earlier books was *We and the Japanese,* a title that assumed the book would be read by Americans rather than Japanese. As work on the manuscript progressed, Benedict came up with the alternate title, *Japanese Character,* which opened up greater possibilities for Japanese and American readers.[15]

Sensitive to the demands of the popular market, Ferris Greenslet, the Houghton Mifflin editor, suggested that the title of the first chapter, "Assignment: Japan," would make an appropriate title for the whole book. Initially, Benedict agreed, but later she requested that it be changed to *Patterns of Culture: Japan.* Benedict preferred to avoid her book's being lost among the many war-assignment type of books on Japan then being published, seeking instead to connect it to her earlier and well-known *Patterns of Culture.* Still seeking to attract and sustain reader interest, she offered to compromise with Greenslet by using the word "assignment" in a new formulation, *Assignment: The Japanese.* As she advised Greenslet, she could not write about "Japan" because she had never set foot on the "sacred Japanese soil." Eventually concluding in 1946 that the term "assignment" had become outmoded or overused, Greenslet suggested an alternative, *Patterns of Japanese Culture,* to distinguish it clearly from Benedict's 1934 book, which continued to attract readers.[16]

Just when it seemed as though the completed manuscript would be published as *Patterns of Japanese Culture,* the discussion between Greenslet and Benedict took a new turn. The editor, obviously thinking primarily of the mainstream American market, now felt that the title sounded too academic and that as a result sales would suffer. Greenslet whittled down a list of alternative suggestions to three choices: *The Curving Blade, The Porcelain Rod,* and the one favored by the press, *The Lotus and the Sword.* Upon receiving Benedict's comments, "chrysanthemum" replaced "lotus." Thus the emblematic title was circuitously chosen, with Benedict's favored option appearing as its subtitle. She, however, had suggested the flower that would create symbolic echoes for future readers in Japan.[17]

Once the title had been decided, Benedict inserted passages related to chrysanthemums and swords into chapters 1 and 12 so that the title appears to spring out of the work itself rather than being added at the last moment. As Benedict wrote about herself (using the masculine pronoun), she pointed out the paradox that the title was intended to convey: "When he [a writer] writes a book on a nation with a popular cult of aestheticism which gives high honor to actors and to artists and lavishes art upon the cultivation of chrysanthemums, that book does not ordinarily have to be supplemented by another which is devoted to the cult of the sword and the top prestige of the warrior." As she then informed her readers, she would write about "both the sword and the chrysanthemum" as a "part of the picture" she would paint of Japanese culture.[18]

Having provided a detailed portrait of Japanese culture in ten chapters, Benedict returned to the image in the twelfth chapter, where the flower became a vivid symbol of cultural rigidity. Drawing on Etsuko Sugimoto's *A Daughter of the Samurai,* Benedict described the Japanese style of carefully controlled gardening. She then wrote, "So too, chrysanthemums are grown in pots, and arranged for the annual flower shows all over Japan, with each crooked petal separately disposed by the grower's hand and often held in place by a tiny invisible wire rack inserted in the living flower." Substituting the wire rack for the sword in the earlier juxtaposition, Benedict used the contradictory images to symbolize the debate surrounding the two conflicting aspects of "the Japanese character." Added late in the production of the manuscript, these passages do not appear in the draft manuscript contained in the Benedict papers at Vassar College.[19] Despite its late appearance in the manuscript, the references to chrysanthemums demonstrated Benedict's sensitivity to Japanese culture in a way that would draw Japanese readers to her analysis decades after MacArthur had ordered its translation into Japanese.

Having been published as an intervention into a specific historical situation, *The Chrysanthemum and the Sword* outlived its contemporary context, perhaps because it revealed the enjoyment that Benedict took in understanding another culture. She had been looking for an unknown country ever since she was a child. Benedict found herself marginalized in American culture because she had impaired hearing and because she was a lesbian. It was this personal self-definition as an "abnormal" that afforded her the empathetic intuitions that gave her contribution to our understanding of culture such enduring value.[20]

Always insisting that Americans seek to understand the cultural choices made by other peoples, Benedict included the following passage in an unpublished draft manuscript that would form the basis for *The Chrysanthemum and the Sword:* "The contradictions which Occidentals see in their character and that of the Japanese are contradictions because we try to see them according to our categories, not according to their own. When we look at them in their terms, we may even come to agree with the Japanese who said to me, 'But the Japanese are so simple.'" Deftly shifting her terms from the usual reference to Orientals, and asking Americans to look outside their own boundaries, Benedict unmistakably endeavored "to make a world safe for differences." She was a "lady of culture" in its most profound sense. She was disadvantaged by gender, sexual orientation, and hearing difficulties and yet empowered by the insights these deviations from the "normal" gave her about difference, both between and within heterogeneous cultures. As suggested by the misidentification of the anonymous visitor to the Hieizan Temple in 1946 that began this essay, the figure of Benedict, like the chrysanthemum and her powerfully expressive words, continues to compel our respect and enliven our imaginations.

9 Ruth Benedict's Obituary for Japanese Culture

C. Douglas Lummis

Ruth Benedict's *The Chrysanthemum and the Sword* has long possessed an almost mysterious power to outlast its critics. Certainly this can partly be explained by Benedict's remarkable writing skill. Set down in marvelously simple and elegant prose, organized with extraordinary clarity, illuminated with wonderfully told stories and brilliant images, the book seems a model of the way one wishes social science could be written. Moreover, given that the research was mainly done during World War II and the book published shortly after, it seems remarkably liberal and tolerant. Perhaps it was the best that American liberalism could have produced under those circumstances. Nevertheless, judged by the criterion that matters the most—whether it helps or hinders understanding of Japanese culture—it is deeply flawed.

While the flaws in the book have been difficult for many Western scholars to see, they are clear enough when viewed from Japan. Devastating critiques of the book's inaccuracies and methodological errors by such scholars as Tsurumi Kazuko, Watsuji Tetsuro, and Yanagita Kunio appeared soon after the Japanese edition of *The Chrysanthemum and the Sword* arrived in Japan. The criticisms that Benedict took the ideology of a class for the culture of a people, a state of acute social dislocation for a normal condition, and an extraordinary moment in a nation's history as displaying typical behavior are by now well known in Japanese scholarly circles.[1] In its tendency to treat Japan as an absolute other to the United States and to explain the complexities of this state-run industrial society with a small number of generalizations about its "culture," *The Chrysanthemum and the Sword* qualifies as a work of what Edward Said labeled "Orientalism." Its view of the Japanese as the "most alien" of peoples, inscrutable to the "Western" mind until unlocked by the "ethnographer's magic," opposed to and incompatible with the people of the "West," had deep roots in the encounter between Asia and that section of European civilization that reached the eastern shores of the Pacific Ocean in the late nineteenth century.

But the book must be located more specifically chronologically. It was written on the occasion of the defeat of Japan and its occupation by the United States, by a person who did her research while working for the government that was engaged in a war against Japan and the subsequent attempt to reconstruct its society to meet Western requirements. Not only did it, unsurprisingly for the time, explain and justify the defeat and occupation, it was also effective in shifting the terms of Japan discourse from a wartime to a peacetime footing, specifically by substituting "culture" for "race" as the key concept to be used for criticizing Japan.

If *The Chrysanthemum and the Sword* was very much a product of its time, it was also deeply affected by the theoretical stance, interests, and obsessions of its author, Ruth Benedict. In addition, it was greatly influenced in a paradoxical manner by the official ideology of Japan's wartime government, especially as communicated to Benedict by her chief informant, Robert Hashima. These influences will be discussed below after assessing the nature and scale of the book's influence as a founding work for what became mainstream postwar Japanology. In particular, though the debt is not always acknowledged, virtually the entire discourse of that branch of Japanese studies called Nihonjinron has been carried out within the framework established by Benedict's book. The debate launched among Japanese scholars over "shame culture" vs. "guilt culture" spilled over into lay society so that the two terms have become established as expressions in ordinary Japanese language. Benedict's book gave birth, in both English-language and Japanese-language Japan studies, to an endless supply of binary "x culture vs. y culture" tools for blunt instrument social analysis.

Why, despite its errors of fact and interpretation, has *The Chrysanthemum and the Sword* exerted such powerful influence? It is primarily because the book is useful, not as an accurate account of Japanese society, but as a work of political literature. The same could be said of other works of anthropology. Long before anthropology was invented, drawing detailed pictures of "Another Country" was a time-honored method of political theory, a method of establishing a "standpoint" from which one's own society could be viewed in a different perspective, thus enriching self-knowledge and making possible self-criticism or self-praise. To serve that function it is not necessary that "Another Country" be a real place. For Plato and More, it was only necessary that their ideal republics be possible; for Rousseau, it was only necessary that Natural Man be logical; for Swift, it was only necessary that his various countries be imaginable; for Augustine, whose City of God is unimaginable, it was only necessary that it be utterable. Like these political theorists, anthropologists have often expressed a not-so-hidden intention to offer overt or covert "lessons" drawn from anthropological knowledge of other societies. What has only recently begun to be noticed is that sometimes the researcher's eagerness to educate led him or her to arrange the culture to fit the lesson rather than to draw the lesson from the culture.

Anyone who doubts Benedict's desire to be a political educator need only read the last chapter of her 1934 book, *Patterns of Culture.* According to Clifford Geertz, "To say one should read Benedict not with the likes of Gorer, Mead, Alexander Leighton, or Lawrence Frank at the back of one's mind, but rather with Swift, Montaigne, Veblen, and W. S. Gilbert, is to urge a particular understanding of what she is saying. *The Chrysanthemum and the Sword* is no more a prettied-up science-without-tears policy tract than [*Gulliver's Travels*] is a children's book."[2] Geertz understood *Chrysanthemum and the Sword* as a piece of Swiftian satire principally about U.S. society. This is an important insight missed by most earlier commentators.

Japan, in addition to being the only country actually on the map that Lemuel Gulliver visited, was also a country in the twentieth century with which the U.S. was engaged in a very intense relationship. It was what Benedict had to say about that country that has been its most important legacy. *The Chrysanthemum and the Sword* established in academic terms the paradigm for postwar U.S.-Japan relations. It depicted/invented Japan as the country the most appropriate for the United States to have defeated and occupied. Of equal importance, it depicted/invented the United States as the country the most appropriate to defeat and occupy Japan. Thus, Geertz is half right: the book is as much "about" the United States as it is "about" Japan. It taught that for the Japanese, being defeated by the United States was quite the best thing that could have happened, and that they should have been—and in fact were—grateful for this defeat. Moreover, the defeat was not a mere accident of power but had a kind of Hegelian necessity. It was Japan's only hope of advancing to a state of freedom.

According to *The Chrysanthemum and the Sword,* Japanese culture contained no concept or spirit of freedom, no principle of liberation. In fact, it contained no principle at all. This is the meaning of describing it as a "shame culture" where people act, not according to principles, but rather according to how they think they will look to others and whether they will be honored or shamed. In 1946 this was a convenient interpretation because it meant that Japan, having just been shamed before the world, would be willing to change itself by importing principles from the outside, meaning from the United States. All this is written in a polite and tolerant tone. What matters, however, is the content.

Benedict's judgment on Japan can be seen in her answer to the question Why did Japan fight this war? Her answer makes no use of economic or political explanations. Japan did not follow the well-known logic of colonial and imperialist powers—seeking markets, resources, investment outlets, and cheap labor. Nor did Japan follow the well-beaten path of tyranny—seeking power, glory, a central place in history. Nor had Japan, in contrast to Germany and Italy, passed over into an extraordinary state of political pathology. Nowhere in *The Chrysanthemum and the Sword* do the concepts of fascism, totalitarianism, or any similar notion appear. To admit the relevance of any of these explanations would be to admit that Japan's behavior was understandable according to ordinary "Western" reason: it was yet another rather extreme and badly timed example of plain old-fashioned imperialism. Determined to show that Japan's behavior was utterly different from anything known in the "West" and understandable to "Westerners" only by means of her "ethnologist's magic," the anthropological method, Benedict did not explain Japan's conduct of the war as found in a crisis in international relations, in Japanese capitalism, or in Japanese politics. Rather, it was "a cultural problem": the war was the inevitable and necessary expression of Japanese culture itself.[3]

Militarist Japan was simply "Japan" for Benedict as it had always been and as it would continue to be unless changed from the outside. This contrasts with her historicized interpretation of Germany. Basing her analysis of the state of German "morale" on British surveys of prisoners of war, Benedict argued that only the generation of men (presumably not women) in their late twenties were solidly Nazi. Benedict claimed that the "Nazi regime" had failed to "Nazify the age group now under 25 as it did the one now 25–30." As for the older generation, "There is no need to discuss the relative non-Nazification of the generation over 30 since the grounds for this are well understood. The fact that Hitler Regime [*sic*] has been of such short duration that there remains a whole older generation who grew up under a different social order, is of great importance in estimating Germany's future." In Benedict's discussion of Japan, however, there is no notion of a "failure of indoctrination" (nor for that matter, of a successful one), no term equivalent to Nazify, no suggestion that "a different social order" may have existed in the recent past. Even the word *regime* does not appear. While Germany's Nazism was a fleeting phenomenon that managed to attach itself to German culture only temporarily and precariously, Japanese militarism was Japanese culture itself. It had existed essentially unchanged from ancient times, and far from being imposed through indoctrination, it had been "voluntarily embraced."[4]

While it is possible that racism played some role in the lower depths of

Benedict's consciousness influencing her interpretation of Japan, it played no role whatever in her theory. Ruth Benedict campaigned against racism and considered anthropology—and in particular her theory of cultural patterns—-to be the definitive refutation of race theory. Moreover, race theory no longer fit the times. While it was compatible with U.S. war propaganda when the Japanese were to be killed, it was inappropriate as an ideology for the postwar occupation under which the Japanese survivors needed to be changed. Benedict's work offers a prejudice more appropriate to the period of occupation and reform and to America's postwar projects of forced economic development and other forms of humanitarian intervention: cultural prejudice. *The Chrysanthemum and the Sword* told its readers that Japanese culture must be changed and explained how it could be changed under the force of the U.S. military occupation. As Benedict wrote in *Chrysanthemum:* "In the United States we have argued endlessly about hard and soft peace terms. The real issue is not between hard and soft. The problem is to use that amount of hardness, *no more and no less,* which will break up old and dangerous patterns of aggressiveness and set new goals" (emphasis added).[5] Benedict's theory of "patterns of culture" has achieved a reputation as a theory of tolerance. Perhaps in some cases, but not in the case of a nation under occupation.

Benedict's *The Chrysanthemum and the Sword* is just as profoundly shaped by her poetic sensibilities as by her political intentions. Benedict came to anthropology from English literature, having graduated from Vassar College in English, taught English at a girls school in California, and published poetry before she entered the graduate program in anthropology at Columbia University. That she received her Ph.D. in three semesters not only testifies to her brilliance but also suggests that she did not undergo a fundamental retraining in methodology. This is supported by her own testimony, that "long before I knew anything about anthropology, I had learned from Shakespearean criticism . . . habits of mind which at length made me an anthropologist." According to Mead, Benedict transferred her sensibilities from literature to anthropology by seeing "each primitive culture" as "comparable to a great work of art" whose internal consistency and intricacy was as aesthetically satisfying as was any single work of art.[6]

In 1925 Benedict, in a New Mexico village, wrote to Mead, in Samoa, "I want to find a really important undiscovered country." Interestingly, she was referring, not to anthropology, but to poetry. Benedict published her poetry under the name Anne Singleton. The lure of the "undiscovered country" set the same person on both a poetic and an anthropological journey. In Benedict's own account of her childhood, she wrote that as far back as she could remember she lived in two worlds, one the world of her family and friends, in which she felt alienated and unhappy, and the other of her imagination, where everything was calm, beautiful, and rightly ordered and where she had an imaginary playmate. "So far as I can remember I and the little girl mostly explored hand in hand the unparalleled beauty of the country over the hill."[7]

The meaning for Benedict of this "country over the hill" is the main theme of her brief childhood memoir, entitled "The Story of my Life." It begins with the remarkable sentence, "The story of my life begins when I was twenty-one months old, at the time my father died." A relative told her what had happened. Her mother took her into the room where he lay in his coffin and implored her to remember her father. This scene was reproduced every March when her mother "wept in church and in bed at night," producing in the young daughter "an excruciating misery with physical trembling of a peculiar involuntary kind which culminated periodically in rigidity like an orgasm." It was this experience that divided her life into "the world of my father, which was the world of death, and which was beautiful, and the world of confusion and explosive weeping, which I repudiated." She resented her mother's "cult of grief" and retreated to the "other world" where her father belonged, identifying him with "everything calm and beautiful that came my way."[8]

This fascination with the calmness and beauty that comes with death shaped Benedict's responses when she encountered the dead. Benedict wrote that as a child she would bury herself in the hay on the family farm and imagine she was in her grave. When she was taken to a neighbor's house where a baby had died, she found the corpse a thing of "transparent beauty" that seemed "the loveliest thing I had ever seen." The same attitude continued into adulthood, as her life story attested. "Even now I feel I have been cheated or unfaithful if I can't see the dead face of a person I've loved. Sometimes they're disappointments, but often not." This theme became deeply embedded in the consciousness of Ruth Benedict and in the poetry of Anne Singleton, who wrote poems that expressed the same feelings: "And nothing dead but is perfected" went the line in one poem that Mead published in her biography of Benedict.[9] As the poem made clear, Benedict continued to yearn for the perfection that death appeared to offer in comparison to the messiness, emotional conflicts, and confusion of living.

Although Mead described Benedict as separating her emotional life, as expressed in poetry, from her anthropological work in the "earlier years," the same sensibility might be discovered inspiring both poems and ethnographic research. According to Mead, American anthropology in those years was a "salvage task." Anthropologists collected "masses of vanishing materials from the members of dying American Indian cultures" that must be pieced together to recover the dead or dying cultures. It is not difficult to see how Ruth Benedict/Anne Singleton could be attracted to this enterprise. Who better than an anthropologist could make a career exploring the country over the hill and contemplating the beauty of the dead under the supervision of Boas, the man she came to call "Papa Franz"? It is not difficult to hear the voice of Anne Singleton in Benedict's most famous passage, the one with which *Patterns of Culture* opens, as Benedict describes how one day Ramon (Benedict's Digger informant) broke in on his descriptions of grinding mesquite and preparing acorn soup. "In the beginning," he said, "God gave to every people a cup, a cup of clay, and from this cup they drank their life." Interjecting that she never learned "whether the figure occurred in some traditional ritual" or was the product of Ramon's own imagination, Benedict wrote that "the figure of speech was clear and full of meaning." Referring to the cup, Ramon told her, "They all dipped in the water" but "the cups were different. Our cup is broken now. It has passed away."[10] The poet and the anthropologist jointly selected this image.

The task of the anthropologist was, as Mead said she learned from Benedict, to "rescue the beautiful patterns," not the survivors. Equally obviously, the patterns could not be restored to living form but only written down. Thus the anthropologist would move backward in time, beginning with the fragments of the shattered cup—some missing, some badly worn—and try to piece together, both from evidence and from sympathetic imagination, the culture pattern as it once existed. The native informants no longer lived this pattern. They were themselves defective as evidence: fragments. What the researcher required from them was their memory of the culture that had once lived.[11]

Although Benedict's analysis in *The Chrysanthemum and the Sword* has been criticized for her failure to learn Japanese language or visit the country, she was only studying Japan as she had always undertaken her anthropological research. As Mead admitted, Benedict "never had the opportunity to participate in a living culture where she could speak the language and get to know the people well as individuals." Handicapped by her deafness from learning languages or being an active participant observer, Benedict relied on informants, the research of other scholars, and documents for her knowledge of the cultures she sought to understand. Remembering her own unhappy research trip among the Omaha, when she had depended on paid informants, Mead described Benedict's research experiences: "She never saw a whole primitive culture that was untroubled by boarding schools for children, by missions and public health nurses, by Indian service agents, traders, and sentimental or exiled white people. No living flesh-and-blood member of a coherent culture was present to obscure her vision or to make it too concrete."[12] Unable to interact with any spontaneity or active involvement in the cultures she studied, Benedict found it easier to discover the patterns she was seeking in the isolation of her own imagination.

Perhaps turning what may have been a necessity into a virtue, Mead pointed out the benefits of Benedict's reliance on a few chosen informants and inability to engage actively with the people of the culture she was studying. "The clarity of her concept" benefited from "the lack of a sensory screen between the field worker and the pattern," that is, from the absence of the actual members of a living culture. Benedict's field letters reflected an amused, patronizing attitude toward her informants. From Zuni she wrote Mead: "Nick and Flora both eat out of my hand this summer. As soon as I go out for water the men begin to come in. One amorous male I have got rid of, dear soul. He's stunning, with melting eyes and the perfect confidence which I can't help believing has come from a successful amour with a white woman." In Cochiti she celebrated another piece of good fortune: "I'm in luck that my old shaman is poor—otherwise he would be frowned on. One of those who rob the poor working girl, you know!"[13] Benedict's letters stressed the separation between herself and the people she relied on to inform her about the culture she was trying to excavate, without considering the dangers that her informants faced.

A footnote to a field report written by Ruth Bunzel, who worked among the same peoples as Benedict, told about the consequences of revealing the "tales and ceremonials" that Benedict extracted from her informants. Bunzel wrote about the death of one of her best informants, who, among other things, told her many prayers in text. "During his last illness, he related a dream which he believed portended death and remarked, 'Yes, now I must die. I have given you all my religion and I have no way to protect myself.' He died two days later. He was suspected of sorcery, and his death was a source of general satisfaction. Another friend of the writer, who had always withheld esoteric information, remarked, 'Now your friend is dead. He gave away his religion as if it were of no value, and now he is dead.' He was voicing public opinion."[14] The description suggested that this informant might have been the same Nick or someone else whom Benedict had also interviewed.

In contrast to her jocular response to the people, Benedict's letters conveyed a different emotional response to the location. Writing to Mead about her last morning in Zuni, she wrote, "Yesterday we went up under the sacred mesa along stunning trails where the great wall towers above you always in new magnificence." Obviously awed by the beauty of the landscape and the vista, she added, "When I'm God I'm going to build my city there."[15] Anthropology gave the poet the opportunity to become God and to design perfect cultures, if only for the dead. For a person who found perfection in death, the ability to discover and write about the aesthetically pleasing patterns of another culture compensated for the difficulties she encountered as a necessary part of that process.

As a complex individual torn between two worlds, however, Benedict did not easily reconcile her different demands and desires. The anthropologist collected information about the customs, rituals, habits, ceremonies, myths, and other institutionalized activities that make up a culture. The poet portrayed them in vivid, dramatic, and intricate detail as a coherent whole. Pattern was her fascination and her trademark. The following entry in her journal, probably written around 1915, thus appears to represent a puzzling contradiction. Writing as the slaughter of the Great War was beginning to penetrate American consciousness and as she was encountering the disappointment of her own recent marriage, Benedict said: "All our ceremonies, our observances, are for the weak who are cowards before the bare thrust of feeling. How we have hung the impertinent panoply of our funeral arrangements over the bleak tragedy of death. And joy, too. What are our weddings, from the religious pomp to the irrelevant presents and confetti, but presumptuous distractions from the proud mating of urgent love?"[16] Ceremonies, observances, funerals, weddings—these are the very stuff of which cultural patterns are made. One might dismiss this as a youthful outburst were it not a constant theme in her journal entries and poetry. It is an attitude that must be described as a horror of pattern. What is the marvelous creature of human genius in the daytime of her anthropology becomes a nightmare in the nighttime of her personal life.

The constantly recurring image in her poems is that of some substance that escapes patterning—breath, wind, mist, water—in contest with the forces of rigidity:

> "How shall we say of this, inductile water,
> It shall be chiseled by the fragile sand?
> Water slips lightly, flawless, from our confines,
> Shaped to no permanent feature, fluid as air;
> Though we stand hewing till the sword is eaten,
> There is no lineament we shall chisel there."[17]

Both poem and diary entry appear to convey a sense of futility about any attempt to impose order on nature. Pattern could not be sustained in a fluid environment.

If Benedict knew that water could not be carved with a sword, she also knew what could happen in winter. Writing about water turned to stone, Benedict mourned for the loss of its fluidity:

> Stone never ran quicksilver in the shade,
> Stone never gathered out of doom a singing,
> Lost now, forgotten, and its dream betrayed.[18]

The beauty of flowing water is tragedy itself. It is not grateful to the winter for transforming it into rigid crystal. It is a choice of how to live with the courage that came from knowing about death and yet choosing to live.

Benedict expressed these sentiments in another poem that showed the same sense of rebellion as she asked, "How should the shred and filament of the air-stepping mist be lovely still, or hush itself to blue against the wintry sky?" Rather than accept that fate, she concluded that it were "best we kissed before the wind, and went as smoke clouds do."[19] For a cultural patternist these are words of rebellion: cultural institutions as prisons of the human spirit. Can a society be built on such ideas? Certainly not, which is why Anne Singleton was a poet, not an anthropologist. But such beliefs can lead to a bitter assessment of one's fellow human beings.

Bitterness can be found in another journal entry. Benedict wrote that in "modern society the majority are lost and astray unless the tune has been set for them, the key given to them, the lever and the fulcrum put before them, the spring of their own personalities touched from the outside." Her entry concluded with an outburst of pure repugnance against "the stench of atrophied personality."[20] The horror of pattern could not be more powerfully expressed in her opposition to conformity. Is it possible to reconcile all these contradictory ideas? Perhaps Benedict's own inability to do so was one of the reasons she wrote under two names—as the poet who celebrated fluidity, and the anthropologist who described pattern. On the one hand, Benedict cherished the image of the beauty of death; on the other, she expressed a horror of atrophy. Atrophy, however, is not death but sickly life, life so undernourished and underused that it is shrunken and decayed.

According to Benedict, culture patterns carry a double meaning. When the culture is dead, its pattern has the same beauty Benedict found in the faces of dead people—the aesthetic closure of something reconciled and finished. For the living, the patterns are a kind of death-in-life, an oppressive, imprisoning force. If the living do not struggle to liberate themselves from them, they will never be fully alive. These "other-directed" ones, as David Reisman was to call them, who live only by pattern and custom, have neither the beauty of death nor the joy of life. They are in a state of life resembling death, a state of atrophy. In her poetry and journal entries, Benedict was talking about her own compatriots.

It was only with *The Chrysanthemum and the Sword* that Benedict transferred this damning diagnosis to the erstwhile enemy country, the shame-culture Japan, enabling her to describe her own country (using a Singletonian air metaphor) as a land of "simple freedoms which Americans count upon as unquestioningly as the air they breathe." Arguing that both *giri* and *on* emerge as central principles of conduct in a shame-based culture, Benedict developed her discussion of these issues in three chapters where she argued that "shame" was "recognized as the root of virtue" in Japan in contrast to the Western "consciousness of sin," which saw it as "the root of virtue." Ignoring, or perhaps unaware of, the rich Japanese vocabulary for expressing guilt, Benedict provided comfort for a corporatized America, fearing for its lost "individualism," that took comfort in imaging the Japanese as a people lacking a conscience or a moral code. Contrary to Pauline Kent's analysis, Benedict made precisely this flawed argument: that in Japanese culture "honor" took the place of conscience, God, and the need to avoid sin in Western ethics, while shame supplied "the soil out of which all virtue grows."[21]

In addition to Benedict's propensities for both seeking and resisting pattern, two other major factors shaped *The Chrysanthemum and the Sword.* One of these was Japanese government ideology, and the other was the uncritical acceptance of that official ideology by Benedict's informant, Robert Hashima. It is common knowledge that after the Meiji Restoration the Japanese government labored to remake Japanese society politically, economically, technologically, and culturally. Eric Hobsbawm and Terence Ranger's *The Invention of Tradition* did not include an analysis of Japan, but surely it ought to be considered as a paradigmatic case. The Meiji elites, using compulsory education, military conscription, institutional reorganization, and many other forms of indoctrination and force, sought to organize the various cultures of the Japanese and Ryukyu Archipelagos—and later the cultures of Taiwan and Korea as well—into a single nation-state under the direct rule of the Tokyo government. The whole apparatus was mystified under the newly organized emperor system and legitimized by means of the "invented tradition" of a modernized version of the ethic of the old *bushi* class. This story has been one of the chief objects of study for historians of modern Japan and hardly needs to be repeated here.[22]

Benedict mistook the existence of an officially fabricated and enforced national ethical system for a culture that grew up naturally because "human society must make for itself some design for living." Perhaps this distinction did not exist or did not matter in the indigenous cultures Benedict had studied before she went to work with the Office of War Information, but failing to take it into account in the case of Japan was a fatal error. The error was understandable, since it is an error that was positively promoted by the Japanese government and passed on, wittingly or unwittingly, by many

Japanese intellectuals. In his 1950 review of *The Chrysanthemum and the Sword,* Yanagita Kunio acknowledged this responsibility when he wrote, "One thing we may criticize ourselves for is that those of us who have tried orally or in writing to explain Japan to the world have often taught falsehoods." Referring to the impact of the Meiji Restoration, he explained how the teachings of "bushido" had "gradually spread throughout the entire society, and in particular came to dominate the field of education. Following the hint offered by Benedict's book, this is a point on which we need to reflect. The life of the bushi class had many peculiarities. To make it the basis for the life of all the people was neither possible nor necessary, and often harmful."[23] An uncritical acceptance of what Yanagita referred to as scholarship "biased toward a class amounting not even to ten percent of the population" had introduced this error into *The Chrysanthemum and the Sword.*

Yanagita's insight that Benedict's errors were grounded in a misconceived self-knowledge among intellectuals in Japan is an important one. Still, identifying these "falsehoods" taught by Japanese scholars as one of her sources does not fully account for her analysis in all its peculiarity and detail. Benedict read just about everything that was available at that time in English on Japan, but her analysis differed from all earlier works. Soeda Yoshiya claimed to have found seventeen points where she seems to have used Nitobe Inazo's *Bushido* as a source, but even if this is true, her analysis is by no means the same as Nitobe's. The core of Benedict's work is her analysis of Japan as a "shame culture" whose central value system comprised a hierarchically ordered series of notions of obligation: *on, chu, ko, gimu,* and so forth, terms that are not part of Nitobe's or any other previous analysis. At one time it appeared possible that Benedict's chief source for these ideas might be the pre-1945 moral education (*shushin kyouiku*) textbooks issued under the authority of the Ministry of Education, through which the state ideology was disseminated in the schools. They do contain most of the words that Benedict analyzed, but they contain a great many other value terms as well: words for cooperation, benevolence, civic virtue, enterprise, mutual aid, self-management, inventiveness, and so forth. Benedict selected a few of these ethical terms and ignored the rest. It is therefore the principles or person who guided the selection in her research process that must be considered. That was Robert Hashima, the only informant mentioned by name in the acknowledgements section of *The Chrysanthemum and the Sword.*

During my interviews with Hashima in Tokyo in 1996 and 1997, he discussed his collaboration with Benedict. At that time I showed him a passage about "the man in the street" and asked him, "Is that you?" He looked at the page a long time, laughed, and said, "I guess so!" More concretely, he said, "Well, as far as providing her with the information, I guess I would say it came from me."[24] It would be a mistake, however, to think of Hashima as a literal "man in the street." Benedict's "man in the street" would be a person who knows the customs and values of a culture simply by being a member of it, not one who has specialized knowledge of them gained through systematic study or who has a personal interpretation of them. Leaving aside the question of whether such an unreflective person exists anywhere, this is certainly no description of Hashima.

Hashima was born in the United States and brought to Japan by his parents in 1932, at the age of thirteen. When he entered school at that time, he knew no Japanese. He also knew little of the official government ideology that dominated the school system during this period. As he recalled his experiences, he was forced to memorize the *kyouiku chokugo* [*Imperial Rescript on Education*]. Hashima was not persuaded by this ideology. As he stated in the interview: "So far as the system goes, I didn't care for it." He decided, however, that he needed to master it in order to survive. He succeeded well enough to graduate from Hiroshima Shihan Gakko teacher's college and to teach school for a while, including classes in "moral education." When I asked him how he was so knowledgeable about all this at the age of twenty-four, he answered, "Well, I went to teachers college there." Hashima thus absorbed the official ideology of "moral education" based on the Bushido.

Advised by an uncle to leave Japan before the war started, Hashima arrived in the United States in 1941 where, ironically, he was sent to an internment camp. There he met the anthropologist John Embree, who got him a job working for the Office of War Information. Hashima, remembering his meeting with Alexander Leighton, said: "Leighton took me to Benedict. And, oh, she's reading [Natsume Soseki's novel] *Botchan.* She told me her assignment was Japan, and she was reading this book, *Botchan,* I remember." Here Hashima related the scene where Botchan throws back the price of a glass of ice water to a teacher who had insulted him. "Dr. Benedict couldn't understand why. So that's where I told her, this is where the *giri,* and *on,* these things start there. Ooooh. She went to Leighton, she says, I want Hashima." Thus, Hashima became the key medium between militaristic governmental ideology and Benedict.

Rather than merely providing information, Hashima interpreted that information. What he was taught in school was not "an ideology," it was Japan itself. He did not like it, but learning it was his assignment, and learn it he did. Underneath his apparent acceptance and his mastery of its details, his interview revealed a deep alienation, one that remains to this day along with the continuing sense that the ideology continued to dominate Japanese thinking. As he explained, older people "seem to go back into this pattern." He believed the Japanese could not "change Japan" unless "the language and the history" was changed. When prodded, he exclaimed, "Get rid of Japanese! Get rid of the Japanese language!" He added that it should be changed to English. Hashima's combination of rich insider information and radical alienation made him the ideal informant for Benedict's assignment to analyze America's "most alien enemy." One can easily see how the deep resentment that must have been instilled into him by his bitter boyhood experiences would harmonize well with Anne Singleton's "horror of pattern."

Hashima was by no means Benedict's only informant, but it seems that he became a kind of touchstone, an authority against which she would test information from other sources. As he said in the interview, "I feel that she—maybe today I kinda feel guilty, but, ah, she would ask for my opinion, what I thought about what these people had said. And she seemed to, if I said no, then she would, you know, maybe change it or something, but, ah, she kinda relied on my opinion quite a bit." Hashima's alienation from Japanese culture may have appealed to Benedict, who obviously felt a similar estrangement from aspects of American culture.

The Chrysanthemum and the Sword is the product of a remarkable convergence of conceptions. Benedict, Hashima, and Japan's wartime militarists—though each for entirely different reasons—all subscribed to the myth that Japanese society was something like a family or tribe, that there were no functional class differences within it, that the ideas of democracy and rebellion were inconceivable within it, that its value system was traditional and was the essence of the national identity—in short, that the system was not the product of state oppression. To be totalitarian and to be Japanese were the same thing.

Benedict's most chilling image was not the sword but the chrysanthemum. For her the sword was "not a symbol of aggression, but a simile of ideal and self-responsible man." This aspect, Benedict conceded, the Japanese could keep. It was the chrysanthemum that represented everything she found horrifying in Japanese culture. The image appeared in a discussion in which the metaphor of gardening was used to illustrate freedom and its absence. In Japanese gardens, Benedict said, nature itself is forced to fit the pattern of culture, its wildness is tamed, and even the pine needles that seem to have "naturally" fallen from the tree are actually spread there by the gardener. As Benedict wrote, "So, too, chrysanthemums are grown in pots and arranged for the annual flower shows all over Japan with each perfect petal separately disposed by the grower's hand and often held in place by a tiny invisible wire rack inserted in the living flower."[25] Here the poet's image of Japanese society as chrysanthemums fixed rigidly on a rack, each flower impaled on a wire, and thus human beings fixed rigidly on a rack, a wire passing through each soul, surfaced in the anthropological text. Once again one can sense a convergence of minds here, for surely this must be very much what it felt like to Robert Hashima, and surely the image described a situation that Japan's military government would very much have liked to achieve.

When confronted with this image of a culture, Benedict's cultural relativity shut down. With regard to Japan, she tried to sidestep the issue by suggesting that the Japanese system was such a violation of human nature that the people would naturally abandon it simply upon being shown the American alternative. As she informed her U.S. readers in an embarrassingly inept version of the Japanese word for democracy, "We must remember, now that the Japanese are looking to de-mok-ra-sie since their defeat, how intoxicating it can be to them to act quite simply and innocently as one pleases." Returning to her chrysanthemum image, she added, "The chrysanthemums which had been grown in a little pot and which had submitted to the meticulous disposition of the petals discovered pure joy in being natural."[26] The trouble is, there is no basis in anthropology—certainly not in Benedict's anthropology—for describing a particular social behavior as natural.

To imply, as Benedict did, that the behavior of the people of one's own country is "natural" was both to fly in the face of her own teaching and to fall into blatant ethnocentrism. Is this the damage war inflicts on the scientific spirit? Benedict hoped that the Japanese would "naturally" change, but as a government advisor she could not leave it at that. In the passage quoted earlier, Benedict made clear that the victorious U.S. government should not shirk its task of using "that amount of hardness, no more and no less, which will break up old and dangerous patterns."[27] There is something a little chilling about an obituary written by one of the executioners. It calls to mind the image of a priest who, when his beautiful funeral ceremony is disrupted by the deceased struggling to sit up in the coffin, smacks him over the head with the shovel and then returns to his speech on how we should honor the life he had lived. It is in this context that Benedict's "respect" for Japanese culture should be understood.

Benedict with two members of the Blackfoot tribe in Montana in the summer of 1939, when she went there to do fieldwork. The Blackfoot, with a tradition of social welfare and individual opportunity, were Benedict's model of a just society. She wrote about them in "Primitive Freedom," *Atlantic Monthly,* 1943. Margaret Mead Papers and the South Pacific Ethnographic Archives, Library of Congress, courtesy of the Institute for Intercultural Studies, New York City.

Mead in Samoan dress, with her friend Fa'amotu, during her fieldwork on Samoa in 1925–26. On the basis of this fieldwork she wrote her first major work, *Coming of Age in Samoa,* a study of Samoan culture and of the behavior of adolescent girls on the island of Tau. It was published in 1928. Margaret Mead Papers and the South Pacific Ethnographic Archives, Library of Congress, courtesy of the Institute for Intercultural Studies, New York City.

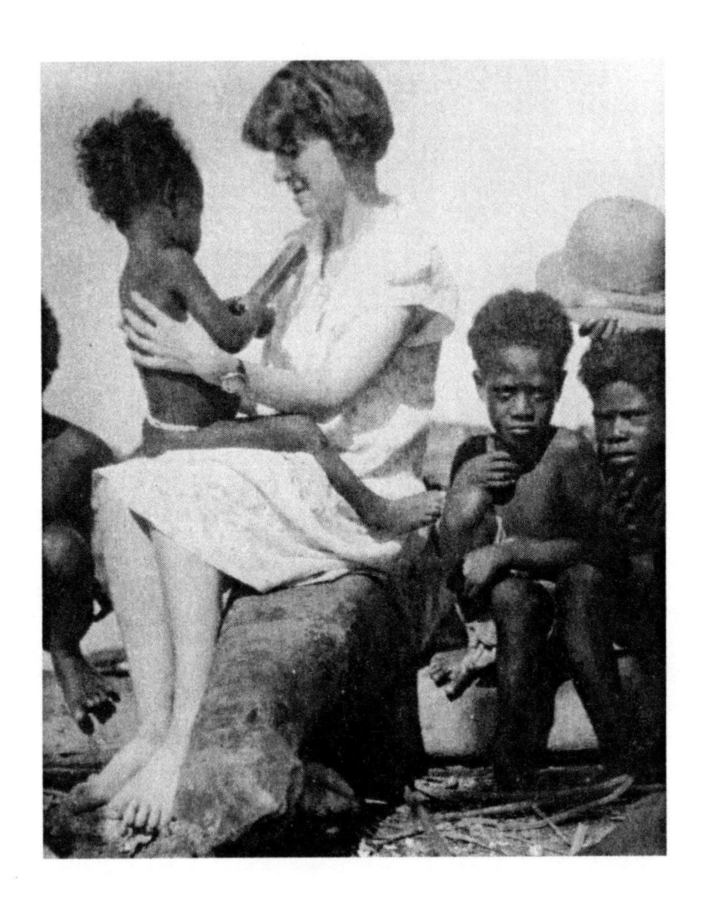

Mead in Peri Village, on the island of Manus, in the Admiralty Islands off the coast of New Guinea, with Ponkiau, Bopau, and Tchokal. These were three of the children she studied for *Growing Up in New Guinea* (1932), her work on the society of Manus and its child-rearing practices. Margaret Mead Papers and the South Pacific Ethnographic Archives, Library of Congress, courtesy of the Institute for Intercultural Studies, New York City.

Mead with Gregory Bateson, writing up fieldwork notes in their thatched-roof hut, during their re-study of the Iatmul tribe in New Guinea in 1938–39. Bateson's *Naven,* based on his earlier visits to the Iatmul, had been published in 1936. Mead wrote about this tribe in *Male and Female* (1949). Margaret Mead Papers and the South Pacific Ethnographic Archives, Library of Congress, courtesy of the Institute for Intercultural Studies, New York City.

Cover of the first edition of *Coming of Age in Samoa,* published in 1928 by William Morrow and Company. Margaret Mead Papers and the South Pacific Ethnographic Archives, Library of Congress, courtesy of the Institute for Intercultural Studies, New York City.

Unattributed photograph of Margaret Mead wearing a barkcloth belt. Reprinted by permission of *New York Times Pictures,* the New York Times Company.

Cover of 1947 Pocket Books edition of *Tales of the South Pacific* by James Michener. Reprinted by permission of Pocket Books, a Division of Simon and Schuster.

Cover of the Dell Laurel Paperback reprint edition of *Coming of Age in Samoa,* published in 1968 by arrangement with William Morrow and Company. Reprinted by permission of HarperCollins Publishers, Inc.

10 The Parable of Manus

Utopian Change, American Influence, and the Worth of Women

Margaret M. Caffrey

Writing about a 1953 field trip to Manus in Papua New Guinea, Margaret Mead intended *New Lives for Old* to challenge the contemporary standard of science and scientific writing— scholarly, impersonal, and detached. In part in response to the impact of anti-Communist witch-hunts, most scholars had decided to avoid action in the world. As Mead later wrote, "A general lack of involvement with wider ethical or political issues was the rule."[1] Mead, however, renewed her engagement with policy questions in her fieldwork and her writing when she researched and wrote *New Lives for Old.*

In contrast to earlier field trips, where she had believed that she should not interfere in a cultural or personal problem, on this visit the people of Manus asked her advice on many issues. "Problems of participation, where one should interfere and where not, take on a different color when the people themselves are as anxious for guidance on Robert's Rules of Order as for quinine to stop malaria." She became involved in "therapeutic participation," preventing a suicide and advising parents and relatives after the birth of a deformed infant. She found herself participating in change as well as documenting it. "I have to combine helping with the school with getting work done," she wrote. As she later explained, "The people treated me as a resource, using competently every skill or supply that I possessed."[2] No longer the detached scientific observer, Mead participated, counseled, and prodded people to make changes. She sought to convey those new roles through the use of a new form of writing appropriate to its subject.

In *New Lives for Old* Mead continued her experimentation with form, paralleling her interest in applied anthropology. By including large portions from *Growing Up in New Guinea*—a more impersonal, omniscient, "objective" account—Mead established her scientific credentials and her professional authority. She framed this within a personal memoir format, placing herself in the fieldwork, using the personal "I," interweaving her own experiences and emotional responses into the text. Her style would evoke criticisms from academics, one of whom panned her writing as "cloying marshmallow prose" and "an increasingly journalistic impressionism which apparently the public welcomes." Mead believed, however, that personal, narrative, and detailed writing would provide a better basis for policy decisions. She republished a United Nations report made up of impersonal data lists and charts with no analysis and no recommendations in an appendix to show her readers that her preferred style addressed "the problem of comprehension and responsible action which faces the modern world," in contrast to the bureaucratic style of the report.[3]

Since Mead wanted to make a difference, she preferred a style that would encourage people to read her work and act on it. "This Manus record is presented as the material which the research scientist owes to the society of which he is a member," she wrote. She offered her readers "carefully collected materials on which considered decisions may be based." Interweaving subjectivity and objectivity, the particular and the universal, the observer and observed, Mead consciously broke with academic convention to achieve her purpose.[4]

Mary Catherine Bateson wrote that her mother saw her restudy of the Manus as a "parable of hope."[5] This was more than just a felicitous turn of phrase. As a literary form, parables are homely stories taken from everyday life to offer guidance or to illustrate a truth or moral. The messages of parables can also contain the promise of utopian change. *New Lives for Old* provided two major parables for her readers: the parable of the Manus as a model for the American people in pledging a new allegiance to a United Nations instead of a United States, and the parable of the Manus as an important model for a postcolonial world. The first was a discourse that oscillated between an American identity built on nationalism and isolationism, and one built on world community; the second debated the end of colonialism, and if so how and on what terms. While speaking within the terms of these discourses, Mead sought to establish a blueprint for utopian change in American society and in the world. This chapter examines these two parables as promoting specific positions within larger discourses on both American identity and colonialism during the Cold War.

Among the eight peoples with whom she had conducted fieldwork, the people of Manus seemed to Mead to most closely parallel American society. "This book," she told her readers, "is offered as food for the imagination of Americans, whom the people of Manus so deeply admire." Although distant and few in number, "their experience is part of our experience as we learn to draw for our inspiration . . . on all that has happened in the world."

Studying Manus, she affirmed, could "throw light on contemporary problems."[6] Without explicitly stating her intention to write a parable, she invited her readers to apply the lessons of her research to their own concerns.

Mead believed that Americans in the 1950s had become confused about American identity and values in the face of a profoundly changed postwar world. Intricately intertwined with this crisis was the issue of utopia. Americans, like the Manus, faced a conflict between desiring an unrealizable fantasy, that is, a false utopia, or seeking to create a valid utopia that was ultimately only attainable through day-by-day, constant effort. For Americans, the false utopia was a "diluted utopia," a retreat to nationalism and isolationism. In a 1949 article Mead argued that the past offered "an inadequate model on which to build procedures in the atomic age." For the Manus the false utopia was a cargo cult named "The Noise." Although ultimately rejected, The Noise became a catalyst for a movement, led by a man named Paliau, that worked toward what Mead perceived as valid utopian change. The Manus would seek to transform their society based on a Euro-American model. Likewise, Mead believe that Americans, after rejecting a return to a utopian past, would find in "our priceless heritage of political innovation and flexibility," the resources for their own valid utopia, assuming a prominent position in the world community striving to be born. As she would write in 1957, Americans would be able to create a "more vivid" utopia without retreating into isolation or vainly seeking a lost golden age.[7]

While favoring utopian change for Americans and Manus—the Manus taking an equal place in the world community and the Americans moving from insular nationhood to world citizenship—Mead wanted to avoid these goals being dismissed as unrealizable fantasies at a time when utopia was often associated derogatorily with Communism. As a consequence, she did not use the word *utopia* in *New Lives for Old* except to describe what had been or should be rejected. This meant using constructions that would allow her to evade this conceptual dilemma. As she told her readers, "From one point of view the whole of history, and very particularly our own American history, can be seen as a struggle between those who seek a utopia here on earth and those who feel that the life of man is made better by ever-changing institutions carefully shaped and daily renewed by human effort."[8] By setting up "utopia" and "reality" as the two poles of binary opposition, Mead avoided having to discuss her criteria for distinguishing between false and valid utopias.

Mead used rhetorical strategies concerning the Manus that both contained subtextual references to utopia and conferred realism on the concept that she had overtly rejected. Mead emphasized the amazing change the Manus had undergone from "nearly naked savages" to twentieth-century people. They "had taken their old culture apart piece by piece and put it together in a new way" to reach for "new goals—a modern Manus way of life." The picture series that opened the book made the same point. The caption read, "From Savage Boyhood to Modern Maturity." The bottom photograph showed a picture of Mead's cook-boys in 1928, with wild hair and few clothes. The top pictured these boys now grown to manhood—with Western clothes and haircuts—making real the utopian vision Mead had discussed for the Manus without explicitly using that term. While Mead emphasized the positive changes—the new ease of male and female relationships and the new democracy—her narrative contained less edifying, "realistic" elements to show that the Manus had not yet created a perfect society. These included accounts of attempted suicide and various conflicts between people.[9] Her overall view, however, depicted the Manus as moving toward human betterment.

The references to "savages" served another purpose beyond stressing the utopian nature of their "new lives." As Michel-Rolph Trouillot has shown, the "savage slot" sometimes moved away from the "utopian slot" as it did in the 1950s, when believable utopias tended to depict people from other worlds or people from the "civilized" future, as Edward Bellamy had done in *Looking Backward* (1887). Mead faced the problem of making the people of Manus seem a credible analog from whom Americans could learn. To do this, she emphasized the change from "savage" to twentieth-century people. Mead showed her keen awareness of the politics of representation in her preface to *New Lives for Old* when she wrote, "The word *native* is used in those contexts where it seems appropriate, when discussing the point of view of the Administration, Europeans, etc., and the term *people* is used when the people of the Admiralties are being referred to as part of the modern world." Intended to show that "all of us" lived "in the same world," Mead used the word *people* precisely because she was not representing them as savages. Indeed, Mead sought to demonstrate the opposite through a rhetorical strategy that engaged the stereotype of the "wild man of Borneo" in American culture and worked to neutralize it, thus also neutralizing it with regard to the Manus.[10]

Mead did not wish to portray the Manus as noble savages or primitives. Psychological portraits showed the faults as well as strengths of individual Manus. She balanced portrayals of the "noble" use of democracy in village meetings with accounts of the lack of sexual pleasure, portraying the faults and strengths of the society as she did with individuals. In these ways Mead strove to portray people American readers could identify with and learn to overcome a paralysis of the American spirit. Mead later referred to "the apprehension and apathy of the dismal 50's." She believed that fear of the atomic bomb, the ongoing arms race with the Soviet Union, and the anti-Communist witch-hunts reflected reactionary retrenchment in the face of change, accompanied by a retreat into domestic life.[11]

Mead agreed with other contemporary critics who decried the conformity and apathy of the decade. Betty Friedan later wrote, "The American spirit fell into a strange sleep, men as well as women, scared liberals, disillusioned radicals, conservatives bewildered and frustrated by change—the whole nation stopped growing up." A historian characterized the winter before Mead left for Manus as "a time of ripe despair for most American liberals." Another scholar stated, "The idea of change—as possibility or threat—was perhaps the summary idea for all the others." The Cold War created a climate of discursive instability that oscillated between an American identity built on isolationism and one based on full membership in a world community.

For Mead the Manus, like the United States, had reached a moment in their history when they had found the past inadequate to build on for the future. The Manus had mustered the courage to step forward, embrace change wholeheartedly, and model themselves on the best other that they had encountered during the war, the Americans. Mead believed that Americans could use the Manus example to go forward themselves into the unknown, using as their models "materials from other cultures," as the Manus had also done. In a time when Americans had come to doubt their "belief that men can learn and change," Mead used the parable of Manus to remind her American readers that they did not have to feel impotent in the face of severe problems.[12] Like the Manus, Americans could successfully change.

Mead desired Americans to cooperate with other peoples to create a new and larger society. The Manus had joined with thirty-two other villages containing a mixture of at least two other New Guinea peoples. They offered a model for Americans and their relation to the United Nations about the possibility of joining together with people greatly different, giving up hostilities, and forming a world community. As the Manus saw themselves as having something unique to contribute to the larger society, so Americans should see themselves as well. Mead, however, faced the discursive problem that, in the context of a dominant American discourse, she was among a minority favoring world government. As Paul Boyer has documented, support for the United Nations and world government had diminished as anti-Communism and nationalism increasingly gained predominance in the Cold War era.[13] Confronting such opposition, Mead framed her parable to seek to change public opinion.

Mead's portrayal of the Manus' striving for democracy reminded Americans of the importance of "the ritual of democracy" at a time when civil liberties were being curtailed at home. At least twice she portrayed a village meeting as reminiscent of New England town meetings. Speaking to readers in the United States, where dissent could mean the loss of a job or possibly prison time, she recalled for her fellow citizens their tradition of dissent, a tradition that strengthened rather than divided, that led people to laughter rather than anger. She also insisted that democracy must be inclusive. "But we have seen over and over again in history," she wrote, "a religion or a political system grown great by its openness, its receptivity to all men, contract and perish in a new throe of cultish exclusiveness. Often, perhaps always, it is reactive to some rival universalistic movement."[14] These words contain echoes of the emerging critique of McCarthyism without stating her opinions so forcefully that she might have alienated her readers.

The second of Mead's parables promoted the Manus as a model for the postcolonial world in the transition from colonialism, the cause of a "running conflict" in the United Nations marked by "long and acrimonious debates." As one foreign policy analyst wrote in 1958, colonialism was "the emotional issue of the decade." He went on to describe the competing discourses on colonialism as ranging from demands for immediate abolition to keeping colonies in perpetuity. The discourse of gradualism and a "leisurely pace" expressed a fear of political violence and cultural chaos that sought to postpone "liquidation day." According to another scholar, "The crucial debate within the Western world" concerned "methods and timing."[15] In *New Lives for Old,* Mead, using her authority as a social scientist, intended to speak to both of those issues.

Within the larger field of discourse about colonialism and its aftermath, support for Westernization was a major current in the thinking of Mead, whose views were shared by most members of her intellectual generation. Social scientists debated whether Westernization would be positive or negative for the formerly colonized peoples rather than whether cultures could reject it altogether. Mead's stance on Westernization was not an uncritical acceptance "regardless of how it was imposed," contrary to claims made by critics who understand anthropology as a social construction, espousing ideas of colonialism, nationalism, and indigenous peoples that Mead could not have encountered in the 1950s.[16]

For Mead, writing in the context of the early Cold War, the postcolonial future held only two major alternatives. New countries could be damaged by the worst of Westernization or benefited by receiving the best of Westernization. Bad Westernization—shoddy trade goods; inflated prices in stores; lack of schools, hospitals, and other positive institutions; the system of dominance and inequality by whites—would create a dystopia of cultural disintegration that made its victims vulnerable to the dreams of cargo cults and Communist "utopias." All of these criticisms appear in *New Lives for Old* in carefully couched form. By 1963, as the constraints of the Cold War eased, Mead would write more critically about the "consciously benevolent" system of hegemonic racial dominance with its enforced inequality in which "the gulf between the level of any white man and any native was fixed and appeared absolutely uncrossable," where "no native sat at the same table with a white man . . . or walked with a white man down the street of a town," using more explicit language that her later critics might have applauded.[17] In the mid-1950s, before the influence of the civil rights movement, she did not feel able to speak as frankly if she wanted to communicate with the American public.

The "bad Westernization" of European colonial influence could also lead to the development of "slum cultures" dominated by Europeans because they were unable to act for themselves. Mead expressed her fear that she might discover that Manus had become such a slum. Before they went to Manus, Mead told Theodore Schwartz, the anthropologist who would accompany her on the trip along with his young bride, "that if Manus turned out to be another cultural shambles—a slum culture, undermined and demoralized as a result of the drastic culture contact and change they had experienced—she would not write about it. What the world needed was a success story." Mead had expected that the Manus, who had experienced only European "dehumanizing and degrading contact-culture, would eventually become poorly educated proletarians in a world they could neither understand nor influence."[18] Associating "bad" Westernization with European colonial administrations, Mead prepared herself and her companions for the possibility that they would find just that in Manus in 1953.

Mead associated the best of Westernization with the United States and its ideals of equality, democracy, and concern for the individual. It was to foster this kind of Westernization that she desired to move colonial policy. Discovering to her satisfaction that Manus could serve as a model for her utopian postcolonial vision, Mead ascribed the result to the encounter with the U.S. armed forces. During World War II about a million servicemen moved through a military base near the Manus.[19] African American and European American troops taught the Manus "the difference between being treated as a 'native,' as belonging to another category of creature, and being treated as an individual." The Americans brought to life the ideal of the "brotherhood of man" that the missionaries had discussed. The Manus "had never seen such vivid illustration before."[20] They watched African Americans being treated with seeming equality as they had never been treated by their colonial administrators. The war experience gave the Manus a model of equality and democracy to which to aspire, bringing an example of "good" Westernization within their reach.

Charles Rowley provided independent evidence of this positive impression of the U.S. armed forces by the people of Manus. He quoted a Manus account produced in 1947 by J. H. Wooton, a staff member of the Australian School of Public Administration, who spent five months in Kawaliapin, the central Manus Island. Wooton's informant complained that the missionaries "did not show us the straight road that would lead us on to your knowledge, your ideas, your language." He told Wooton, "If only you would open your minds to us, we could be brothers. We could sit down together like white and black America. Brother, these are not my thoughts alone. All of Manus is talking thus."[21] Rather than seeing only what she wished to see, Mead accurately depicted the Manus's positive assessment of American culture.

Writing in the mid-1950s, Mead realized that this claim might be rejected as too utopian to be believed. She used rhetorical strategies to emphasize the reality, the factuality, of what she was saying and to defuse the incredulity of her readers. For example, she wrote the chapter on the U.S. Army from the perspective of the Manus. She used language such as, "they describe," "the Manus experienced them as," "from their point of view," "to native eyes," or "the people say," quoting them directly, giving the Manus ownership of these observations.[22] She included evidence about the true state of affairs in the U.S. Army—"treating Negroes as labor battalions was demeaning,"—and reported that Marines had been brought in to quell a disturbance by Negro troops protesting discrimination.[23] She told her readers that the Manus had not recognized the extent of racial antagonisms. They only saw that black and white ate together and that Manus workers ate with them.

After the most optimistic statements by the Manus, Mead added qualifying statements. When she wrote that the Manus told her, "'The Americans treated us like brothers,'" Mead added, "Like other Anglo-Saxon peoples, they were remarkably friendly and sympathetic to 'other people's natives.'" She qualified her own more idealistic statements. Claiming that "the basic American preference" was to "treat every man on his merits," she also referred to the "residue of slavery and the scars of immigration." At times she took the standpoint of open incredulity. She asked the questions a skeptical reader might have expressed: How could the Manus believe Americans were not materialistic, when there was such an excess and emphasis on material goods? She answered by stating that this was what the Manus saw. It was reality to them and reality enough for the Manus to take the Americans as a model.[24] By describing the idealistic visions of American society through the eyes of the Manus, Mead sought to convince her readers to abandon a dystopic model of postcolonialism in favor of adopting a valid utopian one that would encourage Americans to live up to their national ideals.

Contributing to the debate about colonialism, Mead supported fast change over slow, and total change over partial. Although the Australian Minister of Territories, speaking of Papua New Guinea in 1951, stated that "some form of self-government" might be "more than a century ahead," Mead stressed the need to speed toward independence, not cling to colonial privilege.[25] Implicit in go-slow policies was a vision of "savages" unable to cope with the withdrawal of colonial power. Mead's representation of the Manus as twentieth-century people rather than savages worked against the stereotypes of backward primitives and hastened the end of colonialism.

In Mead's analysis, partial change had led to the development of "slum cultures." The Manus of the New Way could "move so fast because they had changed the entire pattern all at once—houses, costumes, ceremonies, social organization, law—making up the new pattern out of the accumulated bits of European civilization which they had learned through the previous twenty-five years." The Manus provided a pattern that could be used by other peoples. "It is most important," she wrote, "that they should change from one whole pattern to another, not merely patch and botch the old way of life with corrugated iron or discarded tin cans, in political peonage in the great cities of the world." While it was difficult to add one new habit to many old ones, it was much easier "and highly exhilarating to learn a whole new set of habits, each reinforcing the other as one moves—like a practiced dancer learning a completely new dance."[26] By taking such bold, ambitious moves, the Manus had avoided the dangers of a prolonged transition.

Finally, Mead emphasized the agency of the Manus, thus stressing the necessity of allowing and encouraging colonial peoples' activism as they moved away from colonial rule. She promoted the necessity of letting indigenous peoples lead in adapting Western institutions and ideas to their needs. She saw Westerners as trapped in "a one-sided picture that something was being done to people." The model of the Manus pointed up "the completeness with which a people may want to change rather than merely submit to being changed." It showed "culture contact as an active choice" for indigenous as well as industrialized peoples. She also pointed out the resistance of colonizers "to share the whole pattern," to accept indigenous peoples as potential equals, while feeling righteous about giving "bits of it," which kept the colonial people subordinate.[27]

Hindsight might suggest that what Mead and others thought of as postcolonialism was actually a form of subtle neocolonialism, substituting a form of Western, particularly U.S., cultural domination in place of an oppressive colonial administration. Mead's parable of the Manus tended to reinforce a model of U.S. imperialism among colonies freed from overt European domination. Lola Romanucci-Ross, who completed fieldwork among other Manus groups while working under Mead's auspices, questioned whether the Manus had changed as totally and completely as Mead believed.[28] At the time, however, her parable expressed the desires of the Manus and other anticolonial peoples for rapid independence on the basis of principles that she hoped would further the process.

While her two parables of the Manus looked forward optimistically to the promise of utopia concealed within her overtly realist text, Mead's ideas concerning women constructed a dystopian countertext. Although *Coming of Age in Samoa* and *Sex and Temperament in Three Primitive Cultures* had contained new ideas for American women, Mead produced no parables or stories suggesting utopian change for women in *New Lives for Old.* This may have reflected her own deep ambivalence during this period or her accommodation to the hegemonic discourse of 1950s America on women. Mead presented Manus women's new equality as utopian in ideal but dystopian in reality. While she formulated a critique concerning women in Manus that could have been applied to 1950s America, she did not acknowledge or pursue a feminist analysis. Her readers received mixed messages about women's place in the new Manus society. Officially emancipated, women had become the equals of men. Some of the narrative showed women testing and enjoying their new state. Women could have nonmarital sexual relations, but sex was related to anger in Manus culture, and men did not expect women to enjoy it. "In the very young marriages there is an appearance of greater tenderness," Mead wrote, "but whenever a disagreement occurs the old attitudes which associate sex and anger seem to reassert themselves."[29] She noted the violence that occurred but did not pause to discuss this issue extensively.

Mead wrote that the New Way offered no public roles for women comparable to those lost under the old system. Instead, a bright woman became trapped in a new domesticity, expected to "keep her house and children in a modern way, and never embarrass her husband by being old-fashioned." In spite of the decree of sexual emancipation, "they still live in a world which in repudiating sex also repudiates women, and which in exalting fatherhood leaves less room for motherhood, except as a sort of delegated fatherhood." A familiar pattern in the unofficial courts of the New Way was to blame women. "If the blame could be firmly affixed to a woman, then any failure of the New Way among men need not be faced so directly." There was a desire in the new situation to increase the Manus population. "Women are exhorted to have children; abortion is heavily forbidden; and that woman who never menstruates because she is always either pregnant or breast-feeding a child is regarded as the most patriotic and virtuous."[30] As Mead moved to portraying a full-blown dystopia for Manus women, she described a pattern that resembled gender expectations in the United States in the 1950s: domesticity, the new participatory fatherhood, the baby boom, Momism, and the repudiation of sexuality while seeming to exalt it.

Despite invoking many of the aspects of gender that would soon become known as the "feminine mystique," Mead did not take the next step and apply this critique to American society.[31] She did not turn this part of her study into a coherent story for Americans. In contrast to her explicit discussions in her earlier books, she provided no sense of warning, no suggested solutions. Though she could see women's problems in Manus culture, she did not explicitly relate those issues to aspects of American culture, except in a muted sub-commentary about the worth of women.

Late in her account of the Manus, while making a point about why colonial people "fail" to progress (a word she used in quotations since it was a common argument in the colonial discourse), she cited examples of other groups who "fail," including women, a comparison group that would not generally have occurred to her colleagues. "Many different kinds of failure and refusal become intelligible," she wrote. She included the example of girls trying to learn physics. Taught by male physicists and teachers "whose full status" they could not hope to attain, "bright girls strangely have no 'ambition,'" because "the desire to learn is blocked by the knowledge that part of the pattern to which they aspire will be denied them." Women could not get jobs in physics, and they did so only with great difficulty in other professional areas in the 1950s United States. Moreover, "the terms in which the participation is denied—physical self-identification as a 'mere woman'"—intensified the problem. "Each such limitation of natural gift and aspiration carries with it a kind of constriction, a denial of the self, which, if once relaxed, provides channels through which great energy can be mobilized and released."[32] Women and other groups, she suggested, could succeed only when society realized its discrimination and created structural openings for them to do so. When that changed, she implied, then both women themselves and society could finally realize the worth of women. Only in this one small example hidden among examples of other forms of discrimination did she directly refer to the constrictions for women, perhaps because of a reluctance to challenge an antifeminist status quo. Mead was a utopian in terms of her views of a changed American identity dedicated to a world community and in her vision of New Guinea as a model for other colonizers and colonized people, but she was not able to apply her utopian vision to the status of women.

In personal terms, as a result of her field trip to Manus, Mead felt a renewed sense of her own worth. The 1950s began as a time of vulnerability for Mead. She and Gregory Bateson divorced in 1950, and her mother died. She herself was entering her fifties, a time in life that marked a loss of status for most American women as they lost youth and conventional beauty. Soon after the end of World War II, Mead and Benedict had launched Research in Contemporary Cultures, a project to analyze numerous cultures through oral interviews and other innovative techniques that involved a significant number of scholars working in groups. Mead had begun it with such high hopes of profound significance, but it had minimal impact. Her own work on child rearing was trivialized as "diaper-determinist." The night she arrived in Manus, she wrote, "I felt almost as if someone—and I was not quite sure who it was, they or I—had been raised from the dead. Someone who, not knowing it, had been dead, and lived again."[33] Her identity had changed from a low-status young wife who had never borne a child, as she had been on her 1928 visit to Manus, to that of an older woman with status—an advisor and consultant to both the leaders and the ordinary men and women of the village of Peri concerning appropriate change. With a renewed sense of her own worth and of her mission in the world, she returned to the United States to write *New Lives for Old.*

IV / Echoes and Reverberations

As authors of the "Ur-texts" of American anthropology, Benedict's and Mead's works have outlasted their creators. *Patterns of Culture* was one of anthropology's most influential texts.[1] *Sex and Temperament in Three Primitive Cultures* continues to be read as a pioneering text of second-wave feminism. However, it is Mead's first major work, *Coming of Age in Samoa,* and Benedict's last, *The Chrysanthemum and the Sword,* that survive as the most provocative of their contributions. These works have also generated the most serious criticisms, perhaps because their longevity produced a sense of betrayal when flaws were found in their analyses.

As the work of a fledgling anthropologist, *Coming of Age in Samoa* is amazing in its ability to attract readers and outrage critics. As exemplified by the debate sparked by Derek Freeman's attack initially published in 1983, scandals have helped to sustain interest in both Mead and *Coming of Age in Samoa.*[2] Mead's association with sexual liberalism and cultural relativity would invite attacks whenever the ideological pendulum swung toward an insistence on defending Western civilization, traditional morality, and cultural absolutes.

Written in the period haunted by the atomic bomb attacks on Hiroshima and Nagasaki, Benedict's *The Chrysanthemum and the Sword* received a respectful reading in the United States as Americans debated the way to treat the former enemy and its imperial leader. Its translation into Japanese, however, gave it a much longer life. Contrasting assessments of the book began a spirited debate that continues more than a half-century after the book's first publication.[3]

Both books still give their readers something that sustains their interest and arouses their passions. Benedict's "gift for intellectual synthesis and popularly accessible writing" achieved its effects as much in translation as in its original English. Mead's enthusiasm for middlebrow American culture and her belief in progress embodied the "vices and virtues" of her writing that continue to raise scholarly suspicions.[4] Angela Gilliam, using a film, *Margaret Mead and Samoa,* as the vantage point through which to view the Mead-Freeman debate, finds that both authors drew from colonialist assumptions about Samoa. In her more positive defense of Mead's portrayal of sexual politics, Sharon Tiffany turns the her attention to the visual iconography of art on the covers of various editions of *Coming of Age in Samoa* as those illustrate the ethnic, gender, and sexual fantasies of U.S. consumers. Pauline Kent ventures into the Japanese debate, addressing the criticisms made by Lummis and other critics to create a sympathetic interpretation of Benedict's intentions. The messages of *Coming of Age in Samoa* and *The Chrysanthemum and the Sword* continue to echo and reverberate as readers encounter texts that have survived their moments of origin and their authors.

11 Imagining the South Seas

Margaret Mead's *Coming of Age in Samoa* and the Sexual Politics of Paradise

Sharon W. Tiffany

> Are we Samoans now to be known as a nation of sex-starved, suicidal rapists? I much prefer my previous reputation as a free-loving orgiast.
>
> Lelei Lelaulu, 1983

In January 1983 the *New York Times* ran a front-page story reporting on the imminent publication of Derek Freeman's *Margaret Mead and Samoa,* which criticized Mead's depiction of female adolescent sexuality in *Coming of Age in Samoa.* Anthropologists returned to a text that many had not read in years as the profession, the media, and the public debated Freeman's allegations as well as those made by earlier academic critics who had not received such public attention. Had Mead "constructed her misinformed account of 'free love-making'" from "fundamentally flawed fieldwork"? Was Mead "duped" or "lied to" by her informants? Did these contrary views represent "paradigms in collision," as Freeman asserted? Were Mead's research findings "preposterously false" and a "confused travesty"? The ongoing debate, sustained by Freeman's subsequent articles and newer versions of his critique, suggested the need for scholars to reconsider the constructed images of pleasure and danger in paradise often associated with popular ideas about Mead's *Coming of Age in Samoa.*[1]

Seventy-five years after the publication of *Coming of Age in Samoa* and nearly two decades since the publication of Freeman's initial challenge, the academic literature concerning the sexual lives of Samoan Islanders continues to grow. Much of this recent work seeks to address the controversy by returning to Mead's ethnography. Some scholars deal with issues of sexuality and aggression, variously based on fieldwork in Samoa or on Mead's published and unpublished materials. Others focus on language and writing style, challenge Freeman's assertions regarding the historical or methodological importance of Mead's book to the discipline of anthropology, or criticize Freeman's methods and assertions. Rather than engage with the debate over the ethnographic text, this essay will examine *Coming of Age in Samoa* and Freeman's critique in visual terms. As I will argue, it is important to recognize that Mead and her publisher initially placed her interpretation of Samoa in the context of the romantic image of the Pacific paradise.[2] Cover art offered a way to attract the reader with visions of paradise whose location and iconography changed to make *Coming of Age in Samoa* appealing to the mythic dreams of successive generations of readers but also exposed it to criticisms for those elements of fantasy. Freeman would become the most celebrated critic of Mead's version of a South Pacific Eden while using visual imagery to undercut Mead's authority and increase his own.

Geographically located in Polynesia, the South Seas in Western imagination encompasses a social, geographic, and ideological space in which men undertake the quest of erotic mastery and self-discovery—recurrent themes in romantic literature incorporated into the colonial experience of possessing distant worlds. One need only mention the famous novelists—Jack London, Herman Melville, Joseph Conrad, Robert Louis Stevenson, and James Michener—to evoke the imagery of tropical abundance, seductive Pacific Islander women, and the thrill of male adventure. The South Seas romance has also informed the anthropological enterprise. *Coming of Age in Samoa* was not the only ethnography marketed with endorsements from well-known experts of the day or intended to appeal to readers' fantasies. Bronislaw Malinowski's provocatively titled *Sex and Repression in Savage Society* appeared in 1927, a year before Mead's book. Malinowski's *The Sexual Life of Savages in North-Western Melanesia,* published in 1929, included a preface by Havelock Ellis, the internationally recognized authority on the psychology of human sexuality. Both would enthusiastically endorse *Coming of Age in Samoa.*[3]

William Morrow's insistence on making the book accessible to a popular audience encouraged Mead to consider the social and sexual agency of the "New Woman" in American society of the 1920s through the lens of a Samoan "Other." Innovative marketing techniques, directed toward a growing urban population eager for consumer goods, were well established in American society by the 1920s. Mead's publisher recognized that *Coming of Age in Samoa,* if effectively marketed, would speak to an audience receptive to the idea of change. This market segment included, of course, the "New Women" who challenged Victorian notions of female sexuality and the proper social roles of women.[4]

Before departing for the Admiralty Islands for another research trip, Mead revised the last two chapters of *Coming of Age in Samoa* in 1928 and asked Franz Boas, her professor and mentor, to write a preface. William Morrow later wrote to Mead that he had committed a "substantial" publicity budget of "nearly $1,500.00 in various forms of advertising and promotion" and that *Coming of Age in Samoa* had already sold 3,144 copies as of December 31, 1928. Havelock Ellis's enthusiastic endorsement appeared as a "bright red band" that adorned the first-edition cover of the book. "That stunt helped the sales materially," William Morrow wrote to Mead on January 11, 1929. Noting the book's "splendid review" in the *New York Times*, Morrow praised Mead and her work: "The basis of our success is the book itself. You wrote something that people could understand and enjoy, and you did it superbly." In Mead's view, rewriting the last two chapters of *Coming of Age in Samoa* had taught her an important lesson that influenced her subsequent ethnographic writing. Such writing had to "make the life of a remote island people meaningful to an American audience." Morrow's advice and her own ability to appeal to American readers enabled her to reach a mass audience beginning with the success of *Coming of Age in Samoa.*[5]

Morrow used cover art to attract readers. Mead's role in selecting or approving the cover for the first edition cannot be fully determined. According to her published comments, she "read proof, and saw a small printer's dummy of the table of contents, chapter one, and the jacket" before sailing for Hawaii, "glad to be on my way to the field in the Admiralty Islands." The first-edition cover, apparently chosen by William Morrow, conveyed the theme of an exotic eroticism, with lovers enveloped in the magic of swaying palms and the light of a full moon. This classic—indeed, hackneyed and stereotyped—image of paradise assumed fresh significance for Mead herself as well as for other women of the 1920s who sought to create independent lives that included the right to sexual pleasure.[6] Indeed, the Samoan woman on the cover of the first edition of *Coming of Age in Samoa* suggests a "partial state of undress" parallel with her modern flapper counterpart, whose signature fashion of liberation revealed bare arms and legs.

By contrast, Freeman's description of the first-edition's dust jacket illustration suggested a different version of paradise. According to Freeman, it showed "by the alluring light of a fuller than full moon, a bare-breasted Samoan girl, inflamed with sexual desire, hurrying with her lover, to what Mead, in her pseudo-poetic language calls a 'tryst' beneath the palm trees. It is exquisitely true to the preposterous fantasies by which she had been hoaxed, and romantic bilge of the first water." Freeman attempted to demolish the notion of a South Seas utopia by questioning the legitimacy of Mead as an ethnographer and as a writer. However, he did so within the ideological boundaries of a constructed paradise familiar to a Western audience. Thus, the fantasy of paradise remains a possibility, even if, as Freeman asserted, it was "romantic bilge" in the case of Samoa.[7]

The artistic construction of paradise on the book's first cover did not appear exactly as Freeman described it, however. Significant details of the scene have been omitted from his account. The young woman, with hair that flows like a wave over her back, was faceless. Wearing a pandanus skirt and a hibiscus behind her left ear, she ran hand-in-hand with her faceless lover. It is important to note that it is she who leads him toward the grove of palm trees on the moonlit beach. The 1928 cover highlighted the evocative image of paradise as sexual playground. Indeed, the dust jacket image with its faceless lovers derived from an Orientalist premise that paradise can exist in just about any remote or tropical place—a place ideally framed by palm trees and a beach. In this instance, the woman's body and dress situate paradise in the South Seas. Shaping the visual image of a South Seas Eden for Western consumption required a bare-breasted woman wearing a grass skirt, just as the veiled woman was the requisite image for visualizing the Middle East. In the case of Mead's narrative, it is the title above the dust jacket drawing that serves to reinforce Samoa as an idealized world of "free love," a specific place where women are not ashamed to show their bodies and to take the sexual initiative.[8]

Freeman's polemical assertions of informants' hoodwinking, combined with his denigration of Mead's mental state as a "chronic state of cognitive delusion," deny the possibility of a pleasurable version of paradise in Samoa. Rather, with a kind of missionary zeal, Freeman posited a misogynist view of Samoan women as liars, as zealous Christian prudes, and as victims who provoked men's sexual aggression. This dark, dystopian narrative is rhetorically linked to repeated comments about Mead's body and brain.[9] Freeman's menacing view of Samoa not only denied indigenous women the right to claim their bodies and their minds, but it also provided a forum for derogating Mead's physical appearance and mental capacity. In other words, women's words and minds, like their sexual behavior, are suspect.

By contrast, Mead's narrative privileged voice and agency on behalf of her female informants while asserting her own legitimacy as an ethnographer. Mead, of course, was aware of the "current daydreams in the Western world" of paradise and the positioning of this dream world in the Pacific of her time. Her use of fictional devices in her narrative, especially in the more "literary" chapter, "A Day in Samoa," combined with the book's evocative dust jacket, clearly situated paradise in Samoa. Islander youth in the South Seas—in opposition to their U.S. counterparts—lived simple lives: Samoan adolescents enjoyed "freedom in sex, lack of economic responsibility, and lack of any pressure to make choices." In contrast to Freeman's forbidding version, Mead's portrayal of Samoan adolescents emphasized pleasure rather than danger.[10] Despite their significant differences in tone and emphasis, however, both narratives assumed a static, one-dimensional social geography that resisted an understanding of the complexities and paradoxes that comprised a human reality. Essentialist portrayals of Islander women's bodies and lives disallowed a careful examination of the dynamics of gender, rank, and ethnicity. Such portrayals also ignored the ways in which women's bodies are appropriated for consumption in a globalizing political economy as the exoticized and eroticized "Other."

The paperbound cover of the 1949 Mentor edition of *Coming of Age in Samoa* reinforced the promise of Polynesian sexuality with its colored drawing of a dancing couple that catered to an audience familiar with visual and literary references to Pacific paradises. An artistic rendering of homogenized vegetation—the conventional marker of paradise, South Seas style—provided a background that highlighted their bodies. This cover attempted to capture the gender dynamics and sexually charged atmosphere of a *siva* that could be danced on many levels of eroticism, depending on the dancers' age and social status, as well as the social context in which the *siva* is performed. The dance of a *taupou,* the ceremonial virgin, with one or more men could be especially provocative. The male dancers' assertive, almost aggressive and charging style contrasts sharply with the sedate steps and elegant arm and hand movements of the woman.

The woman portrayed on the Mentor cover, however, was not a *taupou.* She lacked the elaborately decorated headdress of human hair, barkcloth and pandanus clothing, and other ornamentation associated with this elite status. The woman's facial features appeared somewhat more "Asian" than generically "Polynesian." The male dancer's face suggested Caucasian features that, like those of his partner, are impossible to situate in a particular geographical region or culture. The man's face, framed by a leafy crown worn along the hairline, resembled Robert Flaherty's principal Samoan actor in his documentary film *Moana of the South Seas,* which was filmed in Samoa shortly before Mead's research trip.[11]

Like the woman depicted on the first cover of Mead's book, this dancer also takes the initiative by engaging in an erotically charged gesture as her right buttock and hip nearly touch her partner's left hip. Unlike Moana and the *taupou* Fa'angase in the Flaherty film, however, the dancers who decorate the Mentor cover wear *lavalava,* squares of cotton trade cloth that tie at the waist and hang to knee level. This simple, everyday type of dress suggested that overt expressions of female sensuality did not depend on elite status and the social decorum demanded of ceremonial maidens but rather extended to women of modest social background. This book cover image concurred with Mead's narrative focus on girls and women who, for the most part, were not high status.

Images of women's overt sexuality depicted on covers for the 1928 hardbound edition and the 1949 Mentor paperbound reprint edition ended dramatically with the publication of the 1961 paperback edition of *Coming of Age in Samoa.* There a charcoal drawing of the head and bare shoulders of an exotic-looking young woman with a generic tropical flower nestled in her long, dark hair commands the viewer's attention. Her downcast eyes suggest modesty; her full lips, slightly parted, convey sensuality. The woman on this cover appears as a passive presence who refuses to look directly at her audience, but she is nonetheless aware that she is being surveyed.[12] She entices her viewers but remains mysterious, her thoughts a secret to the reader. In this cover illustration paradise became a place where a compliant female waited to be acted on, as opposed to the woman who ran across the beach with her lover or the dancer who responded to the movements of her partner. This is not a woman of full-bodied Polynesian grace with the insouciant demeanor and the "attitude" of a Gauguin painting, but a small-boned, light-skinned woman combining Asian and Polynesian features, leavened with a hint of Bali-Ha'i from the musical *South Pacific.*

The *South Pacific*-Bali Ha'i association with paradise in Samoa became more explicit on the cover of the Dell paperback of *Coming of Age in Samoa,* published in 1968. The cover depicted an updated version of the South Seas "wood nymph," a small-boned, light-skinned, Caucasian/Asian-looking woman with pouting lips and a silky mane of dark hair carefully draped over a bare shoulder. A large white tropical flower nestled in her hair; a red sarong covered her breasts. In this "soft-primitivism" of an imagined Eden, the lush, verdant landscape that framed the woman's body included the discordant presence of temperate-climate daisies. The wood nymph coyly looks up at the viewer, seemingly aware that she herself has become the spectacle. This is the classic *National Geographic* gaze focused on the non-Western "Other." She signifies the authentic South Seas siren—a woman without artifice who inhabits a static, "natural" world untouched by time and history.[13]

The viewer, located above the subject, is privileged to look down on the woman in this 1968 image, suggesting a relationship of inequality between the one who looks and the one who is looked at. The woman's return look is sideways and upward, while her facial expression, softened by the slightly upturned corners of her mouth, reinforces this colonial positioning of gazes exchanged between the viewer and the viewed. As Lutz and Collins note in their discussion of *National Geographic* photographs, "The smile plays an important role in muting the potentially disruptive, confrontational role of this return gaze." This South Seas wood nymph on the Dell cover suggests what Native Hawaiian activist Haunani-Kay Trask has called a "soft kindness." The woman appears "approachable," suggesting the possibility of intimacy between subject and viewer but on terms set by the viewer.[14] While the tropical-looking world depicted on the cover could be interpreted as a Polynesian Eden, the subject posing in this constructed landscape is not Samoan. Indeed, what is remarkable about the 1961 and 1968 covers of *Coming of Age in Samoa* is the absence of a Samoan woman from the paradise in which Mead and her famous text had placed her.

An earlier proposal for the Dell cover of *Coming of Age in Samoa* suggested that Mead had strong views about issues of representation and objected to the creation of a non-Polynesian image to illustrate an editorial proposal for the book. Asked by a representative of Morrow to approve an illustration, Mead replied thorough an administrative assistant. She refused to give permission because "the girl does not look like a Polynesian; her arms are too thin" and she was dressed in a "Tahitian style garment" in a design that "Samoans do not wear." Mead requested that "the cover be based on a real Samoan photograph" or that no drawing be used. Mead's assistant passed on her demand that "the posture must be right." In order to avoid "these difficulties," Mead suggested that the publisher might "stick to printing and pretty colors, or palm trees and NO people." It is not clear from the correspondence if Mead subsequently approved the illustration for the 1968 edition, but it was evident that she wanted *Coming of Age in Samoa* to appear as serious scholarship at a time when the issue of the sexual objectification of women was becoming a focus of the emerging feminist movement.[15]

Connected motifs of sensuality, culture, and race in the 1961 and 1968 covers of *Coming of Age in Samoa* reflected a shifting of the South Seas idylls to an imagined place beyond the geographic confines of Polynesia. As Mead would acknowledge, "It has twice been my accidental fortune to make ethnological expeditions to islands which . . . have also been current daydreams in the Western world—-the South Sea Islands [of Samoa] in 1925, and Bali in 1936." Noting the efforts of steamship companies to promote Bali as a tourist destination in the 1930s, Mead wrote of their efforts "to make Bali the lineal descendant of the Tahiti of the romantic Twenties, with the slim figure of a high-breasted, scantily clothed girl as the symbol."[16] Despite her own involvement in attracting a popular audience by publishers' use of such romantic images, Mead probably opposed the conflation of Bali or an imagined "Bali-Ha'i" with that of Samoa when *Coming of Age in Samoa* was republished in the 1960s.

In the postwar Pacific of popular imagination, paradise had moved to Bali-Ha'i, the mysterious, forbidden island immortalized by the stories of James Michener and the music of Rodgers and Hammerstein. The romantic appeal of the South Pacific was enormous. Two short stories, "Our Heroine" and "Fo' Dolla'," from James Michener's Pulitzer Prize–winning collection *Tales of the South Pacific* first appeared in hardcover in 1947 to immediate acclaim. Michener described Bali-Ha'i as "an island of the sea, a jewel of the vast ocean. It was small. Like a jewel, it could be perceived in one loving glance." Michener equated the woman Liat with the island of Bali-Ha'i, thereby configuring paradise as a feminized landscape. Indeed, part of the allure of Bali-Ha'i was its young and shapely female form: "[it] was green like something ever youthful, and it seemed to curve itself like a woman into the rough shadows formed by the volcanoes on the greater island of Vanicoro."[17] The paradise of Bali-Ha'i became a reconfigured place in the popular imagination noteworthy for the absence of indigenous Pacific Islander women—like the covers of Freeman's books and Michener's *Tales of the South Pacific.*

The Rodgers and Hammerstein musical *South Pacific* opened on Broadway in April 1949 and closed in January 1954 after 1,925 performances. Songs like "Bali Ha'i" played a pivotal role in establishing the symbolic linkage of woman-as-island-in-paradise. According to music critic Stanley Green, "The haunting sound of the first three notes is almost all that is needed to establish the spellbinding appeal of the South Seas paradise, and the words perfectly match its mystical quality." Four years later the movie version of *South Pacific,* filmed in Hawaii, was released. Transformed in the Western imagination as the ultimate South Seas paradise, the dream world of Bali-Ha'i became the signifier of an idyllic, non-Polynesian landscape, conflated or confused with the actual Bali of Island South Sea Asia—an island that has its own complex history in the Western construction of paradise. As Margaret Jolly has noted, "The erotics of the exotic here work through a series of displacements and fugitive transformations." The shifting of paradise to Bali-Ha'i displaced Polynesian women and ignored Melanesian women, influencing the illustrators who designed the covers of *Coming of Age in Samoa* in the 1960s.[18]

The dust jacket of Freeman's *Margaret Mead and Samoa* contained no photograph or drawing of a human figure, thereby avoiding the issue of bodily representation made problematic by feminism and the emergence of postcolonial scholarship. It depicted a generalized Polynesian-looking design in yellow and brown, thus giving the appearance of solidity and neutrality, academic qualities associated with masculinity and a prestigious university press.[19] The front and back covers served to reinforce the author's purpose of depicting Mead's ethnographic research as suspect. The back cover highlighted the author's assertions of Mead's naïveté and her deception by female informants. It contained five lengthy testimonials, all from men, none of whom were Pacific specialists, representing the fields of physical anthropology, zoology, ethnology, and evolutionary biology. Their words testified to the "hard," masculine work of science, in contrast to Mead's purported gullibility and her narrative, spun from girlish deceptions—the stuff of pseudo-science. The back cover of Freeman's *The Fateful Hoaxing of Margaret Mead* also reinforced this message of masculine rationality overcoming feminine emotion when it appeared in 1999 to continue the attack on *Coming of Age in Samoa.*

By comparison, the cover of the 1984 Penguin edition of Freeman's *Margaret Mead and Samoa* shifted from the disembodied neutrality of the 1983 hardcover dust jacket by highlighting Mead while avoiding the portrayal of Islander women. The cover showed a photograph of a young Mead set against a Polynesian barkcloth design. Mead appeared solemn and serious, her eyes looking unflinchingly at the viewer.[20] The photograph suggests that the "Shangri-La" dream world of Samoa has been transformed into a domesticated enclave of colonial control. Mead's feminine clothes—a long-sleeved dress, neck scarf, anklets, and dainty strapped shoes—sharply contrast with the barkcloth belt she is wearing, which is worn by Polynesians of rank. The belt, symbolizing the elaborate complex of chiefly ceremonialism and status, nearly dwarfs the petite female body that holds it.

At first glance, the photograph suggested a simplistic portrait of Mead as a modern, fashionably dressed woman and ethnographer, draped in the cultural artifacts of her fieldwork site. This "I-was-there" image focused on the incongruity of the artifacts with the subject's overtly feminine dress and pose. Indeed, this visual dissonance fitted Freeman's representation of Mead as too young, too naïve, too small, and too fragile to engage in a man's heavy-duty work of studying the physically and culturally robust Samoans. Alternately, the picture could be interpreted as a woman asserting her authority to engage in a masculine enterprise by conducting anthropological fieldwork on her own, publishing her research, and audaciously promoting herself and her work. The barkcloth belt authenticated the wearer's presence in Polynesia and her legitimacy as a scientific observer of Samoan social life.

Freeman's *Margaret Mead and the Heretic* continued the narrative and visual displacement of indigenous Islander women from paradise. This Penguin edition appeared in 1996, the same year that *Heretic,* David Williamson's play about Mead and Freeman, dramatized the controversy for audiences in Sydney, Australia.[21] The 1996 book title, like the play, portrayed a contest between Mead and the heretical Freeman. The cover connected Mead to her "fantastic" tale of paradise in Samoa. It superimposed Mead's face on Freeman's own photograph of the Manuan islands of Ofu and Olosega, as seen from the island of Ta'u, the site of Mead's fieldwork. The computerized artifice of morphing a close-up, cropped photograph of Mead's head and shoulders onto the geography of paradise merged Mead into the landscape—her ethnographic authority erased, and her voice silenced.

The visual and narrative process of appropriation and displacement continued on the back cover, which contained a quotation from Fa'apua'a Fa'amu, Mead's cherished friend and informant. Fa'apua'a's own words appear to damn her as a prankster and liar who eventually turned against Mead in order to set the ethnographic record straight. In fact, their numerous letters, written in Samoan and English, express great warmth and friendship between the two women as they continued to communicate long after Mead had left Samoa. The text appeared on a shroud-like background of deep black—like a death notice—suggesting the demise of Mead and her Samoan research. Three years later the dust jacket of Freeman's 1999 "fateful hoaxing" narrative depicted a beach scene in which Pacific Islanders are, once again, notable for their absence. Mead has also vanished from the cover image, completing the erasure of her ethnographic presence. Only the physical landscape of paradise remains: palm trees on a sandy beach, the coral reef, and islands in the distance reinforcing the narrative premise of an unoccupied and derelict paradise. Mead, like the girls and women of Samoa, has been rendered invisible.

The commercial success of *Coming of Age in Samoa,* combined with Mead's busy schedule of public-speaking engagements and interviews to promote her research and published work, secured her public status and simultaneously made her academic stature problematic in the eyes of male colleagues. Unlike many women anthropologists of her time, Margaret Mead avoided the erasure of women's writing from the canon of ethnographic literature. Derek Freeman, as the self-described "heretic," undertook a self-imposed quest to remove Mead and her "spellbinding text" from the canon.[22]

Freeman's narrative of sexual politics in Samoa, accompanied by its subtext of biological determinism, sought to restore the legitimacy of men's power and women's subordination. According to Freeman's version of the story, Samoa could not be a "paradise" of free love because male-dominated institutions and the privileging of male-defined ideologies of sex and gender relations constrained the lives of girls and women. In short, lovers cannot rush off to a romantic encounter under a full moon because such provocative behavior, especially on the part of women, subverted the dominant narrative of chiefly and mission authority. By contrast, Mead's story of "love under the palm trees," to use the title of a favorable review of Freeman's initial book, appeared fraudulent.[23]

Mead's and Freeman's texts presented alternate versions of paradise. Freeman's tales of sex and lies bespeak a menacing, alter-representation of Samoa. Book cover images reinforce a sexual politics of ethnographic authority that enabled Freeman to denigrate a woman's ability to observe and write about the lives and experiences of Samoan girls and young women. Freeman's narrative attempted to demonstrate that Mead—the naïve "wanna-be" ethnographer—accepted as "truth" the girlish jokes and deceptions of her informants. The covers of his books reproduced this dismissive story. Paradise, as a site of pleasure, becomes a historical impossibility. There are no Polynesian lovers on moonlit beaches. In Freeman's construction of the other side of paradise, Samoan women, like the "sleeping beauties" and "wood nymphs" of Western image making, become visible only when they serve the interests of men.

Mead, on the other hand, wrote a more complex story that challenged images of compliant South Seas sirens. Most importantly, women are at the forefront of Mead's narrative. *Coming of Age in Samoa* clearly situated Mead, the ethnographer, in the narrative along with her young informants. Moreover, Mead's text conveyed an explicit feminist subtext by privileging women's voices, legitimizing women's struggles to control their bodies, and challenging the structures of male power. In contrast, Freeman's geography of the South Seas had no place for Samoan women or the woman anthropologist who wrote about them. *Coming of Age in Samoa* will survive such efforts to edit women out of their own lives.

12 Symbolic Subordination and the Representation of Power in *Margaret Mead and Samoa*

Angela Gilliam

In 1988 *Margaret Mead and Samoa*, a documentary, presented one side of an anthropological controversy—the "Mead-Freeman Debate"—which had begun five years earlier with the publication of Derek Freeman's book *Margaret Mead and Samoa.* Reversing the typical scholarly framing of the debate as "the good guy (namely Mead) and the bad guy (namely Freeman)," the film took Freeman's side in the debate. The film, like Mead's *Coming of Age in Samoa* and Freeman's critique, reproduced the primary problems of Western ethnography—the dominance of Western voices and the subordination of the objects of their interpretations in image and text. The film's images continued the very eroticization of Samoan women for which Mead's work on Samoa had been criticized.[1]

The academic scandal sparked off by Freeman's original book challenged not only the reputation of Margaret Mead and *Coming of Age in Samoa* but also questioned the ability of the anthropological profession to detect fraudulent research. Originally incited by a *New York Times* report, the controversy focused on Freeman's attack on the validity of Margaret Mead's Samoan research. Presenting himself as a person who had naively gone to Samoa with total faith in *Coming of Age in Samoa,* Freeman repeated his tale of fraud and betrayal on a tour across the United States, punctuated by appearances on talk shows. Framed as a theoretical contest between the famous anthropologist and the obscure Freeman, the controversy became one of many battles fought in the so-called "Culture Wars" between the claims of "scientific rationality" and the cultural relativity attributed to Mead, implicitly understood as a precursor to postmodernist rejections of empirical validity. Conservatives bewailing the "permissiveness" of the 1960s seized the opportunity to attack the woman they blamed for the so-called sexual revolution of the 1960s. Identifying Mead with an assault on traditional values, conservatives welcomed Freeman's criticisms of her work.[2]

Perhaps James Clifford summarized the debate best. Describing "two competing portrayals of Samoan life," Clifford suggested that both Freeman and Mead "configure the other as a morally charged alter ego," despite their professed commitment to science. "What is the scientific status of a 'refutation' that can be subsumed so neatly by a Western mythic opposition?" Clifford asked. "One is left with a stark contrast: Mead's attractive, sexually liberated, calm Pacific world, and now Freeman's Samoa of seething tensions, strict controls, and violent outbursts. Indeed Mead and Freeman form a kind of diptych, whose opposition panels signify a recurrent Western ambivalence about the 'primitive.'" Clifford may have been alluding to the terms of the "Mead-Freeman Debate" as a modern example of the "white man's" or the "white woman's" Samoan.[3] Recognizing that the debate involved the invention and representation of the Samoans to serve their respective author's allegorical intentions, Clifford noted the omission of Pacific island perspectives in both books and in the subsequent debate.

Samoans have recognized the absence of a Pacific ethnography that would incorporate their perspectives. Confronting the faulty constructs deployed by Western observers, Pacific peoples have protested a hierarchical model of analysis that replicated the imperial relationship. Both Mead and Freeman constructed erroneous and one-sided interpretations reflecting their origins in an imperial system whose impact they ignored in their research. As previously argued, Mead's work has created a legacy that Pacific peoples need to confront. Freeman, despite his pretensions to having produced a scientifically verifiable interpretation of Samoan culture, paid inadequate attention to colonial rule. The film presented the same form of symbolic subordination and failed to interrogate the power relationships that helped to construct Samoans as a colonized people.[4]

Viewed from within the anthropological profession, the "Mead-Freeman Debate" contributed to an increasingly reflexive ethnography in which the relationship between observer and the observed became the object of attention. According to Lowell Holmes, the controversy forced anthropologists to "re-examine issues like nature/nurture, proper field methodology, scientific objectivity, and professional ethics." Freeman challenged Western scholars characterized as cultural determinists or "inductivists"—who knew what they were going to find before they looked for it. Invoking the "historical setting" in Samoa, he began a promising initial critique that mentioned the historical opposition of Samoan people to abuses of power by the U.S. Naval administration, which Mead had failed to acknowledge. Freeman, however, inexplicably aborted his argument about the geopolitical situation in Samoa in the 1920s. Instead, he concentrated on refuting Mead by trying to prove that the Samoans were the opposite of the gentle people described in *Coming of Age in Samoa.* "Samoans, as children, adolescents, and adults," Freeman contended, "live within an authority system the stresses of which regularly result in psychological disturbances, hysterical illnesses, and suicide." Presenting "the darker side of Samoan life," Freeman argued for the need to explore biological factors, implying that Samoans are "genetically endowed with a strong tendency to place utmost importance on social rank and hierarchy."[5] Styling himself a defender of the Samoan people, Freeman ultimately used Samoa as the background to promote a theoretical position and an academic reputation.

Despite its title, the film *Margaret Mead and Samoa* uncritically portrayed Freeman's view rather than a disquisition about Mead or the Samoan people. Mead served as a foil, while the Samoans became the background for Freeman and his views. As such, the film continued an intellectual discussion that remained defined as a theoretical contest between two Western anthropologists. This exclusion of Pacific perspectives makes it important to recognize that both the film and the books contribute to a Western-inspired myth of Paradisiacal Samoa—or its debunking—rather than providing insights into the experience of Samoans.

Though the debate became a controversy in 1983, Freeman actually began his critique in 1972, when he called attention to errors in Mead's *Social Organization of Manu'a.* In 1975 Pacific scholar Epeli Hau'ofa lauded Freeman for having criticized Mead's erroneous spelling of the Samoan language and for pointing out that the Samoan language was as deserving of academic precision as European languages. Coming to Mead's defense, Crocombe, in his reply to Hau'ofa, defended Mead's contributions to the anthropology of the Pacific, noting the welcome she had received upon her return to Samoa.[6] Thus the debate began while Mead was still alive, but it did not fully capture public or scholarly attention until the appearance of Freeman's book and its accusations against the most famous anthropologist and her most famous book.

Part of the complexity surrounding the issue is rooted in the history of an ongoing Western intellectual conflict over the relative importance of nature and culture as defining human existence. This has been viewed as the effort to deal with the periodic resurgence of biological determinism as an explication for human difference. In some Western countries, this controversy has been expressed through time as a struggle between those who believe that "biology is destiny" and those who maintain that culture determines how life would be lived by the individual. Mead's emphasis on culture as a primary factor in the development of humankind through time contributed to the struggle against racism and sexism in the United States. By suggesting that a painful adolescence is not necessarily experienced all over the world and may be determined by culture, *Coming of Age in Samoa* helped to defeat the primacy of biological determinist thinking in the United States.

Derek Freeman's apparent desire to revivify the importance of biology has been interpreted by some scholars as an intellectual rationale for racism and sexism, causing them to defend Mead's work as worthy, if somewhat flawed by its uncritical stance toward colonialism. As conservatism appeared to gain ascendancy in the 1980s, such scholars saw Freeman's apparent explanation of violence or crime as biological or genetic in origin as promoting a "new racism." Eleanor Leacock, for example, reminded scholars that physical anthropologists and biological determinists, rather than scholars with knowledge about the Pacific, welcomed Freeman's book with admiring comments on its "dust cover." As George Stocking has argued, "in the minds of many anthropologists, sociobiology is simply nineteenth-century racialist evolutionism reincarnated."[7] Given the importance of culture as a concept to anthropologists who have been influenced by Benedict, Mead, and the other Boasians, Freeman faced a difficult challenge to win this debate.

Sharing with Benedict a commitment to a "liberal, pluralist vision," Mead's *Coming of Age in Samoa* could be described as a "fable of identity" that emphasized the utopian possibilities of human malleability, telling a story about Samoa to provide "moral, practical lessons" for U.S. readers. It neither presented a complete picture of Samoan culture nor did it address Mead's position as an American, hosted by the U.S. Navy, in islands incorporated as a territory of the United States. Mead's initial work on Samoa did evoke many questions, in part because of the unquestioned access to power that she had exercised as an American scholar visiting Samoa in the mid-1920s. That the subsequent debate over the validity of Mead's interpretation did not foreground these questions reflects how difficult it continues to be for anthropologists and other scholars to face the consequences of colonialism and the hierarchies that it produced in colonized societies like Samoa. As a result, at least some of the defenders of Mead's work in the Mead-Freeman debate have offered uncritical support, divorced from a critique of imperialism and unaware of Samoan perspectives. Neither Mead nor many of the scholars assessing her work acknowledged the relevance of the geopolitical situation of Samoa in relation to the other nation-states in the region, particularly the United States.[8]

Perhaps the fact that Freeman conducted his research in Western Samoa, then a New Zealand protectorate, rather than in American Samoa where

Mead had conducted her research two decades earlier, detracted from his ability to incorporate an understanding of geopolitics into his analysis. Instead of exploring the origins of the "darker side of Samoan life" in a colonial situation, he asserted that Mead's most significant error was her emphasis on cultural phenomena. Insisting that Mead neglected "much more deeply motivated aspects of Samoan behavior," Freeman invited the reader to conclude that he was affirming that Samoan cultural patterns are there by nature. As David Schneider has inferred from Freeman's emphasis on the importance of "biological factors in determining human behavior," Freeman was arguing that "Samoans are genetically endowed with a strong tendency to place utmost importance on social rank and hierarchy."[9] Failing to investigate the role of colonial hierarchies, Freeman provided cultural evidence but asserted natural causes for those patterns.

While Samoan critics lauded "Freeman's very solid and complex vision of Samoa," they have recognized that, in the words of Samoan novelist Albert Wendt, Samoan "informants" have been excluded from the debate. That exclusion implied, according to Wendt, "that we don't know enough about ourselves to be able to contribute intelligently to the debate." Wendt's brother, Tuaopepe Felix S. Wendt, expressed even greater concern about Freeman's work: "Some people have expressed the view that Freeman has done us a good turn by finally dispelling Mead's illusion of Samoa. Unfortunately, the more I re-read Freeman's book, the more difficulty I have identifying what constitutes this 'good turn.'"[10] His ambivalence stemmed from Freeman's use of Samoan culture to support his position while claiming to represent Samoan "reality." As Tofilau Eti, Prime Minister of Western Samoa, expressed in 1983, Freeman's and Mead's defenders would benefit from considering theoretical alternatives provided by Samoan scholars.

Moving from text to filmic representation, *Margaret Mead and Samoa* presented Freeman's thesis in a 51-minute documentary format. The film gave Freeman the opportunity to offer his perception of his own personal relationship to Samoa and Samoans, accompanied by footage portraying Samoa and Samoan people on screen. Although interviews with Western academics occasionally challenged Freeman's criticisms of Mead, the information taken in by the eye tended to overwhelm the commentary. Cultural critics have explained that there is a distinction between the act of production and the filmic potential inherent in the moment of reception.[11] At the core of this concept is the question of authorship. Samoans did not produce the documentary *Margaret Mead and Samoa;* the scholarly "talking heads" included only one Samoan, Fanaafi Le Tagaloa. In addition to the experts being interviewed, Mead herself was often foregrounded in these segments with her own words and actual footage or illustrated by photographs of herself with others, including Samoan women. Many scenes placed Samoans in the background—situated within the filmic image in such a way as to give symbolic reinforcement to their textual subordination in Mead's and Freeman's work. This filmic vocabulary subtly reproduced the problems with both Mead's and Freeman's theses and the debate between critics and defenders.

Two modes of Samoan representation accompany the first segment of the film with its attention to Margaret Mead and *Coming of Age in Samoa.* Echoing the romantic version of Samoa that Mead presented in the first substantive chapter of her book, the camera provides twenty shots of "Paradisiacal Samoa." This version of Samoa shows beaches, swaying palm trees, ocean waves, and only a few buildings. It reproduces the exoticized image of Samoa as an idyllic South Seas island landscape. The second mode features Samoan people themselves in either exoticized or eroticized representations. During four scenes the camera lingers for several seconds on the breasts of Samoan women, accompanied by an oral narrative not directly germane to the visual representation of their bodies although somewhat suggestive. At least once, the camera begins with the breasts so that the face and identity of the woman almost seem irrelevant in contrast to the way the "expert" informants are depicted, where the "head" is their primary feature. Another fifteen scenes contain images of non-eroticized nudity or still photographs. Full body depictions of partially nude Samoan women predominate, although one scene depicts nude men relaxing or working in a lagoon. While the camera does not linger on specific body parts, images invited the viewer to share an exoticized image of Samoans posited as "different" from Europeans. The viewer is primarily positioned as a European male accustomed to seeing Polynesian women as sensual and feminine.[12]

Interspersed with these images are shots of several Western scholars making substantive criticisms of Mead or her work or offering critical support for Freeman. This segment includes biographical sequences on Freeman's life, including an explanation of how he received the Samoan title of Logona-I-taga and his subsequent recognition by the council of chiefs. Three of the interviews concern the intellectual context within which Freeman's ideas address broader issues rather than simply the correct analysis of Samoan society. Robin Fox, a noted sociobiologist whose theoretical position is not made explicit in the film, claims that a substantiation of Freeman's critique of Mead would undermine the "liberal humanitarian scheme," which would be proven to be "wrong." Laura Nader expresses disquiet over precisely that outcome by pointing to the historical role the kind of biological determinism Freeman upholds has played in supporting socially conservative, "right-wing" government policies in the United States.

Ian Jarvie addresses the deeper meaning behind the debate that Samoan contributors might have sought to include. Understanding that the European or American vision of the world has political implications, Jarvie suggests that Samoan people, interviewed by an inquisitive researcher within the context of colonial domination, might have told Mead what they thought she wanted to hear. No one made the same suggestions about Freeman, who also conducted his research within a colonial context.

Having concentrated on Freeman's views and those of other Western scholars, the last ten minutes finally enables Samoan voices to be heard. The narrator, possibly an Australian woman, intimates that Samoans voices might add complexity to the debate. Five Samoans join the debate, primarily addressing Mead's work. High Chief Lutali, governor of American Samoa, recalls confronting his anthropology professor about Mead's Samoan ethnography during his days at the University of Hawaii. Critical to Freeman's contention that Mead's work depended on false information is the on-screen interview with an 86-year-old Samoan elder, Fa'apu'a Fa'amu, who affirms that she and Mead "were like real sisters." When referring to the information she gave to Mead, she maintains, through a Samoan translator, that "we [Mead's informants] just lied and lied." The complexity of this scene is not fully discussed within the context of the film, which does not appear to want the viewer to question the authenticity of Fa'apu'a Fa'amu's response. As Felix Wendt asked in 1984, why did Freeman imagine that he was exempt from being duped? Wendt asked in print, but not in the film, whether Freeman was "himself behaving *tau fa'ase'e*—purposely giving incorrect information to mislead the intellectual community."[13] Understanding the significance of the unexamined power relationships in both research processes, Samoan scholars asked questions that the producers of the film Margaret Mead and Samoa did not raise.

Another question not asked about the commentary provided by Fa'apu'a Fa'amu concerns the history of Samoa. Whatever the flaws in Mead's research, the circumstances in the mid-1920s in American Samoa certainly differed by the 1940s and afterward, when Freeman conducted his research in the New Zealand protectorate that would become Western Samoa in 1962. Similarly, the film does not question whether Fa'apu'a Fa'amu might have participated in any discussion or debates about Mead's interpretation since she and Mead became friends in 1925. Naively ignoring the issues of how memory might be conditioned or of the influence of Christian conversion on the recollection of girlhood sexual experiences, the film's producers, like Freeman, do not encourage the viewer to challenge Freeman's assumptions about the validity of this crucial evidence.

The failure of the debate to engage Samoans prevented the asking of essential questions. In commenting on the persistent existence in Samoa of the "Mead industry," Samoan journalist Lelei Lelaulu eventually became convinced that Mead had described herself in the act of viewing another culture rather than the culture itself.[14] The only Samoan scholar interviewed in *Margaret Mead and Samoa* shows this same skepticism. Maintaining that Mead came to Samoa with her own preconceptions and problems, Le Tagaloa explains, "She was just looking for a frame." Through a focus on Samoan "promiscuity," Mead stripped away "our oneness with other human beings." Talking Chief Muasai and High Chief Galea'I Poumele, secretary for Samoan Affairs, share the same doubts about Mead's approach. Inserted into the film's narrative framework that asked Samoans only to comment on Mead's work, these commentators cannot fully establish their intellectual "ownership" of the issues concerning Samoan representation. Left unquestioned, the flaws in Freeman's interpretation do not receive the attention they deserved.

Lurking within the on-screen commentaries by anthropologists Richard Goodman and Tim O'Meara are claims that Samoan contributors would surely have wished to challenge. Their discussion of Samoan aggression, rape, and propensity to violence is a disturbing subtext whose implications become more explicit in Freeman's words near the end of the film. Explaining his interest in Samoa, Freeman describes Samoa as "a place that has assumed immense anthropological and human significance." Justifying his own research, he declares, "I have often said that if only we Westerners can understand the Samoans, then we can understand ourselves." Speaking for the "we" which must have referred to Western scholars, Freeman adds, "And we should be thankful that Samoa was a place where issues of such great scientific importance could be studied and resolved." Freeman ends by professing an "enormous debt of gratitude to Samoans."

The failure of the film to address issues of power and symbolic subordination prevented the Samoans from being asked whether they wanted their society to be the subject of scientific investigation and debate that imprisoned them within a fixed colonial gaze that emphasized "differentness." The interviewer did not ask Freeman to justify his claims. Why do Westerners need to understand Samoans in order to understand themselves? What purpose do Samoans serve in the Western imagination? As a result, Margaret Mead and Samoa, like Freeman's book, continued the same process that had tried to sever Samoans from their "oneness with other human beings."

Had the producers of *Margaret Mead and Samoa* engaged with the work of film critics, the documentary might have included a more self-critical approach. Jane Gaines's elucidation of "looking relations" or Laura Mulvey's analysis of the "male gaze" certainly could assist the viewers of the film in comprehending how the Samoan—especially the female—body has been displayed in this documentary. At issue in the film as in other visual media is "the way these viewing vantage points control the female body on the screen and privilege the visual position (the gaze) of the male character(s) within the film." In those scenes where the camera lingers gratuitously on the bodies of Samoan women, the knowledgeable viewer might recall Peter Worsley's critique of Mead's popular work—namely that she catered to a prurient preoccupation with sex. Despite its pretensions to providing a critique of Mead's work, *Margaret Mead and Samoa* pandered to the same desires whether or not its makers intended to do so. The documentary displayed the same "naturalized patriarchal assumptions" of the "cinematic apparatus" that constructs its spectators and structures "the screen relationship." The film's images evoked "fantasy," "fetishism," and "narcissism" and conveyed the notion that "the woman's image" exists to "be looked at (and to be desired)." Polynesian women especially have found themselves entrapped within controlled inventions of erotic imagination catering to Western readers and viewers. *Margaret Mead and Samoa,* with its acerbic comments from one of Mead's biographers about her attractiveness, clearly reflects the vision of a patriarchal ideology that treats women primarily as objects of male erotic desire.[15]

Reinforcing the eroticization of Samoan women, ethnicity intersected with gender and sexuality within the historical structures of domination embedded in the colonial relationship. The sustained, objectified, sexualized depiction of women contributed to sociopolitical control. The images and discourses that entrapped colonized people shaped the visual politics of the documentary that wove together multiple and reinforcing strands, which blurred the boundaries between erotic fiction, scientific research, and ethnographic allegory. The Samoans depicted in *Margaret Mead and Samoa,* with the exception of those included in brief interviews, have been removed from their historical context to occupy a place in the colonial imaginary. By conforming to documentary and ethnographic conventions, the images appear evidential rather than ideologically constructed. If, however, viewers had read analysts of the gaze, they might understand that photographic images operate as "ideological forms that construct terms of identity and experience for subjects and viewers."[16] The exposed breasts of Samoan women offer visual "proof" of their subordination to viewers who consume these images as evidence of their superiority.

Although women serve as the symbols par excellence of this subordination, the process reflects the status of both colonized women and men. The selection of primarily female images by the producers of the documentary and the overwhelming presence of Western male experts who define themselves sexually as heterosexual complicates the issue. Dominating the terrain of interpretation, these experts produce knowledge about Samoans that occupies the structural interface between the exotic and the erotic in which the bodies of Samoan women operate as signs of the blurring between ethnography and pornography involved in the construction of otherness. Sharing a common discourse of domination, these two ostensibly distinct modes of knowledge satisfy the same "impulses born of desire: the desire to know and possess, to 'know' by possessing and possess by knowing." Structured "hierarchically," the film visually reinforces the colonial experience it fails to acknowledge. A film like *Margaret Mead and Samoa* re-inscribes stereotypical images of Samoans and, despite its "documentary" genre, does not offer an alternative to the Hollywood image of a Pacific nation as Paradise. In the words of Diane Mei Lin Mark, films about Pacific Islands have recycled "exotic scenery and romantic Island Paradise stories" making it "difficult for other stories, real Island stories, to find a place in the market."[17]

Despite Freeman's claims to defend Samoans against Mead's fraudulent depiction of their culture, the film foregrounds him while backgrounding most of the Samoans enhancing his authority. For example, in the scene depicting Freeman in his role of honorary Samoan-designated authority—Logona-I-taga—Freeman's figure and his words occupy the central position in the sequence. Samoans transport and arrange suckling pig and other foodstuffs within the ceremonial house. The camera locates Freeman in a separate space from Samoans—mostly men—who almost seem to be praying and/or listening pensively as he speaks. Although Freeman speaks of his "deep" bond with the Samoans, whom he calls "marvelous human beings," the mise-en-scène renders them subordinate to him. Other than the singing subsequent to Freeman's words, no Samoan says anything. In the last few minutes of the film, in which Samoans do appear, Freeman speaks with the voice of authority. The visual impact of surrounding Freeman with Samoans, promotes a sense of false egalitarianism that masks the fact that nowhere in the film do Samoans express their opinions of Freeman or his "refutation" of Mead's work.

Within the context of the film, Freeman's tears shed on screen are, on one level, moving. In an eloquent expression of communion with Samoans, Freeman reveals a capacity to speak credibly and extemporaneously in the Samoan language, saying, "I trust that we shall all stand firm in truth and love forever." Freeman's "search to belong" to Samoans, as Albert Wendt graciously put it, may well be a factor in his efforts to portray a different picture of the land and people he has adopted. However, Pascal Bruckner aptly warned about romantic "Third-Worldism" and "getting high on paradise," observing that "the passion to be someone else requires first of all that you recognize the otherness of what you are searching for." Bruckner also cautioned that racism could reappear in disguised forms in colonial encounters: "The idea of being open to foreign cultures engendered the monstrosity of differentness, a catchall that inspired not only starry-eyed conversions but a highly sophisticated form of racism." Explicitly addressing people of European descent from highly industrial countries, Bruckner's admonition raises important questions about Freeman's identification with Samoans that neither he nor the documentary addressed. Without more Samoan voices expressing opinion and authority regarding Mead and Freeman, the emotion expressed by Freeman in the film appears manipulative at best.[18] Whether Freeman could be accused of the espousal of a new kind of racism depended a great deal on how the Samoans evaluated his interpretation. Differentness can be used to validate the dominant, the European.

Margaret Mead and Samoa juxtaposed ethnography and film, crossing two frames for representing knowledge and peoples. In the context of the "ethnographic present," the ethnographic filmmaker often violated the reality of the people by an interposition of the narrator or the careful selection of informants.[19] Such depictions reveal the extent to which "subject peoples" are not permitted to voice opinions that suggest authentic intellectual production but rather are made to be the exhibit that accompanies a specific discourse. *Margaret Mead and Samoa* presented a hierarchical exoticization of Samoa and Samoans to create a "visual ethnography of ethnography." The film criticized Mead's methods and portrayal of Samoans, and in turn used classic ethnographic visuals itself by appropriating an eroticized paradisiacal Samoa. In its structure and text, it reproduced the very aspects of Mead's ethnography that many Pacific peoples have found offensive.

Equating Mead with the postmodernism that Freedman disdained as "the most anti-scientific 'school' in the entire history of anthropological thought," both Freeman and *Margaret Mead and Samoa* avoided considering a possible mode of representation that might have subverted the power relations that permeate the film's visual vocabulary. Postmodernism would bring Samoans closer to the emotional core of the quandary about the intellectual perceptions of Samoan culture. According to Robert Stam, "dialogical anthropology" would use "what the group says about itself to frame the observation." The "observations of the observed" assist in transforming the categories of intellectual inquiry that ethnography has conventionally utilized. This is not merely discursive postmodernist analysis in a vacuum, according to Pauline Rosenau, who argues that "postmodern interpretation" centers on "listening" and "talking with" the other. Such a theoretical approach respects the "other's voice." This sort of methodological complementarity confronts the careerism implicated in the construction of the "exotic." As Glenn Jordan has maintained, an authentically "decolonized" anthropology must not only incorporate the new interpretative cultural anthropology but also go beyond it. Noting that postmodernism is often silent on questions of domination, Jordan affirmed that "anthropology concerned with assisting subordinated groups" should continually reveal the relationship between "hegemonic discourse" and "regional systems of power." Without presenting their views on screen about the cultural context in which either Mead or Freeman worked, Samoans remain the "ethnographic other"—a resource for the collector of data to market in order to enhance scholarly prestige.[20]

Although the film apparently places Margaret Mead at the center, as its title suggests, Derek Freeman actually occupies the core of the verbal text, and his concerns dominate the subtext as well. Freeman's cinematic "refutation" of Mead's *Coming of Age* expounds on the errors of her interpretation with extensive detail. As Clyde Taylor has explained, "Every representational sign or image bears one fact that expresses (and perhaps also hides) a cultural-ideological allegory." Adam Kuper recognized that Freeman played the central role, as evinced in the last scenes where "he is shown in his garden in Canberra," then "padding through the woods," and finally, "driving what seems to be an armored personnel carrier through quiet suburban streets." The narrative accompanying these scenes is the only place within the film in which Freeman suggests a "new" model for a science that would merge culture and biology. Acknowledging the importance of the question that Franz Boas assigned to Mead to research in Samoa, Freeman declares that anthropologists must "take into account our evolutionary history and our cultures" combined in an "interactionist paradigm" as the precondition for "a genuine science of our species."[21]

Unaddressed in the film or Freeman's writings is the question of what the "Nature versus Nurture" debate means for Samoans. The silence has been explained by Trinh T. Minh-ha by pointing out that the Mead-Freeman debate is a "conversation of 'us' with 'us' about 'them,' a conversation in which 'them' is silenced." At the very least, the decolonization of visual anthropology requires consultation not only about the projection within the film but also about the opportunity to share intellectual authorship and authority. This required more than mere utilization of Samoan voices to illustrate or "lend authenticity to the story chosen by the film team."[22] Without equal participation within the frame or the ability to criticize Freeman's views as well as Mead's, Freeman's fluency, dress, and emotion mask something far deeper: "going native" merely consolidated the foreign scholar's privilege in a hierarchical situation that dictated what could be said.

Neither the film nor Freeman's writings challenge Mead's contribution to the construction of the concept of the "primitive," perhaps the most vivid evidence of colonial scholarship in *Coming of Age in Samoa,* whose subtitle declared it to be a study of "primitive youth" for "western civilization."[23] Freeman, in fact, reinforced this linkage by painting the Samoans as more violent and emotionally conflicted than Mead's portrayal rather than challenging her use of the term *primitive.* Freeman thus missed another opportunity to "get it right"—an objective to which he gave voice near the end of *Margaret Mead and Samoa*—and to demonstrate an authentic commitment to decolonizing the ethnographic representation of Samoa. As Eleanor Leacock admonished, ethnography must be "historically-oriented, advocacy-linked anthropology, undertaken in active collaboration with people whose cultures are being documented."[24] At the very least, this requires that the roles of the scholar-adventurer and the scholar-star must be interrogated rather than celebrated. *Margaret Mead and Samoa* has failed to address the issues that would enable Samoa to be seen as a nation and a people rather than as an illustration of the Western imperial imaginary.

13 Misconceived Configurations of Ruth Benedict

The Debate in Japan over *The Chrysanthemum and the Sword*

Pauline Kent

Ruth Benedict's analysis of Japanese culture, *The Chrysanthemum and the Sword,* written for the American general public immediately after World War II, led to a wider understanding of the Japanese as a people—as opposed to a previous image of an exotic people, an enemy race, or a fierce enemy who seemed to defy all understanding. Although it can be argued, as C. Douglas Lummis has done, that her methodology and other elements of the study created future problems for the field, Japan studies owes a great debt to Benedict.[1] Fifty years of debate have given *The Chrysanthemum and the Sword* a virtual life of its own as a controversial text.

Japanese critics often bring to their discussion of the book a limited understanding of Benedict's approach to culture and of Benedict's life.[2] As a consequence, extensive insights into "motivations" behind Benedict's works have not been fully incorporated into the Japanese debate over *The Chrysanthemum and the Sword,* with the exception of Lummis, who has incorporated both biography and textual analysis in his work, albeit with serious limitations in his interpretation.[3] As this essay will demonstrate, scholarship on Benedict and *The Chrysanthemum and the Sword* continues to reflect significant misunderstandings of Benedict's motivations and her analysis of Japanese culture leading to a "misconceived configuration." By analyzing some of Lummis's major arguments and the material he used to formulate his interpretation of Benedict, this chapter points to problems with this influential interpretation. It argues for the importance of simultaneously considering Benedict as an author with her own intentions, the work that has fascinated so many Japanese readers for a half century, and the historical context in which she wrote *The Chrysanthemum and the Sword.*[4]

Commentary on *The Chrysanthemum and the Sword* began in Japan with Tsurumi Kazuko, who reviewed the original English version in 1947. Shimada Hiromi has pointed out that most of the earlier critiques focused on the ethics system that Benedict described. After the appearance of Sakuta Keiichi's article "Shame Culture Reconsidered," *The Chrysanthemum and the Sword* became popularly known for its description of Japanese culture as a "shame culture." Many criticisms concerning her sources and research methods arise from, and are limited by, the brief description that Benedict provided in her first chapter.[5] Other criticisms focused on her too-brief and arbitrary use of history to explain modern society, her treatment of society as a whole, her use of "outdated" sources and data, and her attempt to force American or Western values, such as democracy, on the Japanese. Beginning with Lummis's book in the 1980s, criticisms that Benedict's study was based on a poet's intuition, rather than rigorous scholarship, have become more prevalent.

Lummis began *A New Look* by declaring that his discovery of Benedict's interpretation of Japanese culture had been a "catastrophe" because it had misinformed him about Japanese society when he first read it in 1960. He criticized her fascination with the image of death as a thing of beauty, which inclined her to seek the beautiful patterns of "dead cultures" without sufficient regard for accurate depictions of complicated societies like Japan. Spurred on by a sense of betrayal, Lummis sought to explain the reasons why Benedict's work continued to influence Japanese discussions about their culture.[6]

Having set out to explain the reasons why he had been "coerced" into believing the misleading contents of *The Chrysanthemum and the Sword,* Lummis suggested that it had been a product of the propaganda war machine. Surviving the exigencies of its origins, it had retained its ability to persuade because of its appeal as a work of "political literature."[7] It overpowered its readers with unforgettable imagery, a synthesis of poetry and anthropology. According to Lummis, Benedict had artfully shaped the facts to fit her purpose, effectively producing a work of political fiction in the guise of scholarly interpretation.

Lummis cited Benedict's farewell speech as president of the American Anthropological Association, entitled "Anthropology and the Humanities," as proof that her literary studies had clouded her scientific objectivity. In that speech, Benedict talked about the link between anthropology and the humanities because "they deal with the same subject matter—man and his works and his ideas and his history." He quoted Benedict's acknowledgment that "Shakespearean criticism" had been invaluable to her as an anthropologist, giving her "habits of mind which at length made me an anthropologist." However, he failed to quote the introduction to her speech, which declared that "anthropology belongs among the sciences" and added that "it must constantly try to profit by methods and concepts which have been developed in the physical and biological sciences, in psychology and in psychiatry."[8] Too interested in proving his point, Lummis ignored a part of the available evidence.

Lummis's line of attack also ignored the judgment of anthropologists themselves. In contrast to Lummis, most anthropologists believe that Boas oversaw the evolution of anthropology "from an amateur hobby to its maturity as a rigorous academic discipline" by demanding very high standards from his students in their research. At the same time, he allowed them to invest their talents in writing fiction so long as they clearly distinguished it from their anthropological studies.[9] Boas demanded a thorough enquiry of the data used from the time Benedict wrote her Ph.D. thesis "The Concept of the Guardian Spirit in North America" in 1923. The fact that Benedict wrote her thesis in three semesters in no way compromised her anthropological training. She first attended classes given by John Dewey at Columbia before attending the New School for Social Research for two years and then moving to Columbia to study under Boas. When she attended the New School, she studied under anthropologists Alexander Goldenweiser and Elsie Clews Parsons. Through this training Benedict learned not only basic social science methods but also a variety of approaches. Criticisms about Benedict's lack of training are simply unfounded.

Just as importantly, the claim that literary training prevented Benedict from undertaking objective analysis assumes the existence of rigid boundaries between science and the humanities. In response to this sort of arbitrary separation, Arnold Krupat's *Ethnocriticism* outlined the history of the convergence of literature and ethnography as common pursuits of truth. Boas pursued truth by sending his students to the field, ensuring that they adequately understood the language there, and by instigating "an inductive, particularist, and rigorously relativist method" as opposed to the deductive methods of the evolutionists that were based on "ancient prejudice."[10] As Krupat pointed out, the separation of science from literature does not necessarily result in the production of truth. "Science" has often been known to manipulate the "truth" to fit current demands. In relativistic science and cybernetics, a "blurring of genres" has become necessary for the pursuit of a more comprehensive "truth." As Krupat stated, the Boasians kept "their art distinct from their scientific pursuits."[11] Benedict appreciated literature, but this does not mean that her interest in literature and poetry damaged her ability to respect the demands of "scientific objectivity."

The argument that Benedict was really a poet dabbling in anthropology relied heavily on a section of Mead's *Anthropologist at Work* that bore the title "Anne Singleton: 1889–1934." In that section Mead described Benedict's poetry, provided a selection of her poems, presented autobiographical details in "The Story of My Life," and included fragments from Benedict's earlier journals, correspondence with Edward Sapir, and an early paper, "The Sense of Symbolism."[12] Based on this source, Lummis developed his argument about Benedict's fascination with death and the importance of her alter ego, Anne Singleton, the poetic pseudonym through which, he contended, Benedict approached anthropology, using the intuitive methods of the poet. Finding beauty in death, Lummis argues, Benedict valiantly attempted to rescue dying cultures by recording their cultural patterns.[13]

It is not surprising that Lummis focused on the death theme in Benedict's "The Story of My Life," because it begins with the death of her father. The connections he made between this fascination with death, Benedict's poetry, her Anne Singleton persona, and anthropology are much more tenuous. Margaret Mead did state in *Anthropologist at Work* that in the earlier years the principal aim of U.S. anthropology was to collect data on "dying American cultures." Lummis then freely associated Benedict's personal experiences and emotions with the anthropological enterprise that enabled her to make a career of quietly exploring "the country over the hill, and contemplating the beauty of the dead." He went on to ask, "And can it be a coincidence that the man who provided her access to this country—'the world of my father'—was the man she came to call 'Papa Franz'?" He concluded that she tried to create her own undiscovered country through her study of American Indians. Anne Singleton and the anthropologist Benedict fused, causing her studies of cultures to become, in effect, obituaries for dying cultures. Finally, Lummis proved his "hypothesis" with the example of *The Chrysanthemum and the Sword,* which, he argued, should be interpreted as being an obituary for Japanese culture "written by one of the executioners."[14]

More recent biographical accounts of Benedict provide a striking contrast to Lummis's interpretation of "The Story of My Life." Margaret Caffrey set Benedict's biographical sketch into its historical context and her physical development, insisting that "it is important to understand Benedict's own extremely perceptive yet narrowly psychoanalytic interpretation of her childhood and the limitations of that interpretation." She discussed research showing that death can have a lasting influence on young children. In the case of Benedict, these effects manifested themselves in tantrums and later as depression. Benedict experienced the death of her father when she was still at the developmental age when tantrums were normal, but her hearing problem made it more difficult to express her feelings. Benedict interpreted her response in moralistic terms as the actions of a wanton child.

Until middle age, depression or what she called her "devils" tended to overwhelm her, but in later life she managed to achieve a state of serene calmness. Hilary Lapsley similarly emphasized the difference between the young Ruth Fulton portrayed in "The Story of My Life" and the mature Benedict who had obtained professional success and personal stability.[15]

Caffrey demonstrated that attention must be paid to the complexities of interpreting a life history, while showing how it revealed as much about the historical context of Benedict's life as about her character. In *Ruth Benedict: Stranger in This Land,* Caffrey produced a social history as well as a biography, portraying Benedict as a "vehicle for examining the intellectual and cultural history of the first half of the twentieth century." At the same time she dealt frankly with Benedict's "woman-identification" as she explored "Benedict's life as a case study in cultural feminism."[16] Having neither the personal nor professional inhibitions that shaped Mead's analysis, Caffrey explored the significance of Benedict's lesbianism along with her feminism. Benedict emerged as someone who, uncomfortable in mainstream American society, attempted to explore ways to open up society to a greater number of alternatives through an understanding brought about by tolerance.

As a historian, Caffrey understood the prevalence of death for those who, like Benedict, lived in the Victorian age during their girlhoods. Suicide was a common event for those who lived in Benedict's district, a region with an uncommonly high suicide rate.[17] Death was thus very much a part of life rather than a sign of a personal obsession. It would not influence her to identify the Japanese as members of a dead or dying culture equally fascinated with death or suicide. Placed in its historical context, Benedict's concern with death becomes a common experience rather than a predisposition to morbidity.

Lummis explained *The Chrysanthemum and the Sword* as a piece of political literature designed to mold Japanese society into a democratic society that would satisfy American ideological goals. To serve this purpose, Lummis argues, Benedict discovered the patterns that she wished to see. She had displayed this same tendency in *Patterns of Culture* where, in particular, her portrayal of the Kwakiutl was far removed from any reality that had existed for well over one hundred years. Her description was based on Boas's fieldwork data, which included a distorted portrayal of the potlatch. Benedict, in turn, used this faulty data to demonstrate that the Kwakiutl displayed an "unqualified penchant for self-glorification." Lummis's conclusion that Benedict "took no interest in data of that kind because she saw it as irrelevant to her purpose" ignored the fact that the studies that exposed these flaws did not appear until the 1950s, after Benedict's death.[18]

Lummis did make the valid point that Benedict used her anthropology for cultural self-criticism and political education for U.S. readers. Indeed, Benedict actively used anthropology as a means for improving society. Her attempt to make the Japanese as human as possible in *The Chrysanthemum and the Sword* showed her concern to counter the animosity toward the Japanese that had intensified during the war. Aware of the ignorance of Japan among Americans, she began with a sweeping introduction to Japanese history. Her use of Japanese history has been criticized as superficial by many scholars, but Lummis consciously tried to belittle her sources by dismissing them as "standard history books." Sources such as E. H. Norman and George Sansom are, even today, respected and reliable histories, and experts at the time recommended these as the books to consult.[19] Once again Lummis failed to take into account the historical context in which Benedict did the research and writing of *Chrysanthemum and the Sword.*

Lummis objected to the omission of any discussion of economics, politics, power, and class in *The Chrysanthemum and the Sword.* Benedict described the Japanese ethical system, which she considered to be at the core of Japanese behavior, rather than providing a comprehensive introduction to Japanese society. Although Benedict's papers at Vassar College indicate that she had read about economics and Japanese politics, she decided to write a book about Japanese cultural traits that did not discuss all aspects of culture. Perhaps his *expectations* for an all-encompassing introduction to Japanese society lead to Lummis's disappointment, but Benedict was not responsible for failing to deliver what she had not promised.

Having failed to read "between the lines" of the biographical interpretation he used in his analysis, Lummis neglected to consult the 1974 biography of Benedict written by Mead that showed that her perceptions of Benedict had changed over the interval of fifteen years between the two works.[20] This uncritical reliance on Mead's first biographical treatment of Benedict created a superficial portrait that continues to influence Japanese debates about the validity of Benedict's approach and the accuracy of *The Chrysanthemum and the Sword.*

As Caffrey pointed out, the 1959 biography was "of necessity, selective, and this very selectivity suggested things about Benedict, concealed other things, and reflected Mead's own imperatives as well as those of the late 1950s." For example, it failed to discuss Benedict's lesbianism—which included relations with Mead. Caffrey noted that the inclusion of the paper "Child Rearing in Certain European Countries" toward the end of the book reflected the return of American women to domesticity in the 1950s and the age of Momism in which Mead herself was very much active. By emphasizing their maternal interests, it was easier to camouflage their lesbianism.[21]

Lapsley described how both women attempted to achieve an understanding of homosexuality through their anthropological writings but distanced themselves from any personal association with that form of sexuality.

The 1959 biography only briefly touched on the war years when Benedict was carrying out research on Japan and other cultures and the years during which *The Chrysanthemum and the Sword* was being written. Wartime work was for the most part confidential and sometimes secret, and thus Mead was probably not privy to all that went on in the OWI. Mead was also absorbed in her own wartime work, in raising a young child, and in her marriage, which reached a crisis in the postwar period. What Mead chose to mention about *The Chrysanthemum and the Sword* was its "enormous impact on Japan and the U.S.," which eventually led to the Columbia University Research in Contemporary Cultures Project (RCC), directed by Mead after Benedict's death.

The 1974 biography interspersed Mead's observations in the biographical section with Benedict's own words, allowing the reader to gain a more comprehensive and chronological idea of Benedict's life. She added more information to the section on the war years because wartime security was no longer an issue. In contrast to *Anthropologist at Work*'s focus on professional aspirations, *Ruth Benedict* focused on Benedict's struggles as a woman, reflecting the interest in feminism that had emerged by the early 1970s. In contrast to the earlier biography where she too had featured very prominently—perhaps in an effort to establish her own position in the history of anthropology vis-à-vis her relationship with Benedict—Mead now concentrated on Benedict. *Ruth Benedict* made a veiled suggestion of an affair between Sapir and Benedict, perhaps for the purpose of concealing an affair between Mead and Sapir or of deflecting attention from the issue of lesbianism as another indication of difference from the earlier biography.[22] The 1974 biography placed less emphasis on Benedict's poetry, describing Anne Singleton as having disappeared in 1928 as Benedict had become more confident in herself and her anthropological work. Given access to newly released correspondence, Lapsley confirmed this point. Consideration of the implications of these changes might have altered an interpretation that stressed Benedict's persona as a poet as a major influence on *The Chrysanthemum and the Sword.*

Mead demonstrated her awareness that her biographies of Benedict could be seen as incomplete when she encouraged Judith Modell to write about her. Modell began in 1975, following up with her doctoral thesis, making at least some new aspects of Benedict's personality available to Lummis when he began writing his analysis of *The Chrysanthemum and the Sword.* Although Modell's published version of the biography, *Ruth Benedict: Pat-*

terns of a Life, shared some similarities with Lummis's approach, it significantly differed in interpretation. As an anthropologist, Modell chose to interpret Benedict's life in the form of dominant patterns in accordance with Mead's claim that "a finished life can be seen as simultaneous." Rather than relying on only the sharp and irreconcilable contrast between Anne Singleton as poet and Ruth Benedict as anthropologist, Modell described Benedict's life as a constant reconciliation of dichotomies. In *The Chrysanthemum and the Sword,* Modell heard the echoes of the earlier poet. Benedict's portrayal of the Japanese "had more in common with the poet's 'redundancy' than with the psychologist's 'over-determination,'"[23] she wrote, referring to the heavy-handed use of psychology during the war to determine the overall national character of the Japanese. Yet Modell insisted that Benedict's use of imagery and "redundancy" did not destroy her efforts to write "science" but became tools to convey the science more effectively.

Arguing that her awareness of many dichotomies provided Benedict with the motivation to challenge the hard-and-fast customs in society, Modell emphasized Benedict's feminism. Modell posed the first dichotomy as masculine/feminine. She analyzed "The Story of My Life" and the two worlds of father and mother. She included the opposition of life and death as a major theme in Benedict's life, but she delved into other dichotomies: imaginative and practical endeavor; ecstasy and achievement; creativity and attention. Modell described how Benedict was forced to reconcile her poetry with her professional scientific writings as she became more involved in the professional world of anthropology. Toward the end of her career, Benedict was again struggling to validate her efforts to combine anthropology and the humanities in a meaningful synthesis. Modell concluded that Benedict eventually understood "that dichotomy did not have to be eliminated but could in fact be productive, fruitful—the source of creativity in a discipline and a self."[24] Her interpretation contrasted with Lummis's insistence on the unresolved dichotomy between the poet and the anthropologist.

Modell addressed the issue of shame and guilt in *The Chrysanthemum and the Sword* in the context of child rearing, which reproduces culture.[25] She believed that Benedict's Japanese study contributed to the discussions on child rearing and national identity. As a result, Modell evaluated the importance of guilt and shame theories in relation to child-rearing practices and praised Benedict's use of these in her characterization of Japanese behavior. Modell went on to discuss the postwar RCC project in terms of its attention to child-rearing techniques and the influence these techniques have on shaping an individual's behavior within the larger context of national character. Like Lummis, Modell stated that Benedict wrote *The*

Chrysanthemum and the Sword to serve a political purpose, but she described it as pedagogical rather than ideological. Modell argued that "the very idea of pattern allowed her to make the inclusive characterizations that distinguished her anthropology and established for future anthropologists a dynamic relationship between culture and personality."[26] Rather than pattern being somehow related to a fascination with death, Modell stressed its dynamic possibilities as employed by Benedict.

Modell peered inside Benedict's thoughts via the medium of deeply personal poetry and a between-the-lines reading of Benedict's various texts. Her familiarity with literature and poetry deepened the insightful interpretation of Ruth Benedict's life. Moreover, her approach never led her to suggest that poetry in any way marred, detracted from, or distracted Benedict's anthropology. Nowhere did Modell accuse Benedict of writing obituaries for "dying cultures" because of either her personal experiences or her poetry. Instead, Modell described Benedict's contributions to anthropology as soundly based on an evaluation of the data rather than as molded by poetical intuition.

Just as Benedict began her Japan study with her "but/also" introduction of the seemingly paradoxical behavior of the Japanese, those who have written about Benedict since have also used this method to discuss the complexities of her life. Both Lummis and Modell discussed Benedict in terms of dualities. Caffrey noted dualities, but as a historian, she chose to see Benedict as a product of social dualities that emerged in the Victorian era rather than as psychologically prone to a dual personality. As the Victorian period yielded to the modern era, Benedict re-examined the values of her early years and faced many of the paradoxes that Modell and Caffrey explored.[27]

Caffrey traced Benedict's career within the emerging field of anthropology, which had to contend with Darwinism, diffusionism, and racism while attempting to form a concept of culture. Benedict helped to create the Culture and Personality movement and contended with influences of psychoanalysis on anthropology. According to Caffrey, *The Chrysanthemum and the Sword* was not simply an application of earlier theories. Instead, as with Benedict's other works, Caffrey considered it in relation to her immediate preceding work and within the context of the times. Agreeing with Modell that *The Chrysanthemum and the Sword* extended the methodology developed for studying cultures at a distance during Benedict's tenure at the OWI, Caffrey also assessed the book as being a continuation of Benedict's work on racism and synergy and her desire to apply anthropology for the purpose of attaining lasting peace.

While developing her critique of racism, Benedict also carried on her research on synergy, which she developed substantially in 1941 in a series of lectures given at Bryn Mawr College. These ideas would also influence her interpretation of Japanese culture. During the lectures, Benedict explained synergy as "combined action," furthered by "every extension of an area of mutual advantage and lowered by every curtailment of such an area." She sought to identify patterns of interaction that contributed to certain results. The intricacies of the interaction of social arrangements became the subject of her research to identify general sets of interactions that might be developed for the amelioration of social cohesiveness or integration. Benedict defined synergy as applied to culture as follows: cultures with low synergy have a social structure that provides for acts that are mutually opposed and counteractive, whereas cultures with high synergy provide for, and cultivate, acts that are mutually reinforcing. Benedict believed that identifying the mechanisms of high- and low-synergy societies might allow the re-creation of societies where members acted in mutually reinforcing ways, thereby resulting in a culture that was mutually beneficial and congenial to all its members.[28]

In the Bryn Mawr lecture on morale, Benedict brought up the topic of humiliation. "For all societies, good morale can be defined as sustained and mutually reinforcing individual participation in any social activity or crisis." Such a definition of "good morale" became synonymous with high-synergy society. In societies where solidarity and mutual support are honored, there was little need to use humiliation negatively. Where societies employed humiliation to induce good behavior, social mechanisms could be provided to counteract and thereby extinguish the humiliation. In this latter type of society, "children must have experience of discipline and be able to push themselves forward. But when child rearing gives them the requisite hardihood and the cultural forms specify readily available techniques for countering humiliation and insult, these societies are zestful democracies. They are less peaceable than the first type and they certainly do not have Christian ethics, but they carry out great undertakings and are often the most complex and dominant tribes in their area." Rather than referring to Japan in this instance, she pointed to how humiliation functioned as a negative factor in the United States, where it was humiliating just to be a "Negro." Thus, humiliation can produce completely different results depending on how a particular society weaves it into its own social fabric. After pointing out the negative side of humiliation, Benedict acknowledged that it "is one of the strongest incentives known" but "has positive social value only when it is incentive to labor and to learning and to increased social participation."[29]

While analyzing Japanese culture, Benedict investigated the cultural mechanisms that worked for and against the Japanese. Although the term *synergy* never appeared in the text of *The Chrysanthemum and the Sword,* this concept influenced her assessment of the Japanese, whom she thought capable of high synergy and a peaceful society. These ideas, in which humiliation would be absorbed into the concept of "shame," would be applied to Japan as Benedict continued her efforts to identify concrete cultural interactions that could possibly be used to form more cohesive and peaceful societies. Benedict's painstaking efforts to describe the mechanisms of the intricate cultural system should be seen as an attempt to measure the level, and identify the nature, of synergy in Japanese society. Considered in this light, labeling *The Chrysanthemum and the Sword* as a vehicle for U.S. propaganda seems inappropriate. Examining Benedict's analysis in the context of her earlier work produces a more nuanced understanding of Benedict's motivations and intentions.

As Clifford Geertz has pointed out, *The Chrysanthemum and the Sword* is appealing because Benedict successfully painted the Japanese not only as "less erratic and arbitrary . . . but by the end of the book, [as] the most reasonable enemy we have ever conquered." According to Geertz, Benedict's comparison of Japanese cultural idiosyncrasies with comparable American traits disarmed the public. She was able to teach them to begin to understand a people who had previously been viewed as the most fierce and unknown enemy the United States had ever confronted. Geertz described *The Chrysanthemum and the Sword* as Benedict's best book because she had freed herself from "methodological conceits she did not believe." By "methodological conceits" he referred to psychoanalytic theories about the cultural role of child rearing, and in particular toilet training, in the formation of national characters that created an image of the Japanese as compulsive and obsessive. By avoiding the use of such arguments, Benedict did not reinforce previous images of the Japanese as primitive, adolescent, or neurotic.[30] Benedict had learned, during her time at OWI, the value of shunning propaganda for objective data and thus did not convey the "politically correct" image of the day of the psychologically immature Japanese.

As this analysis has demonstrated, *The Chrysanthemum and the Sword* is far from simplistic. Benedict framed her analysis skillfully, taking into account contemporary attitudes and her intense desire for a new world peace that could overcome the death and hatred created during war. The patterns she traced in Japanese culture are easy to understand, but her examination of the mechanisms of the Japanese ethics system is complex. Imagery conveyed a new picture of the Japanese, but this was a necessity when considered against the backdrop of the rampant propaganda of wartime that had divested the Japanese of both humanity and culture. Benedict dismantled stereotypes before replacing them with a picture that proved ultimately to be much closer to the truth.

Like "The Story of My Life," the origins of *The Chrysanthemum and the Sword* are more complex than they first appear.[31] *The Chrysanthemum and the Sword* has its flaws; scholars, including Lummis, have identified many of them. Yet they still find some truth in the complex abstractions of Benedict's patterns. The complexities of its author's intentions make *The Chrysanthemum and the Sword* a work that can elicit a variety of readings. Despite the flaws, it is no doubt this complexity that has ensured that it continues to be read.

V / Re-thinking Benedict and Mead

As figures in the popular imagination and as feminist icons, Benedict and Mead occupy a complicated space, alternately serving as heroines and as subjects of scornful commentary. Mead's very ubiquity and prolixity lead to a more intense focus on her popular and scholarly image as "holy woman" and scandalous transgressor than on the cooler and more distant Benedict, who died before television could bring her into "living rooms, unannounced," as Mead described herself.[1]

With the exception of the ongoing controversy over *The Chrysanthemum and the Sword,* Benedict's writings have usually escaped the passionate attacks directed against Mead's major publications. Benedict's relative immunity to the accusations made against Mead—shoddy research, distortion, fraud, being hoaxed, betraying feminism, and complicity with colonialism—continues in the scholarly assessments of their respective legacies. Comporting herself as the dutiful handmaiden, the Ruth to Boas's Naomi, Benedict fulfilled the expectations associated with her biblical namesake. Her devotion to professional and public duty offered protection against the venomous attacks that accused Mead of publicity seeking, promoting sexual permissiveness, and self-aggrandizement. Remaining at her post at Columbia, Benedict's dignified, maternal lady stayed within her academic home.

It is important to recognize that part of Benedict's appeal lies in the juxtaposition with Mead. Constructed as a feminist heroine consigned by deafness, emotional distress, and a strong sense of duty to perform the "Martha" duties of a maternal academic housewife, Benedict displayed the emotional characteristics that engage feminist sympathies. Painfully building an academic career despite discrimination and masculine hostility, her eventual success as an academic is a tale of feminist triumph. Her escape from an unhappy marriage and her living with female partners made Benedict a doubly feminist heroine.[2] Mead, on the other hand, moved through life more optimistically, making a less satisfactory feminist heroine for those interested in victimization. Mead's persona appeared almost masculine in her discarding of husbands, her sexual freedom, and her ability to command the public stage. In contrast to Benedict's roles as Ruth or Martha, Mead's liberated "Mary" fits less well in a narrative emphasizing patriarchal oppression. Rewriting their relationship as feminist hagiography, however, should be resisted. Although they sometimes disagreed, together Benedict and Mead crafted those works that appear under one or another of their names.

A careful assessment of their scholarship more evenly distributes the honors that have been more readily, if stringently, granted Benedict, while Mead wears the dubious distinction of "popularizer." Called by Benedict to aid in making the public understand the relevance of anthropology to addressing social problems, Mead deserves no more criticisms for her efforts as a public intellectual than does her mentor. As Nancy Lutkehaus argues in her contribution to this volume, Mead deserves credit for her willingness to cross the private/public dichotomy between the academic and the public sphere. Judith Modell revisits her earlier interpretation of Benedict's emphasis on pattern to illuminate the intellectual context in which she wrote. Virginia Yans explores the underlying reasons for the criticisms that Mead has often received from her academic colleagues, while defending the positions she took.

14 Margaret Mead

Anthropology's Liminal Figure

Nancy Lutkehaus

> "How does it feel to be famous?" This is the question that children like to ask when they hear I am coming to dinner. My first answer is that you get a very skewed view of the world; it makes you feel that people are so much nicer than they actually are.

> My friends like to tell me about the things that my male colleagues tell their classes, the snide innuendoes that they are said to make. . . . My colleagues who imply that I couldn't know so much about the private lives of other people unless I had participated in it represent an equally authentic part of the culture. . . . It is a culture in which the success of a woman in his own field damages a man's esteem.

Mead wrote these two statements for a draft chapter of her autobiography titled "Vicissitudes of Public Life."[1] They reveal contrasting aspects of her experience of being famous: on the one hand, the kindness extended to her by total strangers because of her celebrity status, and on the other, the petty gossip, backbiting, and sexism of colleagues—especially male colleagues resentful of her fame. Although the chapter did not appear in the published version of *Blackberry Winter,* the two statements pose a paradoxical question about Mead and her status as an anthropologist. Why was she denigrated by so many within her profession while she was revered by the public? Mead suggested an answer: her tremendous popular success may have been a major source of her colleagues' denigration.

Indeed, Mead was a liminal figure in anthropology—someone who during the course of her lengthy professional career remained "betwixt and between" the world of academia and the world of celebrity because of the role she constructed for herself as a mediator between the world of academic anthropology and the public—and the ambivalent response she received from her anthropological colleagues as a result of her success in this role. The most salient aspect of liminality, as Victor Turner—who first used the term in the analysis of ritual process—has pointed out, is that of being on the threshold between one status and another, of being "betwixt and between" two different social positions and their assigned roles. Mead was a liminal figure in anthropology because she was betwixt and between the "high culture" world of academic anthropology and the "low culture" world of popular or mass culture. She was also a liminal figure because her success was based on a gender-bending mixture of character traits, personal style, and subject matter—elements of which have been associated with men rather than women. With the posthumous revelation of Mead's bisexuality and her romantic and sexual involvement with both men and women, her liminal status is even more strongly confirmed.[2] In other words, Mead was an anomalous being, by her own choice neither fully of one world or the other, not staying firmly in one place but delighting in her multiple positions.

My interest in the relationship between Mead's popular acclaim and her scholarly denigration began in the early 1970s when I was an undergraduate major in anthropology at Barnard College hired to work for Mead at the American Museum of Natural History. While taking courses from Stanley Diamond and Marvin Harris—both outspoken in dismissing Mead as a lightweight—I organized Mead's busy schedule of meetings and lectures.[3] I became acutely aware of the contrast between the disdain for Mead expressed by many in the anthropological profession and the acclaim she received from the public as well as from institutions and professionals not associated with academic anthropology, including officials at the United Nations and various nongovernmental organizations, members of Congress, later President Carter, talk show hosts, and readers of her monthly column in *Redbook.* If the term "public intellectual" means writers and thinkers who address a general and educated audience, Mead certainly succeeded in gaining and keeping an enthusiastic group of readers and supporters.[4]

Although there have always been anthropologists who have championed Mead, far more prevalent have been those who have criticized her, especially individuals from elite academic institutions, and men in particular. In the first instance, Mead has been criticized for flaws in her scholarship, beginning with the attacks on her first popular ethnography, *Coming of Age in Samoa,* from its first appearance in 1928 to Derek Freeman's infamous attack in 1983, as well as for her subsequent work such as *Sex and Temperament in Three Primitive Societies* and *Male and Female.* Her championing of culture and personality theory and the study of cultures at a distance also drew sharp critical attacks from fellow anthropologists. Critics focused attention on the aspect of Mead and her colleagues' theory of child development that posited a connection between national character and child-rearing practices, especially the so-called "swaddling hypothesis" derisively referred to by critics as "diaperology."[5]

In addition to attacks on Mead for weak theory and deficient scholarship, anthropologists have also criticized her writing and personal style, her politics, her choice of subjects for research, and her use of the popular media. While acknowledging that Mead's writing style was "felicitous," Clifford Geertz also characterized it as "undisciplined, loose-limbed, and improvisational," and as "saying seventeen things at once and marvelously adaptable to the passing thought."[6] Both he and Peter Worsley were critical of Mead's tone in *New Lives for Old.* Geertz criticized the book for its "exaggerated self-consciousness." Worsley criticized her romantic and lyrical "rustling-of the-wind-in-the-palm-trees" style of writing—a phrase he picked up from Evans-Pritchard's earlier critique of *Coming of Age in Samoa.* Such criticisms verge on sexism.[7] In his essay on Benedict, Geertz not only praised Benedict's writing style, in contrast to Mead's, but accused Mead of constantly associating herself with Benedict as a means of self-aggrandizement. Furthermore, he implied that Mead's success was based less on innate ability than on bald ambition.[8]

Other professional anthropologists scorned Mead's use of the popular media. They felt that she used the media to promulgate her personal opinions and that she published her opinions as though they represented anthropologically (read "objectively") derived insights. For example, in reviewing *A Way of Seeing*—a collection of *Redbook* articles—Alice Kehoe contended that she could not accept Mead's claim that her writing was connected by an "anthropological viewpoint," not just a personal one. Kehoe concluded that Mead was not an objective observer but rather "a voice of American liberal humanism," implying that Mead had no right to claim to speak for the discipline of anthropology.[9] Paul Bohannan, in a review of *Some Personal Views,* another collection of *Redbook* articles, said exactly the opposite. According to Bohannan, "Mead managed . . . to be very much of an anthropologist in a fresh context." Furthermore, he said, "I know of no situation that in any way compares in scope to her bringing anthropological points of view to a wider public."[10] The contrast between Kehoe's evaluation and Bohannan's simply underscores the ambivalence with which Mead was viewed within her own profession.

Another dimension to the criticisms of Mead's popularization of anthropology via mass media is the fact that Mead championed the use of film and photography as research tools and as means to disseminate research findings. As film scholar Alison Griffiths recently pointed out, during the twentieth century visual material was not accorded the same status in anthropology as written material, being regarded more as a means of illustration than as scientific or scholarly work. Interestingly, Griffiths argues, the denigration of visual material in anthropology is a result of the close relationship between film and popular culture that developed with the advent of cinema at the turn of the nineteenth century.[11] Indeed, only very recently—in November 2001—have criteria been set up by the American Anthropological Association to evaluate the production of a film or video as a scholarly work.

Many anthropologists have criticized Mead's politics. Especially during the Vietnam War period there were those who disliked Mead's opposition to the censure of several anthropologists accused of having aided the CIA in Thailand and Cambodia. This reaction to her was in spite of the fact that Mead also opposed the war. She claimed that she did not support the censure move partly because evidence against the anthropologists had been acquired illegally through the theft of documents from their university offices.[12]

Finally, the topics such as childhood and child-rearing practices; adolescence; male-female relations; and attitudes toward sex, marriage, and the family that Mead chose as the focus of her research and writing have been denigrated as "female" topics, and thus of lesser value, in the male-dominated field of anthropology, which has favored studying such topics as political, economic, and social organization. While some anthropologists, such as Robert Levine, have posthumously honored Mead by designating her as the founder of the anthropology of childhood, anthropologists in general still accord the study of children less regard than other topics. Since the popular press has often been eager to exploit topics relating to sexuality, especially with regard to non-Western cultural practices, many professional anthropologists have been all the more eager to distance themselves and their profession from these topics—and to attribute Mead's popular success to her catering to her readers' more prurient interests. Thus, Worsley wrote of Mead that "though she tackles serious questions, there is little doubt that she has played upon the fact that many of her readers have a less than scientific interest in sex and in her work, aroused by titles and headings suggesting all sorts of salacious possibilities." Malinowski, however, did not receive the same criticisms from Worsley, despite the provocative title *The Sexual Lives of Savages.*[13]

Finally, as Mead herself points out, some of her male colleagues suggested to their students that Mead could not possibly have known so much about the private life of other people unless she had participated in it herself. Turning this alleged accusation into ethnographic evidence, Mead offers a typical (anthropological) Meadian response. These comments, she says, represent an authentic part of her culture, an example of "a culture in which the success of a woman in his own field damages a man's esteem." It is this point about celebrity and anthropologists' reactions to Mead's popular success that are particularly addressed in this chapter, along with its relationship to Mead's gender. For as Mead herself was well aware, male academics in general, and male anthropologists in particular, seemed to resent her fame as an intellectual and her popular acclaim because she was a woman. Her success in their arena, she surmised, diminished their self-esteem.[14]

The analysis of the ambivalent relationship that anthropologists (and many other academics) have to celebrity and popular renown reveals aspects of a cultural process that Robert Foster long ago identified via the example of Mead and anthropologists' reception of her popular writing for *Redbook* magazine. Specifically, it demonstrates the practice whereby American academic culture works to preserve the hegemony of the semantic categories of "high" as distinct from "low" or "popular" culture. This is true despite the current lip service paid to the postmodern blurring of these very categories and the embracing among many academic disciplines, including anthropology, of the study of popular culture.[15] This hostility reveals a gendered dynamic at work within the culture of anthropology—and perhaps within academia more generally—whereby certain academic styles and modes of intellectual work are gendered "male" and others "female," with "high" culture being associated with male, and "low" culture being identified with "female," and "male/high culture" being valued more highly than "female/low culture."

It is useful to approach this dichotomy through a consideration of anthropologists' reactions to Mead at specific historical moments. For Mead, like the subject of totemism for Claude Lévi-Strauss, is a cultural phenomenon that is "good to think." It is good to think analytically and self-reflexively, that is, about the image of the anthropologist as professor and intellectual, the processes of the production and circulation of knowledge, and the relationship between anthropology as an academic discipline and anthropology as portrayed by, and presented in, the media.[16]

Members of the "high" culture—that is, academic anthropologists—ostracized Mead for having abandoned the critical distance that should characterize the anthropological observer. They criticized her for writing "science fiction" rather than science because she often used a popular vernacular rather than the specialized language of academic anthropology, because she addressed topics that were not always considered to be within the purview of the anthropologist, because she wrote and spoke about these topics without having performed the requisite amount of research or fieldwork that were hallmarks of the anthropologist, and because she wrote or spoke in venues that were not considered sufficiently professional.[17] In other words, activities such as writing in a colloquial vernacular, publishing in mass media, and appearing on radio and television talk shows associated Mead with various aspects of popular culture—especially mass culture. By stepping outside the boundary that divides high culture from low or popular culture, Mead was a transgressive, hence, a liminal, figure.

The distinction in Western discourse between "high" and "low" culture started as long ago as the second half of the nineteenth century with the rise of modernity in Europe and the United States. It reflected a concern among the arbiters of Western culture—both conservatives and progressives—that mass production, and later mass media, was corrupting the world of so-called traditional high culture.[18] Modern popular culture attracted the scorn of individuals of both ideological persuasions because it threatened the standards of high culture on the one hand and because it overwhelmed "authentic" low culture (folkways and customs) on the other.

In America, as Lawrence Levine has shown, the late-nineteenth-century use of the terms "highbrow" and "lowbrow" reflected an upper-class concern that the influx of new immigrants would mean the loss of a traditional "high" culture in the United States, the so-called "culture with a capital C" that the educated elite associated with (predominantly) northern Europe music, art, and literature and with Matthew Arnold's notion of "the best that has been thought and known in the world."[19] Education, especially a university education, both instilled knowledge of the best in culture and replicated the social hierarchy associated with this cultural hierarchy of values.

As theorist of contemporary culture Andreas Huyssen has cogently argued, "The gendering of mass culture as feminine and inferior had its primary historical place in the late nineteenth century, even though the underlying dichotomy did not lose its power until quite recently." He goes on to say that "the universalizing ascription of femininity to mass culture always depended on the very real exclusion of women from high culture and its institutions."[20]

Professional anthropologists, as representatives of the academy and thus as members of the subculture known as professors, protect a set of disciplinary academic standards that make them arbiters of proper—that is, professional—anthropology, as opposed to amateur or "popular" anthropology, of "science" rather than "science fiction," of the "high" culture of academia rather than the "low" culture of mass media. As Henrika Kuklick pointed out in her history of British social anthropology, its practitioners felt the need to distance themselves from those nonprofessional individuals who spoke to the public about their adventures among, as well as their observations of, so-called "primitive" tribes. Thus, professional anthropologists began to write for their own scholarly journals and to speak at meetings convened by professional societies.[21] Peer-review of books, articles, and later, research projects, served to delineate what was "good" anthropology from "bad" anthropology, and who was practicing each.

A similar process was at work in the United States. Beginning with Boas and his shift in focus from working as a curator at the American Museum of Natural History—an institution devoted both to research and to public outreach—to teaching at Columbia, universities became the most important site for the professional anthropologist. Although Mead taught anthropology at Columbia and other universities, she was never a tenured professor. The American Museum of Natural History, with its dual functions of research and public education, remained her professional home. Even at the museum there were scientists and administrators who criticized her and her fellow curator, anthropologist Colin Turnbull, for the popular acclaim their work received. Having learned from personal experience, Mead warned Turnbull of the potential ostracism he might receive from his colleagues when he published work for a popular audience. It may be significant that Turnbull was known to be a homosexual, a fact that displeased some of his colleagues at the museum—an institution traditionally associated with an extreme masculine image, from the statue of Theodore Roosevelt astride his horse outside the museum to the dioramas of African beasts inside the museum that hark back to the era of colonial exploration and hunting safaris. It may also be impossible to sort out feelings of sexism against Mead and homophobia against Turnbull from the professional disdain they received from colleagues for writing for a popular audience.[22]

By the second half of the twentieth century, as Russell Jacoby pointed out in his discussion of the "academization" of intellectuals in the United States, the independent author and critic who wrote for public consumption for his or her living began to be displaced by the private intellectual, the university professor. What this meant in terms of the high/low distinction with regard to anthropologists and other academic intellectuals was that they became even more insular and inwardly focused, writing and speaking more and more for peer-reviewed journals and for tenure and promotion committees that listened to the comments of other academics. For Mead, it meant that precisely at a point in her career when her work shifted to the study of cultures at a distance and the comparative analysis of her own society and when she increased her appearances on radio and the rapidly developing medium of TV, many of her peers disregarded or denigrated the time she spent giving public lectures, being interviewed by the press, appearing on television, or writing for *Redbook* magazine as nonprofessional work.[23]

There is another sense in which Mead was caught by the high/low dichotomy in anthropology. Along with fieldwork and attention to the taken-for-granted conditions of everyday life, a cross-cultural comparative perspective has historically defined sociocultural anthropology's complementary value with respect to other disciplines as well as its appeal to the public. Moreover, twentieth-century anthropology has had two important styles of comparative analysis—a positivist style, for example, the practice of objectivism and generalization, and an interpretive style. Rena Lederman identifies these two different styles of cross-cultural comparison as anthropologists' "local version" of C. P. Snow's "two cultures" by which he identified the differences between practitioners of the sciences and the humanities as constituting "two separate cultures." What Lederman found useful about Snow's "two cultures" tag is the notion of an oppositional relation between different ways of knowing as well as Snow's central concern in his essay with "the public reception of increasingly specialized knowledge."[24]

Within anthropology's "two cultures"—the positivist/objectivist style of comparative anthropology versus a reflexive/interpretive anthropology—Mead has been characterized as a "humanist" heir to Franz Boas's historical particularism—hence, associated with the practices of interpretation and reflexivity—rather than as a "scientist," in the terms of Marvin Harris or Napoleon Chagnon, concerned with objectivism, empiricism, and the search for general "laws" of culture and society.[25] As anthropology has struggled with its identity as a social science, Mead's particular brand of humanist anthropology—while eliciting, as Worsley contended, "a warm liberal-humanist appeal . . . from progressive and open-hearted readers"—has often been seen as detrimental to the discipline's aspiration for "scientific" rigor; thus, as Worsley argued, producing "science fiction" rather than scientific fact.[26]

The anthropologist is both professor and fieldworker. As such, he or she occupies multiple roles in and outside of the academy. And as Mead herself noted, drawing on the breadth of time she had been a professional anthropologist, the image of the anthropologist held by outsiders to the profession changed several times during her life: "There was a time," she wrote in a draft chapter for her autobiography titled "Being an Anthropologist in the 1970s," "when the anthropologist was greeted by the native peoples as the only breed who really understood them, and excoriated by colonial officers as having picked their brains and used their knowledge of the people to write books, but today anthropologists are excoriated for having been friends of any government, including our own, which is indubitable proof that they are on the side of imperialism, the establishment, and the status quo."[27]

As Micaela di Leonardo's recent work on the popular images of anthropologists in American culture as well as filmic representations of anthropologists, such as the deceitful and bumbling cultural anthropologist in *Krippendorf's Tribe,* demonstrates, the American public—at least as represented by the creators of mass media (film, advertisements, cartoonists) ascribes to a set of images of anthropologists that do not necessarily accord with the self-image that most anthropologists tend to hold of themselves. These mass media images of the anthropologist range from the basically benign (although prevaricating) and comically inept Professor Krippendorf to the more respected "translators of the sacred" (or "technicians of the sacred," as di Leonardo labels them) and shadows of forgotten imperialism, to the more maligned images held on one hand by politically conservative individuals who see Mead as having brought "barbarians to the gates," and by leftists who now identify anthropologists as evil imperialists, handmaidens to colonialism and the imperialist project of the West.[28]

To a certain extent, anthropologists' responses to Mead have been shaped by their concern to present a particular image of the anthropologist to their fellow academics as well as to the public at large—an image that Mead did not embody. Thus, another way in which Mead was a liminal figure in anthropology relates to the contrast between the public persona she came to embody and the self-image of the professor adhered to by many in the academy, unconsciously if not consciously.

Media scholar Dana Polan contrasts media images of professors with Edmund Wilson's description of Christian Gauss, his favorite professor at Princeton in the 1920s. For Wilson, Gauss was a riveting lecturer, a radiant inspiration to his students, causing them to feel the aura of historical events or characters. Polan describes Gauss as a figure of mediation whose role was to relay a knowledge that does not originate with him. Polan labels this concept of the-professor-as-mediator the "transmission model" of knowledge. An individual personality is irrelevant to the dissemination of knowledge, becoming a totally disembodied figure, simply a conduit for information.[29] At its extreme, the transmission model emphasizes the spiritual purity of the academic as the conveyor of truth. At the very least, it disassociates the professor from his or her students and establishes a social distance between the academic and the public at large—hence, the image of scholar in the ivory tower. Significantly, this image of the professor is also the image of a man. At some unconscious level, professors of anthropology—in their role as academics—aspire more closely to the "transmission" model of the professor than to the public's various stereotypical images of the anthropologist as adventurer, spiritual guru, and more recently as charlatan or buffoon.[30]

Today's world of celebrity includes female "academostars" like Judith

Butler, Gayatri Spivak, and Jane Gallop, who have perfected a striking personal style of self-presentation. As Martin Jay has pointed out with reference to the personal style and performative aspects of contemporary female academic "stars," this combination of personal style and public performativity calls into question "the distinction between high and low culture, which tends to inform conventional educational practice as well." He recognized that there are gendered dimensions to this high/low distinction and that female academics seek "to subvert by deed as well as rhetoric the domination of male models of cultural superiority."[31]

Mead's appearance, with her signature flowing red cape and the large staff or "thumb-stick" that she carried, might not seem so unusual to contemporary eyes. In the 1960s, when she first adopted this personal style of dress, it contrasted sharply with the more conservative tweed-jacketed image of the college professor and its female counterpart, a sheath dress or suit. What was most disconcerting—and hence, most subversive—was her ability to combine both masculine and feminine capacities. Long before it was as commonplace as it has become, Mead was both a mother and a full-time professional working woman. She could write "through the body," emphasizing her own sensate experiences as a woman, but her speaking and writing also exhibited a strong "masculine" pragmatism.[32] Mead challenged the rigid dichotomy between male and female roles and between masculine and feminine patterns of behavior. Thus, there is an additional sense in which she was a liminal figure as well, as she transgressed her assigned gender role.

To many anthropologists, however, Mead's self-consciously cultivated public persona was at best somewhat silly, at worst, melodramatic and self-aggrandizing. Likewise, her willingness to speak "from every podium to which she could gain access" has been interpreted as further evidence that she was primarily interested in winning and maintaining fame rather than spreading the message of anthropology, or, as public speaking often functioned for Mead, gathering further knowledge for herself of what people in various corners of the country or the world were thinking and doing.

Despite the recent move among anthropologists to extol the virtues—indeed, the necessity—of embracing the popularization of anthropology, they are careful to delimit what they see as "good" versus "bad" popularization. Jeremy MacClancy distinguished Benedict's "good" popularization (she was interested in changing people's values) from Mead's "bad" popularization (her motives, according to MacClancy, were self-interest). Hence, although the popularization of anthropology may be heralded as necessary to ensure the continued vitality of the profession, there are still boundaries to be guarded and standards to be maintained.

The limits now seem to rest on the contrast between the anthropologist as media figure/celebrity and the anthropologist as intellectual/professor. According to Daniel Boorstin in his well-known work on images in twentieth-century America, there are no longer true heroes, valued for what they have achieved or accomplished, but only celebrities. Originally, celebrity was a condition, the condition of being much talked about; but now a celebrity means a person. A celebrity, according to Boorstin, is an individual who is well known for his or her "well-knownness," not for some substantive achievement or demonstrated skill.[33] If Mead has achieved the status of "Grandmother to the World," and achieved this title through her own self-interested pursuit of it, as some anthropologists attest, her very celebrity was cause for fellow anthropologists to denigrate her.

Ironically, at a time when there is a movement within anthropology and history toward a recuperation of Mead as a positive image, among the general public there now exists a more ambivalent reaction to her name than when she was alive. As a result of the publicity generated by Freeman's attack on Mead's Samoan research, much of the public now associates her name with controversy. A reader of *Natural History* magazine—a medical doctor—responding to a recent article I wrote about Mead for the magazine wrote: "Although *Coming of Age in Samoa* had considerable cultural impact when it first appeared, it has been exposed by Derek Freeman as a 'shabby confabulation.' . . . If she had any lasting influence, it was because her popularity assured that both the lay public and professional anthropologists would learn how easily fieldwork can go astray when used in the service of one's politics and preconceptions."[34] Unaware of the lengthy rebuttals anthropologists have presented to Freeman's claims because those have occurred largely within disciplinary confines, this member of the public sees Mead's legacy and that of anthropology as tarnished and diminished.

There has been and still is an antinomy in academic anthropology between "high" and "low" cultures—whether the dichotomy distinguishes the academic from the popularizer, the academy from the media, or anthropology as a science from anthropology as part of the humanities—and hence a suspicion of individuals who cross the boundaries between the two, particularly when that individual is a woman. This distinction exists despite a growing sense among many anthropologists of the need to create a public anthropology, that is, an anthropology that engages the public.

The media's positive response to *Darkness in Eldorado*, journalist Patrick Tierney's critical account of anthropologist Napoleon Chagnon's research on the Yanomami, as well as their earlier embrace of Freeman's attack on Mead, demonstrate that academic infighting provides good entertainment—and ends up reinforcing negative stereotypes of both anthropologists and academics. But this is not further reason to retreat from engaging with the media; quite the opposite. What Mead has shown is that anthropologists and other academics ignore the media and the wider public at their discipline's peril. If academics abrogate their responsibility to convey to the public the breadth and depth of issues of current concern, they risk leaving a vacuum that the media will fill, often to their dismay.

15 "It is besides a pleasant English word"

Ruth Benedict's Concept of Patterns Revisited

Judith S. Modell

In 1934 Ruth Benedict's *Patterns of Culture* used a word that has attracted and annoyed commentators in and out of the discipline of anthropology for the past seventy years. "Patterns" had been used in American anthropology before, and it can be found scattered through Benedict's earlier writings. Other anthropologists had used "patterns" in ways that resembled her usage to refer to language, habits, or the relationship between individuals and culture. What then was different in 1934? Benedict expanded the scope of the word to refer to an array of anthropological issues, from method and theory to ameliorative purposes. Benedict used "patterns" to refer to cultures and individuals in culture, ways of seeing culture, and "truths" that she sought to teach her readers. Setting an intimidating goal, Benedict characteristically couched her aims in "a pleasant English word," as she described "patterns" in a letter to her editor.[1] By making "patterns" a "household word" without formally defining it, Benedict helped to create both the confusion and the lasting impact of the word.

In her book *Patterns of Culture*, Benedict allowed the meaning to emerge over the course of her narrative—which moved from introductory chapters on the discipline of anthropology, through descriptions of three distinctive cultures—the Dobus, the Zuni, and the Kwakiutl—to concluding chapters on the theme of the individual in culture. A passage in the third chapter came closest to a definition. There Benedict argued that "a culture, like an individual, is a more or less consistent pattern of thought and action. Within each culture there come into being characteristic purposes not necessarily shared by other types of society. In obedience to these purposes, each people further and further consolidates its experience, and in proportion to the urgency of these drives the heterogeneous items of behavior take more and more congruous shape." Referring to "the form that these acts take," she explained that it could only be understood by "understanding first the emotional and intellectual mainsprings of that society." A few pages later Benedict added: "All the miscellaneous behavior directed toward getting a living, mating, warring, and worshipping the gods, is made over into consistent patterns in accordance with unconscious canons of choice that develop within the culture."[2] These sentences contained the several dimensions of Benedict's use of the word.

As used by Benedict, patterns referred simultaneously to an arrangement "on the ground," to patterns perceived by an observer, and to a condition that could be judged by aesthetic and moral criteria. The first can be rephrased as "patterns in culture," and the second as "patterns in the eye of the beholder." The third meaning is aesthetic and introduces value judgments based on coherence and intricacy, aspects that are inseparable from the political connotations of the concept. The first meaning, patterns in culture, is by and large the conventional meaning of the word in anthropological writings; the second and third dimensions distinguished Benedict's usage from the conventional and gave her concept its staying power.

The immediate origins of "patterns" lay in anthropology and the development of Boasian approaches to culture throughout the 1920s.[3] Benedict added to these approaches her knowledge of literature and art, supplemented by an intense correspondence with Edward Sapir about the poetry and the discipline they shared. A more general exchange of ideas across the Atlantic undoubtedly made Benedict alert to British as well as American literary movements. Since only some of the influences on her choice of the word can be documented, a part of this analysis assumes Benedict's familiarity with developments in literature and in art during and after World War I in her own country and in Europe. Of particular significance in these developments was an emphasis on order and form, a confidence in the insight expressed through concrete imagery, and a faith in the power of art to alter human character.

Other scholars have examined the aesthetic dimension of "patterns" but have not fully explored how complexly aesthetic the concept was.[4] The very imprecision of the term reflected Benedict's attempt to incorporate revisionist statements about anthropology, fieldwork, and the anthropologist's obligation to "the modern world" into one concept. The meanings she drew from aesthetic theory expanded the concept of patterns from its purely typological and its less purely relativistic references. Benedict's use of "patterns" as explanatory and evaluative partially explains its persistence in the discipline. Unpacking these meanings will do justice to an often-irritating concept and also shed light on its pertinence to contemporary anthropological issues.

Personal and intellectual influences led Benedict to choose—with uncharacteristic certainty—the word "patterns" for the title of her 1934 book. Her choice reflected the process through which she found her way into an academic discipline while continuing to develop a sensibility that was grounded in literature and art. Close relationships with colleagues, particularly Edward Sapir and Margaret Mead, paved the way into the discipline. Both influenced Benedict's accommodation of the poet to the anthropologist. If Mead nurtured her self-image as a professional anthropologist, Sapir forced her to consider what she might lose if she suffocated the poet in the anthropologist. While all three colleagues wrote and shared verses in the early 1920s, Sapir continued to treat Benedict primarily as a poet well beyond the years in which she shifted her commitment from verse to anthropology. At times equally deferential to the opinions of Mead and to those of Sapir, Benedict simultaneously used their viewpoints as a springboard to achieving her own, unique, voice in the discipline. During the 1920s, as Benedict responded to Mead's advocacy of the discipline, she also attended to the insistent reminders from Sapir that "poetry" would keep her spirit alive.

Achieving confidence in her own anthropological voice enabled Benedict to immerse herself in the creative innovations of the 1920s that were manifested in the aesthetic of modernism. Through this immersion, Benedict's "patterns" developed beyond Sapir's favored term "form," the "configurations" that Alfred Kroeber gave to anthropology, and the colloquial use to which Mead usually applied the term.

Three biographers have interpreted the story of Benedict's entry into anthropology.[5] Each of them has discussed the specific influences of Franz Boas, Sapir, and Mead on Benedict in the crucial decade when, dissertation in hand, she moved forward as an anthropologist while simultaneously maintaining her poetic persona, Anne Singleton. To the evolving interplay between poet and anthropologist must be added the impact of Benedict's field trip to the Southwestern Pueblos, a trip that stirred the poet as much as it inspired the anthropologist. Out of the cauldron of personal change, exposure to a stunning landscape, and absorption in a modern aesthetic came "patterns" as exactly the right word to sit on the title page of her first book.

Margaret Mead and Edward Sapir provided the primary filters through which Benedict reconciled her future in anthropology with her past in poetry. Sapir is the more important, not only because his notions of form and harmony developed in parallel with Benedict's notion of patterns, but also because he challenged her in ways that Mead—the comfortable and comforting "arm chair" of Benedict's description in her diary—did not. Letters from Sapir forced Benedict to explore her decision to be an anthropologist, to strive to preserve her sensibility as a poet and writer, and to justify her intention to serve humanity—despite Sapir's cynicism. Sapir urged her not to relinquish poetry, art, and literature.

Understanding her friend's ambition, Margaret Mead took another tack, mentoring her teacher into the discipline of anthropology. Mead insisted that Benedict be rigorous in her approach to culture, the central concept of Boasian anthropology. The rigor Benedict achieved, however, benefited from the self-conscious assessment of her choices on which Sapir's letters insisted. Throughout the 1920s, Mead assured her friend that her role as a professional anthropologist could satisfy her poetic sensibility. At the same time, Edward Sapir insisted that Benedict remain acutely aware of the tensions involved in bringing poetry and anthropology together. These tensions give Benedict's writings their peculiar dynamism and intensity.

During the decade of her professional apprenticeship, Benedict published two major articles: "The Vision in Plains Culture" and "The Concept of the Guardian Spirit in North America," both based on the research she had undertaken for her doctoral dissertation. Each of these differently reflected the influences of Boas, Mead, and Sapir, while introducing the aesthetic method that is unique to Benedict and that would mature into *Patterns of Culture.* Representing Benedict's first attempt at adapting Boasian principles to the field she had chosen, the articles examined the then much-discussed topic of diffusionism, with an emphasis on sharply delineated, distinct, culture traits as receiving their character and meaning from their contexts. In different contexts, the "same" trait looked different, obscuring the common origins proposed by diffusionist theory. Rather than extending the more conventional discussion of the routes along which traits spread, Benedict turned diffusionism into an account of what happens to traits when they land in place.

Benedict's writing in "The Vision in Plains Culture" and "The Guardian Spirit" foreshadowed a stylistic technique crucial to the maturing of her anthropological oeuvre. First, she reiterated a set of details to provide evidence for the potency of a shaping force in culture; and second, she marshaled examples of striking contrasts in order to "prove" diversity in cultural arrangements. Benedict's approach began an argument that continued in *Patterns of Culture,* as her examples demonstrated the "compulsiveness" of existing "drives" within a culture. In an incipient form, Benedict grappled with the notion of a "purpose" in culture that was not simply a sum of individual goals but that gave to individual goals their creative potential and significance. The "purpose" in a culture shaped every trait to fit a unique design.

"The Guardian Spirit" introduced another theme that would receive further development in *Patterns of Culture.* Here the reader gains a clear sense of Benedict's urge to "save humanity" and to understand anthropology as a humanitarian discipline rather than as simply an intellectual exercise. "That is your faith," Sapir wrote in 1925, "and an inhuman absorption in a purely intellectual pursuit will never satisfy you."[6] Benedict's unyielding support of diversity, couched in vivid descriptions of particular manifestations of the guardian spirit, ended in a plea for improving existing social arrangements.

Between finishing her dissertation and writing her first book, Benedict experienced anthropology's rite of passage, fieldwork. Her delighted response to the Southwest Pueblos gave her a future chapter of *Patterns of Culture*, while the aesthetic impact of the Southwestern landscape brought together the concerns she had been exploring in journals and diaries, autobiographical sketches, sonnets, and anthropological papers.[7] Her perceptions of the ceremonialism and formalism of the Pueblos, their celebration of religious awe, and the smooth fit between personality and custom pervaded her portrait of the Zuni in the ethnographic triptych of *Patterns of Culture*. Whatever the combination of chance and self-conscious interpretation, Benedict gained the confidence that she could commit herself to anthropology without relinquishing all that she had invested in her persona as Anne Singleton.

The year *Patterns of Culture* was published, Benedict finished a version of her life story, a last draft of fragments she had been composing while writing the book. In one of these fragments she provided another clue to "patterns": "It is curious to see how the basic patterns of our lives hold from babyhood to decrepitude. All the tale could be told if we could set down the simple theme at the more and more significant times in which we work it out in our different stages of growth." Benedict constructed the narrative of her life around contrasts, purposes that sharply diverge. In a published version of her life story, she puts the contrast in terms of recognizing "two worlds . . . the world of my father, which was the world of death and which was beautiful," calm and quiet, and the "confusion and explosive weeping" that represented the world of her mother.[8] In these contrasts, Benedict saw the propelling theme and motivating energy of her own life.

Patterns, her life story revealed, arrange the prevailing "worlds" of a life with more or less coherence and, Benedict would later add, with more or less beauty, harmony, and intricacy of form. The fragment also suggested the predictive quality of a pattern, its endurance over time, and its ability to contain (and organize) an increasing number of new elements. Benedict apparently believed that by knowing the pattern an individual could move with assurance into the future and change as circumstances demanded. Simultaneously, she adapted the principle that "all the tale could be told if we could set down the simple theme" to her anthropology. In *Patterns of Culture* she expressed the conviction that knowledge of the "intellectual and emotional mainsprings" of a culture could guide its future development.

Whether or not the fragment represented Benedict's interest in life history, as Mead assumed, it certainly captured the back-and-forth movement between autobiographical exploration, anthropological inquiry, and poetic experience that marked the 1920s for Benedict. Coming at the same time as her first book, it offered another version of the significance of patterns, not only as an organizational grid but also as a dynamic and driving force. As Anne Singleton began to disappear from the pages of poetry journals, the poet brought into her anthropology an awareness of the significance of a driving impulse toward form in cultures. This would eventually lead her to advise her editor at Houghton Mifflin that the word *pattern* must be in the title of her book: "I have turned over in my mind some fifty titles for the book, and I find I have the strongest possible preference for a title as exact as possible under the circumstances." She asked Ferris Greenslet, "Would you consider 'Patterns of Culture'? 'Patterns' has been used in the sense I have in mind and it is besides a pleasant English word."[9] That decision evolved during the course of composing her book.

By August 1932, Benedict wrote Franz Boas that she had produced 60,000 words of her book on "cultural configurations."[10] A year later, "patterns" had replaced "configurations." Was it only that "patterns is a pleasant English word" that would help her to reach a wide audience on both sides of the Atlantic? While this was surely a component—a possibility Benedict sought to assist by insisting on British spelling in the published version of *Patterns of Culture*—there were other motivations to her choice. She wanted to sound a distinctive note in anthropology that would distinguish hers from other prominent practitioners in a new field. In this quest, she considered Mead a handmaiden and a helpmeet, less a challenge than a confidante. Other anthropologists of the period offered sharper corners against which she honed her identity as an anthropologist.

Edward Sapir's tortuous movement through the discipline, with its combined messages of "change" and of "relativism," influenced Benedict perhaps most of all. The linguistic anthropologist at once envied and doubted his friend's trust that aesthetic principles could engender an anthropology that blended commitment and objectivity. "I think," he wrote, "I shall end life's prelude by descending into the fastnesses of a purely technical linguistic erudition."[11] Sapir's last letters to Benedict at the end of the 1920s warned her against losing her poetic sensibility in a science, even a humanistic science. The letters arrived just as she was gaining confidence that she could bring the two worlds together in her anthropology. Benedict, seeing the temptation and wishing to avoid the futility of descending into anything "purely technical," resolved to continue her efforts to combine poetry and science.

The confidence that she could succeed in her quest became evident in three articles published between 1929 and 1932. Although not drafts in a literal sense of the word, the three articles immediately preceding *Patterns of Culture* tested the ideas that culminated in the theoretical and the ethnographic chapters of that book. "The Science of Custom," "Psychological Types in the Culture of the Southwest," and "Configurations of Culture in North America" presented part of the approach that the concept of "patterns" would incorporate. As the three articles anticipated, the concept of patterns ultimately induced Benedict to substitute it for "configurations" not only rhetorically but also as a method, a mode of presentation, and a political agenda.

"The Science of Custom," a short article, appeared in 1929 in the middlebrow *Century Magazine.* Benedict defended the usefulness of studying cultures and pointed her readers in the direction of social engineering. "We hope, a little, that whereas change has hitherto been blind, at the mercy of unconscious patternings, it will be possible gradually, in so far as we become genuinely culture-conscious, that it shall be guided by intelligence." Addressed to a general public, the article made a strong argument that every person should become aware of the "lenses" of cultural assumptions and recognize the "pressure" of custom on behavior. In this article Benedict wrote for a thoughtful reading public who possessed common sense and an interest in improving the conditions under which they lived. Advising her readers that no culture was "so good that it needs no revision," Benedict asked: if culture exerts such a strong force, what are the chances for change?[12] She also pointed to another problem: if culture "molds" the person, what are the opportunities for individual creativity and for freedom? Left unanswered in the article, the questions motivated the analysis and interpretations in *Patterns of Culture.*

Benedict published "Psychological Types in the Cultures of the Southwest" in 1930, based on a version of a paper she had delivered two years earlier at a conference. There she presented the dichotomy that would become a centerpiece in *Patterns of Culture:* the Nietzschean distinction between Apollonian and Dionysian to characterize contrasting cultures. "The Southwest Pueblos are, of course, Apollonian" she wrote, referring to the personality and the "purpose" that drove Zuni culture into its distinctive form. Returning to the vision quest, she explained that "the objective performance" between the cultures of the Plains and the Southwest remained "much the same," but "the significance is utterly different" because the quest is thoroughly molded by the "emotional background" of Zuni culture. By the end of the article, a plethora of images of mourning, orgy, and controlled ceremonialism proved not only the existence but the purposefulness of a "fundamental psychological set" "which has bent to its own uses any details it imitated from surrounding peoples and has created an intricate cultural pattern to express its own preferences."[13] Diffusionism had yielded to a new and more complex understanding of "patterns."

In 1932 Benedict added one more piece to her design by publishing "Configurations of Culture in North America." The notions of "drive" and of "compulsion," the dynamic aspect of patterns, had begun to emerge. She wrote, "Another and greater force ['than the mere ringing of changes upon some simpler underlying human response'] has been at work that has used the recurring situations of mating, death, provisioning, and the rest almost as raw material and elaborated them to express its own intent. The force that bends occasions to its purposes and fashions these to its own idiom we can call within that society its dominant drive."[14] The ideal cultural goal became coherence, and the ideal culture was coherent and integrated. Yet, as *Patterns of Culture* would argue, the "force that bends occasions to its purposes" did not eliminate the role of individual creativity.

"Configurations of Culture in North America" confirmed the importance of a notion of integration for anthropological studies of culture: "Such configurations of culture, built around certain selected human traits and working toward the obliteration of others are of first-rate importance in the understanding of culture."[15] Of equal importance was the compulsiveness of "dominant themes" in shaping cultures. This "compulsiveness," Benedict concluded, was missing from the concept of configurations. So configurations gave way to "patterns," the basis for her analysis of culture in the book.

Yet implications of the word "configuration" remained central to *Patterns of Culture.* Indeed, these implications would become the best-remembered feature of the book. The portrayal of an Apollonian culture, sharpened by contrast with a Dionysian culture, became the core of the book for many readers. The connected point, that a culture has a distinct personality, only reinforced the impression of a configuration, enhanced by the vivid description of its character—measured and calm in the instance of the Apollonian, and given to excess in the case of the Dionysian. The descriptive adjectives continue to stick in a reader's mind, "tags" for the cultures to which they are applied. The choice of adjectives recalled the use of psychological types that Benedict had tried out in the article and intended to surpass in the book. While she wanted her argument to be easily accessible, she did not want it to be misunderstood. The book is not about culture personalities, she insisted in a letter to Mead, who was then engaged in her own fieldwork. In 1932, in the heat of an August summer, Ruth wrote to her friend: "I've turned over titles and titles. I want the title of the book to clearly indicate that my competence is in anthropology, nothing else. That is, I don't want any psychologizing title." As she made clear to Mead, Benedict did not want the title to suggest the book was about psychological types or about the strikingly different personalities of diverse cultures. She took words from philosophy and psychiatry to describe but not to provide static portraits—"procrustean tags"—but to evoke the energy of drives and motivations in a culture. Culture patterns "really mean," she wrote in a blurb for Houghton Mifflin, "inner drives and determinations."[16] The energy as well as the content of Benedict's articulation of concepts in the three articles owed a good deal to the non-anthropological ideas she incorporated into her view of anthropological theory, method, and goals.

"Patterns" fit the post–World War I world in which Benedict became an anthropologist. The war had driven her back to school, altered the world forever, and liberated a way of thinking that Benedict embraced with a paradoxical mix of eagerness and caution. A decade after finishing her dissertation, she wrote to Mead from the family farm in Norwich: "I don't write verses anymore, but in my present mood I can well do without them."[17] Benedict really had not abandoned "them," but her engagement with poetry had taken a different turn. This turn is crucial to understanding the non-anthropological aspect of the choice of "patterns."

In January 1932, when she was just embarking on the composition of her book, Benedict wrote to Mead from New York City: "Did you like *The Waves?*" She referred to Virginia Woolf's novel published the year before. "And did you keep thinking how you'd set down everybody you knew in a similar fashion?" She continued, "I did. I suppose I'm disappointed that she didn't include any violent temperaments, and I want my group of persons more varied." In a long paragraph, Benedict continued to speculate about the British novel to her friend doing fieldwork in the Pacific. "Well, what is its theme? People have suggested the strangest assortments. To me of course it's just about life's being a wrapping and wrapping oneself in one's own cocoon. What you can spin is all you have to work with, and the result is altogether dependent on that."[18] The image of a cocoon, like the "basic patterns" of a life, tested Benedict's faith in change. She would develop that idea into a theoretical point in *Patterns of Culture.*

Benedict's 1934 book bore a stylistic resemblance to Woolf's novel. *The Waves* presents a society through the voices of six characters, with a seventh—Percival—conjured by the others. Benedict's book presented diversity through the "voices" of three cultures and conjured the presence of the silent fourth, American culture. Both the novel and the anthropological study depended on concrete, reiterated, and sharply drawn details to convey large philosophical and humanistic issues. Both prepared a program for the future, beyond the recent terrors of a world war. As she confessed to Mead, however, Benedict objected to the lack of a violent temperament in *The Waves* and to the stultifying constriction of a "cocoon." While influenced by the structure and style of *The Waves,* Benedict relied on other sources to develop her thesis more fully.

The rest of the Bloomsbury group offered further resources to the anthropologist. Essential to the revolution in art and literature that followed the Great War, the ideas expressed by the tight-knit Bloomsbury group suited Benedict's sensibility and gave her a way of negotiating the two worlds of poet and of anthropologist. As Sapir reminded her—not always in so many words—she needed to immerse herself in poetry, in art, and in imaginative works in order to develop her distinctive anthropology. In the years before writing *Patterns of Culture,* she read widely in the contemporary literature, poetry, and critical essays of an explosive postwar era. Rather than merely obeying Sapir, she was being true to her own life's pattern; a quick response to the visual and an appreciation of precise moments of perception did not vanish when Benedict became the student of Papa Franz. The condensation of emotion into image and of an elaborate passion into a pure prose could be found in the writings of the Bloomsbury group.

Benedict evidently knew the novels of Virginia Woolf. No equivalent proof exists for her knowledge or appreciation of other Bloomsbury writers. Bloomsbury, however, was a set of ideas and experiences as much as it was a group of individuals. The ideas most likely to appeal to Benedict appeared in consummate form in the art criticism of Roger Fry. According to Woolf, Fry spoke for the rest of the Bloomsbury set, while a biographer of the group called Fry "standard Bloomsbury." Moreover, Fry's penetration into the meaning of art and the significance of artists would have appealed to a person like Benedict, who saw her father in a portrait painted by El Greco.[19] Fry's opinions, pointed and clearly phrased, introduced Bloomsbury ideas to a wide audience. The art critic's essays appeared in journals Benedict certainly read—*The Dial, Atheneum, The Nation*—but the most persuasive argument for a shared discourse and intellectual orientation lies in the substantive similarities between *Patterns of Culture* and Fry's collected essays in two volumes, *Transformations* and *Vision and Design.*

The criticism Fry wrote concentrated on the concepts of "form" and "design." Through those concepts, he developed a theory that incorporated purpose into form and that gave energy or drive to the process of achieving "harmony." Fry, too, wrote about artists with an absolute conviction of the importance of art for an intelligent and culture-conscious "transformation" of life. These were the elements Benedict added to "patterns" so that the word did not mean what it meant in the corridors of Columbia, did not mean "configurations," and did not exactly mirror Sapir's notion of form.

As early as 1907, Fry claimed that a new way of seeing could alter the very shape of civilization. Changed perceptions, he wrote, create a "new reality." Ten years later, now sharpened by a conviction that the war had wiped away the residue of old modes of expression, he wrote about the opportunity that chaos provided for rebuilding civilization. The war had put the last nail in the coffin of an outmoded Victorian world, with its heavy narratives and tedious factual renderings. Form would be the focus of modern art, not content. The quality of design became the measure of a work of art, not the story told or the lesson offered. Modern artists and poets replaced the "sentimentality" of content with the significance of form. In 1917, claiming that a reinvigoration of aesthetic principles provided a foundation for transforming society, Fry described "modernism" as "the re-establishment of purely aesthetic criteria in place of the criterion of conformity to appearance—the rediscovery of the principles of structural design and harmony."[20] In subsequent essays, he claimed that attention to harmony, balance, and form provides the route to an enlightened improvement of civilization.

According to Fry, form referred to the shaping that gives significance to the elements of content. Furthermore, form lies in the world, not in the idea of an artist or a writer; form, Fry wrote, emerges from the visible and the tangible. In Fry's essays, form also meant order as opposed to chaos, specifically in terms of an elegantly designed composition that demonstrated the imaginative creativity of the observer. These several meanings would appear in Benedict's "patterns." In her 1934 book, Benedict used "patterns" to refer both to the traditional meanings and to the meanings she adapted from a context in which form and design contained a message about art and reality.

Although it is perhaps not possible to demonstrate that Benedict deliberately borrowed ideas from Roger Fry or from writers who represented a post–World War I modernist movement, it is clear that she shared an intellectual orientation with Bloomsbury and other literary and artistic movements. The shared orientation encouraged her to relate anthropology to "modern" problems. The fact that the concepts and the politics voiced within an English literary group fit with the direction of Boasian anthropology made the flow of influence easier. A textual analysis of "patterns" in *Patterns of Culture* shows the extent to which she included in her concept the assumption that art, literature, and poetry are primary sources of social renewal.

The beauty of the word lay in its intertwining of meanings. In Benedict's most conventional use, patterns meant that the elements in a culture are ordered and reiterated, but she went beyond the conventional by incorporating a notion of how patterns "come about." Her emphasis on the purpose or drive that motivates a cultural pattern introduced a dynamic element that she had begun to formulate in earlier articles. For Benedict, evidently, the notion of consistency and coherence implied "coming about"—evolving according to a "simple theme," as she had written in her life story. While she continued to associate "pattern" with the distinct character of a culture as a "type," her inclusion of an emphasis on process transcended the purely typological aspect of the word.

Benedict's use of patterns included a claim about the consistency and integration of the traits that constituted a culture. She exploited the word's aesthetic connotations to insist that such integration was not only functional and systematic: it actively fashioned a style or a spirit. Elements were molded in the service of a unique "ethos." As her prefatory quotation revealed, the achievement of a style or ethos became a matter of cultural urgency. "In the beginning God gave to every people a cup of clay, and from this cup they drank their life," she quoted from a Digger Indian chief she had interviewed. The image of a cup represented a cultural pattern containing, holding together, and carefully molding the culture and the members of a specific cultural group. The image of a cup suggested the life-sustaining aspect of patterns. Without the "form," the contents spilled out and disappeared. The words "they drank their life" moved her concept toward its political meaning and emphasized the necessity of coherence and integrity to building a viable life.

The image of a cup missed, however, the dynamic or purpose-driven dimension of patterns. As an image, a "cup" did not convey the compulsion that Benedict injected into the concept. Culture patterns, she wrote, emerge in obedience to a purpose, a forceful master composing separate items into a cultural whole. According to Benedict, such "composing" referred to "the same process by which a style in art comes into being and persists." She explained to her readers: "What was at first no more than a slight bias in local forms and techniques expressed itself more and more forcibly, integrated itself in more and more definite standards, and eventuated in Gothic art."[21] Like the development of an art style, the development of a culture pattern was "unconscious." A culture did not decide on its characteristic purpose. Once in existence, however, that purpose determined the shape and the meaning of all elements. Patterns, Benedict wrote to an editor at Houghton Mifflin, did not refer to "a catalogue of items" but to the "drives" that molded them.

Benedict's reference to Gothic art as an explanation for the origins of culture patterns came from a book that exercised an important influence on the modernist movement, Wilhelm Worringer's *Form in Gothic.* Worringer defended the importance and the manifestations of "unconscious canons of choice." He argued that form was inevitable and that style represented a yearning toward consistency. Moreover, form represented the "basic attitudes toward the universe" of an era. "Every age is strongly biased in favor of the artistic activity which corresponds to its particular will to form."[22] None of this reflected conscious intention. Though "will to form" had an anthropomorphic sound, Worringer did not attribute determination or decision to the process. Form was a force beyond the content or concerns of individual actors.

An expression like "will to form" acquired a special significance in the aftermath of a horrendous world war. Seeming to counteract the chaos recently experienced, the yearning toward form and consistency represented a desire for an order and integrity that had social, political, and aesthetic ramifications. The achieved "style" not only represented but also actually cohered the attitudes of an era or, as used by Benedict, a culture. In both usages style made harmony out of disarray. The nonintentionality of the process was equally important. The style was achieved above and beyond the deliberate decisions of particular actors—it was inevitable, not instrumental. Style, in Benedict's interpretation of the term, was greater than the sum of individual creations, and patterns were more than an accumulation of individual occasions. Internal demands for consistency that transcended particular moments propelled a culture pattern, like a style in art, into existence.

Benedict's explanation of consistency—the distinctive character of a culture—like the art historian's explanation of style, rested on a notion of inevitability. Form happened relentlessly, pulling every trait and every "act" in its path. Ruth Benedict explained the distinctive character of a culture by the strength of its purpose and the compulsiveness of its drive toward form. Taken into her discipline, this assumption provided her with a way of comparing cultures and also prompted her to delve more analytically into the mechanisms by which form is achieved. In *Patterns of Culture* she referred to the need to understand the "emotional and intellectual mainsprings of that society," adding another layer to the meaning she gave patterns. If pattern became evident in visible and tangible traits, these traits resulted from an ongoing adjustment between the intellect and the emotion.

Influenced by *The Waves,* Benedict's *Patterns of Culture* drew the intellectual and emotional mainsprings out of the details of social life. In her novel, Woolf evokes the profound intellectual and emotional mainsprings of an individual life through the descriptions of day-to-day moments her characters offer. Through her three ethnographic "characters," Benedict also returned to the primary contrast of her life, the "intellect" of her father and the "emotion" of her mother. Left undefined in the anthropological work, the meanings of "emotion" and of "intellect" appear in the synonyms Benedict used. "Drive" indicated the emotional sources of a cultural pattern, while "purpose" became synonymous with the intellect. In both cases, the synonyms added the element of motivation or thrust that was crucial to Benedict's interpretation of a cultural pattern—the push toward form and coherence. Including this double-edged push, patterns meant more than the formal organization of traits (configuration) and more than repeated ways of doing things. As Alfred Kroeber acknowledged in his review of the book, his colleague had succeeded in leaving behind the "typological" connotations of configurations as used by other anthropologists.[23] By adapting the "will to form" of Worringer, Ruth Benedict broke open the "tags" to show the motivating sources of a pattern.

Like the Bloomsbury group and modernist American poets, Benedict believed in the power of the precise detail to convey both the motive and the character of a culture pattern. A precise detail, like an image or metaphor, communicated the complex process through which coherence evolved out of the intellectual and emotional mainsprings of any life, whether individual or cultural. Benedict also knew that the anthropologist must refine her senses in order to grasp the pattern as it is revealed in the hodge-podge of "items" that comprise a culture. This conviction issued in the second aspect of patterns, its methodological dimension. As Alfred Kroeber wrote in his 1923 textbook, *Anthropology*, "This gift of seizing character, with its suffusion by insight, admittedly partakes as much of the faculties of the artist as of those of the scientist."[24] Kroeber uncannily countered a criticism of Benedict's book ten years before *Patterns of Culture* appeared. He was not as critical of art in science as other reviewers of her book would be, and his comment suggested a way of assessing "patterns" as a method for studying diverse cultures that Benedict would find quite sympathetic.

The gift of seizing character, or achieving insight into the spirit of a culture, constituted a major component of Benedict's fieldwork. The approach did not violate the developing canon of Boasian fieldwork, with its emphasis on empiricism and its appreciation of the "immersion" advocated by the British anthropologist Bronislaw Malinowski, but Benedict took liberties. She expanded Papa Franz's prescriptions with her own interpretation of the artist's method. On this point, Fry's essays become especially pertinent. His accounts of the work of individual artists and of the development of a "grand style" in art constituted a program for modernism that applied equally well to Benedict's approach to the study of cultures. In his essays, Fry described the artist's method in language and with an argument that anticipated the method in Benedict's "patterns."

In a 1919 essay called "The Artist's Vision," Fry wrote: "As he [the artist] contemplates the particular field of vision, the (aesthetically) chaotic and accidental conjunction of forms and colors begins to crystallise into a harmony." He explained further: "It is the habitual practice of the artist to be on the look-out for these peculiar arrangements of objects." Being "on the look-out" serves as a fair summary of Benedict's method as outlined and carried out in *Patterns of Culture.* "When one is mastering the language and all the idiosyncrasies of behaviour of an esoteric culture, preoccupation with its configuration may well be an obstacle to understanding," she writes. "The fieldworker must be faithfully objective." For her, fieldwork involved discarding a priori assumptions and waiting for the crystallization of observed "occurrences" into a harmony. Benedict made the point even more strongly twelve years later in *The Chrysanthemum and the Sword;* in her observations of Japanese culture, "hundreds of details" fell "into over-all patterns."[25] Like Fry's artist, Benedict's anthropologist, in the field or not, immersed herself in the chaos of impressions and observations and waited for pattern to emerge.

This is not as passive as it sounds. Both the artist and the anthropologist have to learn how to observe, how to be "objective," and how, as Fry put it, to render vision into form. "The fieldworker must be faithfully objective," Benedict wrote; and under the influence of contemporary aesthetic theory, she meant more than gathering data without the intrusion of self. "Objectivity" did not so much eliminate the self as give it a place. Once again, Benedict's implicit point can be illuminated by the explicit point made consistently by Fry throughout his essays. Not "self" but subjectivity is to be excluded from modern art, poetry, and fiction. The self remained to grasp and present precise details, but without the imposition of personal feelings or assumptions. Fry delineated the relationship between observer and observed perhaps most elegantly in his biography of the impressionist Paul Cézanne. For Fry, Cézanne radically revised the way the artist sees and thereby profoundly changes reality.

In many ways an extremely subjective document in the author's identification with the subject, Fry proclaimed objectivity to be the "true" artistic method in Cézanne. Fry outlined Cézanne's growth from youthful immaturity to full flowering as an artist. First imposing his own moods on the landscape, Cézanne learned to extract form from landscape and find in this a correlative for his state of mind.[26] The visible form might represent the artist's internal subjective state, but it did not develop from that state. This is the difference between the "pathetic fallacy," in which the artist creates a world to suit her or his feelings, and an interest in things in themselves because they reveal the truth of an emotion or perception. Fry's essays rejected the romanticism and sentimentality of nineteenth-century artists who imposed their moods onto the landscape.

A rejection of sentimentality suited Ruth Benedict's anthropology. For her, working in the field was a Cézanne-like effort to see intensely without looking intentionally. Like Fry's heroic impressionist, Benedict pressed on the "externals" and the "things out there" for the sources of spirit and of integrity. The task then involved a faithful rendering of what appeared before one's eyes and a patient waiting for the organizing principle—or better, the dual mainsprings—of a culture to become clear. Such patience bespoke a trust in the capacity of distinct elements to provide information in and of themselves. Patiently recording, the observer—now artist—discovered meaning in an accumulation of details. For Benedict, anthropological method lay inextricably in presentation, in choosing the right image or phrase to represent a world. "One line of insight would have done more than all those lines of description," wrote Virginia Woolf about Arnold Bennett, complaining about his discursive, heavy-handed message to his readers.[27] The call of modern writers and artists, the effort to render "reality" in one precise perception, became Benedict's gift to anthropology.

Absorbing her Anne Singleton voice and Bloomsbury ideas about form, characters, and images into her prose, Benedict injected "writing" into anthropological method. Through the concept of patterns, she, like Fry, offered a way of conveying a world through a finely drawn image. An inquiry into culture, from this point of view, was only complete with its "drawing"—or, in Virginia Woolf's phrase, when the anthropologist found her "one line of insight." Benedict achieved her goal surpassingly well in the ethnographic chapters of *Patterns of Culture,* where detail upon elegant detail provided insight into three strikingly distinct cultures. Like the images that make a poem or the lines and shadows that constitute a portrait, Benedict's carefully composed details evoked the realities of Zuni, Dobu Island, and Kwakiutl cultures. From those details emerged the intellectual and emotional mainsprings of a complex life that had produced them.

Benedict's method of presenting precise details of belief, behavior, and personality upheld her conviction that patterns actually existed on the ground. Anchoring patterns to reality, the details proved her conviction: "patterns" were neither an abstract concept nor the imposition of an observer's framework on the world. Like Roger Fry, Ruth Benedict trusted the observer to glean the composition from the world and to reproduce this reality for a reader or spectator. Fry used different words to insist on the reality of "form" in landscape or still life and to argue that the artist evoked and did not simply construct a reality. In his writings, the concept of form referred to existing patterns, whereas the concept of design referred to the artist's process of representation. Design was the "aesthetic construction" of form, he wrote in discussing "the transformations which forms undergo in becoming parts of aesthetic constructions." In essence, he asserted, there was no "truth" without the artist's representation. The role of the artist is to transform vision in order to provide "truth" to the viewer. As an offshoot of his rejection of nineteenth-century style, Fry dismisses "fidelity to appearance" in favor of design in any art.[28]

Design combines insight and style, vision and the rendering of that vision. At once mystical and familiar, "design" represents the burden the art critic puts on the artist. Both noun and verb, Fry's concept condensed his theory that "form" existed and at the same time must be extracted and reproduced by the artist. Benedict's word "patterns" is a noun, but her text implied a process of designing as fully as did Fry's concept of design. Like "design," patterns urged the anthropologist to realize that the meaning of elements comes from the arrangement the writer provides to those elements: the fit of a Zuni death ceremony with Zuni child-rearing practice gave meaning to each trait. Like the images in the sonnets she wrote as Anne Singleton, Ruth Benedict placed the details of Pueblo culture into a rigorous, almost stanzaic form that mimicked the anthropologist's process of discovery.

As method, Benedict's rendering of vision into form suggested the poet and not the disciplined fieldworker. In his review of *Patterns of Culture*, Kroeber wondered whether the main methodological lesson Benedict taught her readers was the importance of intuition, of an acute perceptiveness, and of an ability to transform the perception into a persuasive alignment of images. Recognizing the value of intuition in any inquiry into a culture, he did not condemn this approach. Unlike negative reviewers of the book, Kroeber credited Benedict for respecting the self-presentation of a culture and for refusing to impose her own framework on the way of life in front of her. The book showed that the anthropologist's design, like the artist's, issued neither from an accumulation of facts nor from an a priori assumption. Design served to convey reality.

Fry struggled in his essays with the relationship between form and reality. With an abiding belief that the artist can alter reality by transforming modes of perception, Fry also discriminated in his essays—he was a critic, after all—between design that succeeded and design that failed to produce a changed perception. Fry looked for criteria for judging the persuasiveness of a composition, and he found those criteria in intricacy and in complexity.[29] A similar measure can be applied to *Patterns of Culture,* where the persuasiveness of Benedict's highly composed portraits, delineating the "form" of each culture, depends on the intricacy with which she composes the observed cultural traits. Had he been given the opportunity, Fry might well have reviewed the ethnographic sections of *Patterns of Culture* from the point of view of how well they evoke the "realities" of three cultures—and been positive. With a modern sensibility, Ruth Benedict pushed anthropology from its effort to replicate reality through faithful documentation into an attempt to convey the experience of "reality" through design.

Patterns of Culture calls into question the science of ethnography for some of its readers, just as Margaret Mead's *Coming of Age in Samoa,* where she evoked the reality of Samoan life through the snapshot image of one day, continues to do. Mead's work, like Benedict's, confounds the reader who expects "science" to exclude imagery and descriptive narrative. Fry made the opposite point. For him, science and art share a mode of presenting truth through the elegance and intricacy of form: "In both the mind is held in delighted equilibrium by the contemplation of the inevitable relationship of all the parts of the whole, so that no need exists to make reference to what is outside the unity, and this becomes for the time being a universe."[30] Fry compared art to science as part of his effort to claim that art, as completely as science, can transform a civilization. The world, Fry claimed, is not the same after a great artist renders his vision any more than it is the same after the invention of a printing press or a cannon.

One cannot leave the subject of Benedict's patterns without considering this last shared orientation. For Benedict, the endeavor of anthropology involved portraying the realities of diverse cultures so that no person would be left unchanged after exposure to the portraits. She had maintained the faith that Sapir challenged her on. She would use her anthropology to "save humanity," and patterns played a key role in achieving that goal despite Sapir's questioning of that motive. "Must everybody contribute his share toward the saving of humanity? That is your faith, and an inhuman absorption in a purely intellectual pursuit will never satisfy you," Sapir had written to Benedict in 1925.[31] Addressing the poet in whom he read a humanitarian impulse, Sapir predicted that she would not be satisfied in anthropology alone. Ruth Benedict's humanitarianism was satisfied within her discipline, however, when she found a way to merge the poet and the anthropologist. Patterns, like order and form, incorporated a view of human potential, the possibilities for change, and the mechanisms for redesigning social arrangements.

Benedict's *Patterns of Culture* achieves its political ambitions at several levels. In its first and last chapters she preached a conventional liberal position of tolerance for others and the need for critical intelligence about one's own way of life. In the three ethnographic chapters, style and tone make the point, conveying a message that combined cultural relativism with a call for the self-conscious assessment of the culture within which anyone makes his or her way from birth to death.

In a 1920 essay Roger Fry wrote: "Even the most casual spectator, passing among pictures which retire discreetly behind their canvases [that is, English nineteenth-century paintings], must be struck by the violent attack of these forms."[32] This in itself striking statement appeared in an essay on El Greco that, given her fascination with the Spanish painter, Benedict may well have read. Fry's language suggested both the violence of war and the appropriate response of shock. Being "struck" by a "violent attack" of forms recalled the horrors of the previous five years and asserted Fry's interpretation of art. At their best, great paintings strike the viewer a blow from which he or she does not survive unscathed. The attack came from the form rather than the content. If an El Greco strikes the viewer, it is because the formal arrangement of features, not fidelity to appearance, brings a shock of recognition. The design of details evokes empathy, alters perception, and leads the viewer to a changed understanding of the world. The more intricate the design, Fry continued in his account of the transformative power of art, the more impact the painting has on viewers.

Fry refused to separate his aesthetic theory from an interpretation of the way "civilization" ought to be improved. Great art gains its power inasmuch as it compels a viewer to re-view the world of her or his familiar experiences. For readers of *Patterns of Culture,* Fry's faith in the way a painting transforms the world sounds an echo. Like Fry and his Bloomsbury cohort, Benedict did not preach intricacy for its own sake, but because intricacy intensified the impact of a "picture" on the viewer. This would be her route to bringing the "saving of humanity" into her anthropology. Shunning Fry's language of attack, Benedict believed as strongly as he did in the ability of a "picture" to jolt and disturb the viewer—or, in her case, the reader. Like Fry, Benedict believed that the "jolted" individual, the most casual spectator, was the source of change in any civilization.

The aesthetic theory built into "patterns" relieved her of the inclination to preach that she shared with Mead. Under the influence of Fry, Bloomsbury, and the imagist movement in poetry on her side of the Atlantic, Benedict relinquished the voice of the schoolmarm to let her readers be "shocked" into new perceptions. Not through being told, scolded, or nudged would readers, like spectators before El Greco, achieve a new way of seeing, but through feeling the impact of a representation. The aesthetic Benedict adapted from the post–World War I milieu gave her a means of conceptualizing change that inserted the individual into the equation. It would be the "beauty" of an integrated pattern, not the details of a particular culture, that struck the reader and prompted her or his re-vision of reality. Improvement came from a view of cultural integrity and not from the awareness of a distinctive custom. If Mead drew a lesson from Samoan adolescence for her American audience, Benedict accomplished a different task when she presented a truly integrated, tightly knit pattern—regardless of content—as the goal for a humane social order. Her reader was the "casual spectator" that Fry addressed.

Roger Fry banished "content" from good painting for a generation of artists and art critics. Content, he disparagingly remarked, turned art into illustration. The illustrator, he explained in one essay, "manipulates the sentiments" of his audience through an emphasis on content. Stories in pictures impose an idea on the viewer and thereby dampen critical engagement. The true artist does not intervene between the audience and its experience of the work and instead allows each viewer to be struck in her or his own way. With form predominating, the viewer is free to follow the insight of the artist. Applying this to Benedict's ethnographic chapters exposes the extent to which their impact likewise depends on form and not on content. For while one may remember details of Kwakiutl life, what really strikes a reader is the form: repeated iterations of a megalomaniac culture. The same can be said of the other two chapters: "Apollonian" and "paranoiac" are the functional equivalents to "form" in Fry's sense of the word.

The anthropologist diverges from the art critic when she considers the importance of form, or patterning, to the living of a life. Benedict shared the cultural relativism of her teacher, Franz Boas, and found Fry's aesthetics amenable to this position, granting value, as it does, to any intricate pattern, regardless of content. She also used her concept to compare one culture with another. The intricacy of form, then, gains meaning not only for the observer (reader) who is struck by it but also for the individual in a culture. Intricacy is a measure of cultural coherence and vitality; moreover, in a patterned and integrated culture, individuals find fulfillment and creativity.

In *Patterns of Culture,* Benedict began her journey toward reconciling cultural relativism with a conviction that some "patternings" are better than others. In the descriptive and in the methodological modes, "patterns" remained neutral, applying to the inquiry into and representation of a culture. In the political mode, "patterns" measured a culture by the worth and satisfaction it granted individuals. This criterion, implied in the ethnographic chapters, supplied the foundation for the advocacy of enlightened change in Benedict's theoretical chapters. In 1934, still uncertain of how to blend cultural relativism with saving humanity, Benedict depended on the multivocal connotations of patterns to carry both respect for each culture and judgment of any culture.

The ethnographic chapters of *Patterns of Culture* brought home the value of an intertwining of all traits within a culture. A molded, well-formed design is presented as a "good" in its own right, regardless of content. "Our cup is broken now," the Digger Indian chief says poignantly; the whole fabric of life has disintegrated. Yet one must ask, would Benedict have regretted the breakdown of any "whole," regardless of its content? The inconsistencies reviewers pointed out in *Patterns of Culture* are related to this question. Throughout the book, Benedict praised intricacy as a value in its own right. At the same time, the descriptive terms for each pattern subverted that point. Who can read of a paranoiac culture, filled with suspicion and witchcraft, with the same response as reading of an Apollonian culture, where no person raised hand or voice to another? Intricate as both patterns may be, the result for an individual in the culture—and for the reader of the book—is not the same. In the chapter discussing the Kwakiutl, Benedict again returned to the value of intricacy in a startlingly positive few sentences about the displeasingly aggressive culture: "Kwakiutl life is rich and forceful in its own terms. Its chosen goal has its appropriate virtues, and social values in Kwakiutl civilization are even more inextricably mixed than they are in Zuni."[33] Intricate as it is, the reader must think, the pattern is not as good.

The inconsistency is inevitable, given Benedict's efforts to combine intense description—the precise image—with a commitment to enlightened social change. Like Fry and other modernists, Benedict believed in the power of an image and in the ability of changed perception to alter reality. Unlike the art critic or the poet, the anthropologist did not trust to image alone; and while Anne Singleton stayed, a voice of urgent political commitment swept the poet along. This becomes most apparent when the silent character, the "Percival" of Benedict's book, appears on stage. As many have pointed out, *Patterns of Culture* is primarily about the culture that does not have a chapter of its own: American culture. Introducing this conjured character, Benedict intended, as Fry might have said, to "attack" the perceptions of her countrymen and open their eyes to the mainsprings of their culture-pattern.

Fry's startling verb obscured his subtle account of the nature of an attack by form. This account slows down the process of transformation, asking the spectator to travel the road of the artist. "The spectator in contemplating the form must inevitably travel in an opposite direction along the same road which the artist had taken, and himself feel the original emotion." The artist prepares the way in his choice of images, his rendering of those images, and his "design" of the whole. Responding to the presentation, Fry continues, the viewer goes back to the artist's initial vision and reflection. In the best of any art, the artist "has revealed us to ourselves in revealing himself." In the best of anthropology, Benedict claims, the anthropologist reveals to the reader her own culture through the reflections of other cultures. And so, like Virginia Woolf in *The Waves,* Benedict creates a striking portrait of an absent American culture through the images displayed by the three presented cultures.

Through the deliberately indirect evocation of the pattern of American culture, Benedict permits her readers to take their own steps to awareness or, adapting her words, to culture-consciousness. As Fry articulated and Woolf accomplished in *The Waves,* an arrangement of precise details draws the reader along the path of the writer's vision. There is a virtue in the non-obvious, for it demands that the reader follow the artist through the process of transforming vision into "form" or comprehension. Benedict attempts to do the same thing when she presents American culture in a non-obvious way, through the details of Pueblo, Dobuan, and Kwakiutl lives. At its best, like the encounter with Percival, the reader suddenly "meets" American culture in a way she or he has not before. This is the transformation of perception that has more enduring impact, according to Benedict, than a thousand words of fact.

Benedict did not leave so much up to the reader's perception as did either the art critic or the novelist. The illustrator slips into *Patterns of Culture,* telling the story of a peaceful culture, a suspicious culture, and an aggrandizing culture, so that American readers do not mistake the type of spirit each represents. Her confidence that the reader would be beneficially attacked by form is not as great as that of the British art critic, though it was greater than that of her colleague Edward Sapir. In her doubts, Benedict returned to the model Mead set, an explicit pointing of lessons. But she never succumbed entirely to Meadian anthropology, resisting the "illustrative" style epitomized in the last half of *Coming of Age in Samoa.*

Patterns of Culture, the product of ambition and of a pull between the two worlds of writer and of anthropologist, has flaws. Benedict's decision to use "patterns," the pleasant English word, solved some problems while creating others. As a concept, patterns offered a radical way of simultaneously conceptualizing the "nature" of culture and the "method" of cultural anthropology. The concept also glossed over the gaps in Benedict's anthropology, especially the gulf between cultural relativism and her urge to "save humanity." It would also earn her the criticism of skeptical anthropologists like Sapir, who used the term "piffle" to describe Amy Lowell's poem "Patterns." The decision to reconcile the poet and the scientist might not meet with approval from all the readers of her book.

Yet "patterns" goes a longer way to reconciling cultural relativism with standards of judgment than the negative reviewers recognized. While serving as a mode of judgment, intricacy and complexity of form allowed for the distinctiveness of content. The Kwakiutl, with their aggressive "occasions," exemplified a tightly formed and integrated culture, good in that sense. In the culture of the Northwest Coast, every item of behavior, belief, and personality had significance; nothing was superficial. The striking quality of the pattern, for the moment, overrode the distressing content of its "spirit." Further upholding the cultural relativism of her concept, its methodological dimension demanded respect for the slow emergence of a pattern. Benedict's warning against bringing a priori assumptions to the field gestured to the distinctiveness of any culture pattern. The political dimension caused the problems as Benedict, holding fast to a vision of cultural relativism, steadfastly maintained her conviction that an awareness of diverse cultures must lead to an intelligent improvement of society. As a criterion, "patterns" offered a mode of judging cultures that at once embodied Boas's commitment to the uses of anthropology and diverged from it. More subtle than Margaret Mead's addresses to her "own" society and more optimistic than Sapir's retreat into the formal domain of language, Benedict's patterns brought her into the heart of the anthropological debate that continues to rage some seventy years after the publication of *Patterns of Culture.*

In its ambitiousness and in its contradictions, the concept of patterns incorporated core issues that continue to challenge the discipline. Addressing what is currently phrased as a battle against the "subjective" in anthropology, in which the observer imposes her scheme or frame of reference on the other, Benedict's "patterns" insisted that insight into, and representation of, a culture depended on objective observation. The fieldworker who is obliged to intuit the spirit of a culture is equally obliged to discard her or his prior assumptions. Addressing what today appears in anthropology as a reliance on the testimonies of others, Benedict's concept insisted that observed occasions will gradually and inevitably yield the "truth" of a culture. "These occasions, whether of marriage or death or the invocation of the supernatural, are situations that each society seizes upon to express its characteristic purposes."[34]

And like today's emphasis on writing as a crucial step in anthropological inquiry, Benedict's "patterns" insisted that the anthropologist choose precise images and arrange them precisely. The anthropologist, like the artists in Fry's essays, transformed the raw material of vision into a striking form rather than engaging in tedious documentation of one fact after another. With her absorption of the modernist ideas characteristic of the Bloomsbury group, Benedict brought her word "patterns" into a world of postmodern anthropology and proved its enduring aptness to the dual goals of anthropology: to represent another reality, and to acquire a critical perception of one's own reality.

In a ringing anticipation of contemporary debates, Benedict believed that anthropology could guide the members of a society into self-assessment through the enhancing of culture-consciousness. More like the art and literary critics she read than like her close colleague Mead—the speaker of the pundit's voice—Benedict bequeathed to a later anthropology a conviction that its audience should include the ordinary citizen, who could journey to a radically transformed vision of the world.

In a world just leaving a war behind and entering an era whose cataclysms were becoming evident, *Patterns of Culture* provided a beacon of hope for those suffering from the Depression and from fears of Nazism, Fascism, and other impending conflicts. For the individuals who felt disastrous choices had been blindly made, Benedict optimistically offered the chance to choose with awareness of cultural constraints, thus escaping the bonds of the "unconscious canons" that drove societies. In the twenty-first century, as the fear of another cataclysm sits high on the horizon, the lessons of *Patterns of Culture* are just as necessary as they were in the 1930s.

16 On the Political Anatomy of Mead-bashing; or, Re-thinking Margaret Mead

Virginia Yans

This is an essay about Margaret Mead. It is also an essay about anthropology, a social science conceived by Mead's generation as a tool to be used responsibly for human benefit, now reconceived by her postmodern critics as an instrument of domination. Alongside these two narratives, two other tales unfold. One reveals how anthropology's internal disputes stoke the culture wars smoldering outside the discipline. The other is a tale of matricide and suicide: Mead and scientific anthropology are the victims.

Our narrative is tangled and sometimes perplexing. Consider the major characters: Mead, who died in 1978, still appears as a central figure. Her opponents are unlikely allies—postmodernists and postcolonialists, Marxists, feminists, and disciples of Karl Popper, the philosopher of science. Their attacks on Mead are startlingly vehement, often gratuitously nasty.

Clearly, Mead's current symbolic reincarnation serves important purposes among the living. As an outside observer and a historian, I ask: why? Why is so much negative attention devoted to Mead twenty-five years after her death? To cite E. R. Leach's words regarding the interpretation of symbols, the Mead critiques do not seem to be "immediately explainable as a rational response to a given situation" nor do they "relate directly to the observable facts."[1] As a historian, I am interested in understanding the "symbolic" Mead and the "observable" Mead, situated within the historical context and archival remains of her life.

Anthropologists have generously attended to the Popperian "scientific" critiques of Mead and of cultural anthropology. I focus here instead on the postmodern critical perspective, now widely diffused within anthropology and often adopted by postcolonialists, feminists, and neo-Marxists.[2] I am concerned with postmodernism's ahistorical, anachronistic interpretation of Mead and the history of her discipline. I argue that dedication to specific anti-science and political agenda, professional ambitions, and resentments about Mead as a symbolic Founding Mother motivate this anachronistic critique.

Whatever position they ally with, the Mead critics' disregard for historical analysis is disturbing. Nowhere do critics of Mead's work, such as *Coming of Age in Samoa* or *Sex and Temperament in Three Primitive Societies*, engage in any careful, deep, or comprehensive reading of Mead's ancillary published writings. Archival, even published sources that might complicate their vision of Mead or contradict their own political or scientific agendas are used selectively or ignored. Derek Freeman has been roundly and deservedly criticized for his failure to examine Mead's archives and even some of her published texts on Samoa, materials that contradict his claims of her scientific inadequacies.[3] But postmodern critics also fail to locate the Mead canon within the matrix of her related private, unpublished—and sometimes even published—writings, all of which she made available for future researchers' use, and which have been located at the Library of Congress for some time.

In the words of the historian of Enlightenment science Lawrence E. Klein, postmodernism invites "anachronism because it interprets aspects of the past by reference to what they are alleged to have led to." As Klein and others have suggested, their purpose is to "define a history of modernity against which critique can then be launched."[4] The ghost of anthropology past—Mead and the ancestors—can then be blamed for anthropology's failures. In this way, today's anthropologists can absolve themselves of their guilt by association with the imperialist project.

Klein observed certain inconsistencies within the postmodern analysis of the past: "Notwithstanding Jean-François Lyotard's identification of the postmodern with 'incredulity toward metanarratives,' postmodernism," he notes, "has its own grand [historical] narrative. . . [concerned with] genealogies and lineages, ancestors and their legacies." Margaret Mead, as cast in this metanarrative, comes under attack as a direct descendant of the Enlightenment scientific legacy and of its political utopian social agendas. And Klein is critical of postmodernists who are committed to diversity, nonetheless overlooking the diversity of the Enlightenment's "intellectual cultures."[5]

Klein's criticism applies equally well to postmodernism's interpretation of Mead and her generation of social scientists. While in Bali and New Guinea in the 1930s, for example, Mead developed methodological strategies that not only questioned Enlightenment realism but also anticipated postmodernism's highlighting of fieldworker subjectivity.[6] Furthermore, she operated within a pragmatist philosophy of science that proposed its own critique of Enlightenment authority.[7] Understanding Mead and her generation of social science peers, then, requires substantial revisions in contemporary postmodern criticism of Mead and of its narrative concerning the culture and politics of twentieth-century social science.

Let us take a look at some specifics of the postmodern assessment of Mead's scientific project and its politics. The postmodernists associate Mead with specific Enlightenment ideas about science that they call into question, namely that science can be objective, that it can discover and represent an external reality, and that it can be used to better the human condition. They understand science differently: it is "an authoritarian, 'totalizing' project that impedes rather than promotes truly democratic and egalitarian values."[8] Science is an instrument of Western domination; scientific realism is a form of "crypto-fascism." Fieldwork, anthropology's signature methodology, and the fieldworker's ethnographic texts are monologues by and for the powerful. Works by Mead and other anthropologists are judged accordingly.

The postmodern metanarrative views Margaret Mead and other anthropology ancestors as imperialism's puppets or even as complicit racists supporting the imperialist project. The postmodernist anthropologists' most favored allegations concerning Mead's racialist complicity are her failure to properly acknowledge power differences between white fieldworkers and colonial subjects and her arrogance in allegedly proposing to speak for the natives. It is assumed—not demonstrated or substantiated—that all anthropologists furthered colonial interests; the corollary that colonized peoples were therefore subordinated to, or that they acceded to, these interests remains unexamined and undemonstrated.[9]

Attention to the historical context that produced postmodern theory may serve to clarify its political content and anti-science stance. After 1968, when the French student movement collapsed, French intellectuals conceded that they were unable to break the structures of state power and turned instead to subverting the structure of language. As one commentary explains: "The student movement was flushed off the streets and driven underground into 'discourse.'" An anti-authoritarian movement, postmodernism called into question all forms of systematic thought and political theory, including the Enlightenment's deification of Reason and Truth, its political and social agenda, and any other political program (including Marxism) that aimed to change society as a whole.[10]

Eschewing universal truths, postmodernism offers only a relativist politics, at best a rhetorical "moralizing and political commitment," but no praxis, no humanistic agenda, no critique of the social and economic conditions causing human alienation.[11] Yet as Richard Rorty observed of the literary postmodernists, "They view themselves as 'subverting' such things as 'the humanist subject' or 'Western technocentrism' or 'masculinist binary oppositions.' They have convinced themselves that by chanting various Derridean or Foucaultian slogans they are fighting for freedom." In fact, some postmodernists understand their critique itself to be political, even an antecedent to revolutionary change.[12]

One could interpret postmodernism otherwise. Distinguished philosopher Jürgen Habermas has gone so far as to characterize it as "destructive," its critical emphasis aborting the production of knowledge. Scholars from other disciplines have called postmodernism anti-humanistic, anti-democratic, and, because of its attack on reason and "normative values," an enabler of conservatism, mysticism, religious intolerance, and even Neo-Nazism.[13] Consider, for example, the discipline of anthropology wherein an internalized, hypertrophied self-scrutiny has silenced the social and utopian purposes to which preceding generations of social scientists (including Mead, the Boasians, and the Marxists) believed their scientific knowledge could and should be applied.[14]

Nonetheless, when French postmodern theory arrived in the United States in the 1970s, it offered a promising alternative both to disillusioned Marxists and to those who once favored social science participation in policy and research.[15] The failure of Communist experiments in the Soviet Union and elsewhere, American working class disinterest in such experiments, and anger at the actual and alleged participation of social scientists in counterinsurgency efforts in Latin America and southeast Asia suggested the poverty of existing strategies. Many Marxists and feminists whose scientific, humanistic, or universal human rights agendas would seem to contradict it, for example, embraced much of postmodern theory and politics. Within the academy, postmodernists came to regard Marxists—and, of course, social scientist reformers and government consultants such as Mead—suspiciously. The rift between what Zygmunt Bauman has called "legislative" and "interpretive" intellectuals—those who, like Mead, believed their knowledge had some social purpose and that purpose was to transmit knowledge—grew deeper.[16]

Consider the Marxists first. Twentieth-century Marxist anthropologists were critical of Margaret Mead. They had long argued that Ruth Benedict, Mead, and other psychological anthropologists accepted the significance of the human subject without paying attention to the social conditions and architectures of power that shaped that subject. Marxists understand Mead's embracing of psychological anthropology as a bourgeois and typically American preoccupation diverting her and her audiences from the real issues of structural and economic oppression. While rejecting Mead's liberal reform, social engineering, and "technical assistance" strategies (what I describe later as her pragmatism, though they do not understand it as such), the Marxists expressed a preference for radical, take-to-the-streets social protest and for social analyses that challenge existing power structures.[17]

Their most outspoken contemporary Mead critic, the self-styled political economist Micaela di Leonardo, presents an unrelenting attack on Mead in her book *Exotics at Home.*[18] Revealing herself as a creature of her time and her generation, di Leonardo, despite her protests to the contrary, embraces postmodernism's privileging of a negative adversarial posture above the positive construction of knowledge, of textual analysis above fieldwork, and discourse above social agenda, as well as the postmodernist use of metanarratives of anthropology's past as a means to condemn anthropology. Even though di Leonardo offers a biting assessment of postmodernism, her major work of Mead criticism, a critical discursive analysis of Mead's canonical writing, is a postmodern artifact—one in which she substitutes rhetoric, the stock in trade of the postmodern critique, for a reading of history and historical sources. For her historical analysis, di Leonardo relies completely on published works and some selected historical syntheses. Her interpretation of Mead and anthropology, for example, relies on a short essay by Eric Wolf in which Wolf chastised his anthropology colleagues for their failure to confront power in American society. Mead, on this interpretation, would be characterized as a "progressive" reform liberal who fully supported corporate capitalism.[19] Marxist historian Norman Markowitz disagrees with di Leonardo's interpretation of Mead, and his interpretation is grounded in thorough historical research:

> There was a distinction of great importance [from the 1920s through the 1950s] between the "humanitarian social liberals" which Mead certainly fits, and the "corporate social liberals," the organization men. Mead did not try to use the people in Samoa to advance advertising or to develop social structures for counterinsurgency, and she was not simply a professionally ambitious academic. She had a framework; she was trying to educate large numbers of people in a progressive way.[20]

The contrast between di Leonardo and the ancestors she claims from the post–World War II era such as Eric Wolf, Sidney Mintz, and Eleanor Leacock is striking. These individuals had their differences with Mead, but instead of an obsessive preoccupation with interpretation and criticism, they set out to represent the history of working-class, peasant, and colonized peoples whose fate, they hoped, would be altered by revolutionary transformations.[21] By and large, they examined Mead's work with more open minds than does di Leonardo. Eleanor Leacock, for example, held a Marxist critical stance toward Mead's Samoan studies, yet after a careful search of Mead's work, Leacock acknowledged that even though they were not her primary concern, Mead had not neglected Samoa's economic and class issues.[22] Leacock and other radical anthropologists of her day shared Mead's engagement with the Enlightenment's humanistic and progressive social agenda. Although they were Marxists, they engaged in the "enlightened discussion of different points of view" advocated by twentieth-century American pragmatists and by Mead herself. For example, one searches di Leonardo's work in vain for statements of the kind Marxist anthropologist Marvin Harris made acknowledging the part that Mead and other Boasians played in undoing an earlier imperialist ideology, that of scientific racism. Mead, in turn, engaged with the work of the Marxists and recognized its significance.[23]

Is it possible to unite Marxism's realism, materialism, and political agenda with postmodernism, as di Leonardo and other neo-Marxists attempt? I doubt it. Di Leonardo's adversarial position, linked to her adoption of the postmodern strategy, leaves no room for "conversations" between scholars. Readers might sympathize with di Leonardo's plaints against exploitation of "the primitive," but is her method likely to change their situation? She concerns herself with the deficiency of Mead and anthropology's "representation" of the primitive, not with the reality of the lives these individuals lead.

Certain feminists who adopt a postmodern position are mired in yet another set of contradictions when they analyze Mead, Benedict, and other early woman anthropologists. Feminists have conflicting metanarratives. Wishing to add women to anthropology's pantheon, some feminists acknowledge Mead as an originator of social constructivist gender theory and as a defender of the Enlightenment's equal rights program.[24] On the other hand, feminists who identify with postmodernism or postcolonialism feel obliged to demonstrate Mead's blindness to, or complicity with, colonial and class interests. Both of these metanarratives coexist within contemporary feminist thought. Some unsettling contradictions result. Thus, while di Leonardo calls Mead a "limousine liberal" and condemns her for an alleged soft approach to sociobiology,"[25] she stops short of calling Mead a racist; instead, she hurls most of her thunder at Mead's alleged feminist failures.

Feminist historian Louise Newman, whose writings aim to expose the racism of white liberal feminism (her own feminist metanarrative of condemnation) takes a different position. She claims that Mead's career represents the outcome of the "history of Western liberal feminism" and its "theoretical claims that Western societies are superior to non-Western ones." She contends that "implicit in [Mead's] work was the belief that Samoa was a flawed society because it restricted the freedom of its women."[26] Mead never said such a thing, and to interpret her as having done so is simply wrong. Newman offers absolutely no evidence from the published Mead texts (on which she otherwise relies), and none from archival materials (which she totally ignores) that Mead believed the United States was superior to other cultures.[27] Because Mead believed in the decidedly Western Enlightenment concept of freedom of choice both for men and for women, Newman reasons that she considered "primitive" societies without such traditions to be inferior. But why does a preference for freedom of choice imply racism or inability to criticize racism, or even an assumption concerning Western superiority, as Newman insists it does? The logic of Newman's argument is flawed and problematic.

Newman's analysis, like the postmodern position with which she sympathizes, is interpretive, not factual. Moreover, she seems to believe that "re-readings" and "re-interpretations" of texts (in this case Mead's) "are as important as the original anthropologist's" and that these re-readings will change the relationship between feminism and imperialism.[28] I refer readers back to the postmodern political strategy, the claim that textual analysis itself will lead to social change, a claim that stands in stark contrast to more pragmatic proposals for curing racism and other social ills.

What are we to conclude? For di Leonardo, Mead is neither a feminist nor an outright racist. For Newman, because Mead is a feminist, she must be a racist. Such conundrums are the inevitable outcome of works motivated by anti-Enlightenment metanarratives of condemnation. They are anachronistic in the sense that Klein characterized anachronism, that is, they interpret the past "by referring to what the past allegedly led to."

My plea for a critical examination of anthropology's past is more than a plea to convert readers to historical praxis. I am, just as Mead was, concerned with the social responsibility of intellectuals. Postmodern discourse, typically conducted in a coded language and impenetrable prose, acts as an effective barricade to any unconverted outsider or laymen who might wish to understand its social program. But as I have suggested, there is no social program. "Social amnesia," said the historian and social critic Russell Jacoby twenty-five years ago, accompanies historical amnesia. Social amnesia's symptoms manifest in the forgetting and repressing of "the human and social activity that makes and can remake society." Under its spell, Jacoby noted, social analysis degrades into "slogans that serve more to sort out one's friends and enemies than to figure out the structure of reality."[29] This seems to be exactly where postmodernism has taken us over the past quarter-century.

A return to a social science engaged with public culture and a critical reading of the past (including anthropology's past) that is grounded in "the structure of reality" or, as Rorty would have it, "a "conversation about" that reality, would provide a cure for these ethical and historical blind spots.[30] We might investigate, as does George Stocking, the dean of historians of anthropology, how anthropologists of Mead's generation actually understood themselves. We might consider what they hoped to do as socially responsible scientists and citizens.[31] As twenty- first century academics, we might marvel at their belief that what they thought and did could make a difference. We might also explore why today's postmodernists encourage us to think so darkly about these issues and ancestors. And we might turn some of our energies from destructive criticism to learn something about our past, about the world we live in and how we hope to improve it.

A review of the historical situation of Mead and her generation may serve to undo some of the postmodern narrative's misreadings of Mead and her discipline. Mead and others of her generation worked in different historical circumstances and fought different culture wars from those we are now witnessing. In "the wake of the Moscow trials, the Nazi-Soviet Pact, and Stalin's support for Lysenko's destruction of Soviet genetics," these social scientists were "opponents of a generalized 'totalitarianism' present in Communism as well as in fascism"—and later, of course, in McCarthyism.[32] They saw their world "filled with 'prejudice' and with efforts to 'impose certain opinions by force.'" Against these evils they believed that one must affirm "free inquiry," "open-mindedness," and democratic values. Science (far from the "crypto-fascism" the postmodernists describe) was understood as an analog to democracy. Furthermore, they "denied absolutes," admitted "uncertainty," and were committed "to the continuing viability of the idea of democracy [and science] as a way of life."[33]

Mead and many of her social science peers believed that reason and science—referring to an attitude as opposed to work in laboratories—could inform public culture. For example, Mead ended her World War II morale-boosting treatise *And Keep Your Powder Dry* with a plea to Americans to keep "faith in the power of science." Science could get things done; it could solve the problems of society. And Mead and her peers believed that they, as scientists, could and should participate in a public culture engaged in discussing these problems.

Many social scientists, including Mead, took John Dewey's emphasis on free inquiry, liberal reform, and conversation or dialogue as a means to resolve social problems as the agenda for public intellectuals. James T. Kloppenberg described the early-twentieth-century pragmatists as envisioning "a modernist discourse of democratic deliberation in which communities of inquiry [or competent experts] tested hypotheses in order to solve [social] problems."[34] Unaware of this genealogy, some anthropologists, even her admirers, understand Mead's strategies and her belief in scientific expertise to be simply naïve. As one recently commented: "She thought if the right folks were given unlimited phone time, they could change the world."[35]

These public intellectuals, moreover, conceived of their work in ethical terms. In 1944 Mead's contemporary, Mark A. May, director of Yale's Institute of Human Relations and a distinguished social scientist of his day, wrote of the "morality of science," which he hoped would spread throughout the world, a code that would direct ordinary citizens to live by the scientists' codes of conduct of "honest, free inquiry, the code of critical, interactive, evidence-based, universalistic, antiauthoritarian, and hence 'scientific' conduct." Mead, May, and others of her generation aimed with others of her generation to use science as "an agent of cosmopolitan liberation" for peoples everywhere.[36]

Mead and her generation of Boasians were not, as postmodernists make them out to be, simply apologists for the West. In fact, they were often critics of the West. "Ideologically," as David Hollinger argues, the cultural relativism that Mead and others practiced, was a "critical device fashioned for the purpose of undermining the authority of aspects of a home culture." They reflected "on the possible implications for 'Western Civilization' of what had been discovered about other cultures, especially 'primitive' ones. That is what made anthropology in the cultural relativist mode a major episode in the intellectual history of the twentieth century, rather than simply another movement within a discipline."[37]

Despite her shared projects with other social scientists, Mead was unique. She was a *famous* social scientist. She was at once one of the most successful—and most criticized—of modern anthropologists. Though standing only five feet, two inches tall, Mead had an imposing, and for some, an overwhelming, presence. She projected an "air of monumentality on the lecture platform"[38] Her certainty about most things, including her notions of what anthropology is, rankled her contemporaries and continue to irritate those who came after her. Some of her colleagues jokingly called her "God the Mother." "We can't tolerate a monumental woman figure," admitted Lenora Foerstel, one of Mead's critics.[39]

Mead is inextricably linked to the public face of anthropology. From the late 1920s, when her popular work on Samoa earned public attention, those outside the profession viewed her as anthropology's *charge d'affaires.* Even those who were annoyed by her folkish wisdom and belittled her popular audiences acknowledge the significant public presence she brought to anthropology. That voice, particularly her access to the media, gave her tremendous power. It was an unusual power for a woman (or even for a man), and it was a power molded from the alchemy of charisma and expertise. Mead held an exceptional position as a public intellectual; it is understandable that many would envy her, particularly at a time when academics, once thought of as intellectuals, must have perceived their own declining influence and public relevance. One anthropologist noted that among anthropologists there was "a feeling of discomfort when Mead's name is mentioned; discomfort that she is popular, and then discomfort that the popularity should bother them."[40]

Mead's position as anthropology's Founding Mother did not result from a disciplinary consensus but from public appreciation and her own intentions. She actively cultivated her image as a motherly wise woman. She radiated "the air of authority that comes from many years of telling students, anthropologists, and other people things for their own good."[41] Her thumb staff and cape expressed in sartorial form her authority as an elder. She relished her role as advice-giver to worried parents who read her columns in *Redbook* and *Parents* magazines. She acted as an elder to her fellow anthropologists, many of whom have noted that she claimed that position to the end of her life, dying in December 1978, when her "family" of anthropologists assembled in their annual scholarly convention. Mead's historical position as a Founding Mother of a discipline is a situation with which few other academic disciplines have had to contend.

Indeed, this unexamined ambivalence toward Mead as a symbolic mother explains some of the irrational responses to her and to her work. Her critics (some of them Mead's actual or figurative mentees) proceed without much awareness that they seem like angry, even envious children who, despite their protests, sometimes fail to separate themselves from the mother figure and in fact seem to unconsciously imitate her.

In *Exotics at Home,* for example, di Leonardo continually reinserts Mead into the narrative both literally and imitatively. She complains that Mead "threatened to take over" her book, hovering and haunting her dreams, "throwing her cloak over my books and papers and following me."[42] She mimics Mead's well-known, impatient dismissiveness of those with whom she disagreed; she also assumes Mead's self-appointed role as privileged authorial voice and American cultural critic. Moreover, the titles of books written by several Mead critics bespeak identification with Mead. The subtitle to Lenora Foerstel's and Angela Gilliam's edited volume, *Scholarship, Empire, and the South Pacific,* more accurately describes its contents—Pacific anthropology as practiced by Mead and others—than does its market-savvy main title, *Confronting the Margaret Mead Legacy.*

Foerstel, a former Mead protégée, attacked her mentor's fieldwork methods as patronizing and racist, even though she participated in that fieldwork as a consenting adult researcher some twenty-five years previously. Foerstel speaks of herself as a fieldworker in the third person. She thus distances herself from her own actions and places full responsibility on Mead, as if she were wholly dependent on Mead and not separate from her. "I am somewhat shocked," Foerstel reports, "at the submissive compliance with which the villagers accepted their intrusive anthropologists. If villagers complained, we certainly did not hear about it, and perhaps for this reason, we did not question our own behavior."[43] Foerstel does not report what the natives had to say, for the natives have no voice and no agency in her written commentary. Were the natives mute? Why is it that Foerstel and di Leonardo feel entitled to speak for them?[44]

Another Mead protégée, Lola Romanucci-Ross, conveys the shared powerlessness of both men and women confronting Mead's maternal presence, describing an incident when she and Ted Schwartz, her co-fieldworker and husband, accompanied Mead on an expedition and realized too late that they had forgotten their film. To cover their indiscretion, the two arranged a secret return trip to the field sites, this time shooting the pictures Mead believed they had already taken.[45]

Co-workers who imagined themselves powerless, behind-the-scenes tricks, protestors against Mead's influence appropriating her name for their book titles—these appear to portray the actions of rebellious, envious children.

If one patiently follows the footnote tails, almost all criticism of Mead's politics leads back to Foerstel's and Gilliam's work. Despite questionable interpretations and selective use of evidence, most critics cite such them as authorities.[46] These authors have the hybrid political agenda discussed earlier in this essay: they declare their indebted to Marxist and postmodern anthropology.[47] As Pacific postcolonial anthropologists, they have a special interest in condemning both American imperial interests in that region and their predecessors, including Mead. They also share the postmodern program of calling Western scientific authority into question. To their credit, they enhance their credibility by including New Guinea contributors to their edited volume. They are bitter that Mead's public opposition to nuclear war never included opposition to nuclear testing in the Pacific. But they are badly mistaken when they characterize Mead as a militarist and as ambivalent about race. Let us look at the historical record.

Marxist Peter Worsley summarizes Foerstel's and Gilliam's assessments of Mead's politics in his Introduction to their book:

> Margaret Mead's theoretical posture of concern about the danger of global war never prevented her continuing in the service of the U.S. State Department and the military long after World War II was over. . . . Mead's posture of theoretical opposition to nuclearism contradicted her denunciation of U.S. labor unions and

> others who opposed nuclear power and nuclear weapons, and it never resulted in her participation in anti-war movements.[48]

Evidence available to Foerstel and Gilliam and other Mead critics contradicts these statements. Margaret Mead never served the U.S. State Department, although she and a small "army" of social scientists did work as World War II government consultants on homeland morale, nutrition, and intelligence for the National Research Council, the Office of War Information, and the Office of Strategic Services during World War II. Mead opposed nuclear proliferation; she opposed the Vietnam War; and she lent her name to anti-war efforts.[49]

Foerstel and Gilliam claim that Mead's "close involvement with the American military" began in 1925, when the U.S. Navy aided her fieldwork in Samoa. Mead did acknowledge that she could not have done her work without the assistance of the navy interpreters, transportation facilities, commissary, and her residency in a Ta'u island naval dispensary. But her associations with the navy were not always useful; upon her arrival she received a "reasonably frosty reception" from the navy captain who was acting as governor of American Samoa, and she got into more difficulty with him when she asserted that her youth gave her an advantage over him in learning the Samoan language.[50] Moreover, to characterize the fragile 23-year-old Ph.D. candidate (whom Benedict, Boas, and Sapir worried about because of her emotional and physical fragility) as an agent of imperialism and a military collaborator is laughable.

Foerstel and Gilliam also discuss a 1960 Peace and Disarmament Resolution presented at the annual meeting of the American Anthropological Association as further proof of Mead's alleged military and pro-government position. They use a 1960s Mead quote to imply that she was a militarist: "One of the very serious misunderstanding[s] that exists in our search for new forms of national protection is the persistent mistrust with which many liberals view all members of the military. It ill behooves those social scientists who are working on military problems to aggravate this distrust in any way." In fact, Mead was the author of the original disarmament resolution: its purpose was to encourage "individual anthropologists" to "seek appropriate opportunities within their professional competences to develop ways and means to ensure the survival and well-being of our species" and to use "scientific competence in research on the conditions of peace and disarmament." The quote cited by Foerstel and Gilliam as evidence for Mead's militarism comes from Mead's response to two hawkish anthropologists who feared that anthropologists' involvement in peace organizations would result in their being placed on the U.S. Attorney General's subversive list. In the Mead statement that Foerstel and Gilliam quote out of context, Mead (who often tried to reconcile disparate positions, to create bridges between opposing sides) was trying to avoid strained relations between the military and anthropologists in the interests of working toward the goal of peace and disarmament.[51]

Consider also the 1971 Thailand controversy of the Vietnam era, cited by Foerstel, Gilliam, and other Mead critics as another demonstration of her collusion with American military efforts. Again, they fail to fully investigate the evidence, which demonstrates a different reality. In 1971 many anthropologists insisted that the American Anthropological Association should condemn any fieldworker's participation in studies that could be used to support American counterinsurgency activities in Southeast Asia. Mead was asked to chair the Ad Hoc Committee to Evaluate the Controversy Concerning Anthropological Activities in Relation to Thailand. The committee, after reading six to seven thousand pages of documents, presented its report on November 20, 1971. They found that some anthropologists had engaged in research that aimed at ameliorating the "lot of downtrodden or neglected people, research that may have diminished chances for insurgency in Thailand" but concluded that "the activities there were well within acceptable ethical behavior."[52] According to D. L. Olmstead, a committee member and later editor of the *American Anthropologist,* "The few official reports we had concerning working with the CIA, army, etc., proved, in our view, to be devoid of substance." At the annual meeting, however, "the report was characterized from the floor as the work of pro-war types (far from the truth about any of us) and it was suggested that a vote against the report was a vote against the war, an irresistible proposition in those days."[53]

Feelings ran high at this meeting, Olmstead continued, as they did at many other professional association conventions during the Vietnam era. After a distinguished senior anthropologist (probably Eric Wolf) got to his feet waiving a purloined document allegedly proving that an anthropologist had assisted a government agency to develop anti-guerilla weaponry, the assembly voted to reject the Ad Hoc Committee's report. The audience jeered at Mead, who was reportedly moved to tears by their reaction.

Mead and other committee members believed that the anthropologists, using documents stolen from Michael Moerman, a UCLA Thai specialist, had demonstrated "reprehensible," "unscholarly" behavior, a discredit to the discipline. When interviewed later, Eric Wolf described Mead's views as "one-dimensional," but added, "I think she acted out of what she thought was right."[54] The archival record shows that Mead tried to protect anthropology from McCarthyite tactics, whether these tactics were practiced by the Right or the Left, by persons within their profession or outside of it. Considered from a historical perspective and their own prior experience, Mead and her colleagues were anti-McCarthy and anti-Nazi activists seeking to protect an atmosphere of free inquiry. This, not counterinsurgency, promilitary, or pro-war sympathies, motivated them.

Once again trying to effect a compromise, Mead suggested that in order to avoid polarizing the association, the committee should rewrite its Thailand report to suit the majority. Committee member D. L. Olmstead saw it differently, concluding that the majority of members of the association, who had not read the documents, were acting as "an irrational mob." They wanted to choose "an amanuensis who could write them a report not constrained by knowledge of the documents."[55] Moreover, Mead's archives and the archives of the American Anthropological Association reveal that the citizen anthropologists and members of the association allegedly involved in Thailand's counterinsurgency were not involved at all. The anthropologist named as a government collaborator in the case of the anti-guerilla weaponry, for example, was a case of mistaken identity. It turned out that he had the same last name as the actual author of a study on aerial reconnaissance that he was supposed to have written.[56]

In earlier years, the World War II U.S. bombing of Hiroshima and Nagasaki severely challenged Mead's belief that democracies could be trusted to employ science solely to improve the human condition. Disillusioned, she terminated the formal government associations she eagerly sought in the war against Hitler. She turned instead to university research institutes and nongovernmental organizations such as the United Nations, where she promoted population control and women's rights. She helped to establish New Directions, a citizen's lobbying group promoting human rights abroad. She chaired the National Council of Churches committee on the ethical implications of fast-breeder plutonium reactors, a committee that was a thorn in the side of energy industry interests.[57]

Indeed, Mead's postwar scientific projects defy simple political pigeonholing as a collusion with the government that some critics, reading back into the past from the perspective of the Vietnam era, assign to them. Mead was not a consistent supporter of U.S. policies. She publicly made it known that she opposed the use of nuclear weapons. She openly opposed Civil Defense efforts to encourage private purchases of bomb shelters and mentioned the unmentionable: private distribution of bomb shelters would result in class differences in access to them.[58] After demobilization, several kinds of working relationships developed in the United States to manage the "uneasy partnership" between government and science. The National

Science Foundation, founded in 1950, would offer a structural solution to potential conflicts between the ethics of scientists and the interests of government. In the meantime, scholars accepted government funding for their university-based research projects. The Research in Contemporary Cultures project at Columbia that continued after Benedict's death was carried out with funding from the U.S. Navy and, later, from the Rand Corporation (for Columbia's Russian Studies project). At the time, the Office of Naval Research was "the chief agency used by scientists to obtain federal support for basic research in the universities." Notwithstanding government funding, the relationships between government and scientists remained fluid, sometimes unfriendly, and the scientists cautiously guarded their autonomy.[59] Government, moreover, could be hostile and suspicious of the social science community, particularly those conducting research on Communist countries or those associated with leftists.

During her tenure at the Columbia Project, Mead worried that she might be subject to an FBI investigation because of her cooperation with leftists in efforts to establish an experimental integrated school.[60] What finally emerged out of Mead's Russian studies, in collaboration with Geoffrey Gorer, was the much-criticized "swaddling hypothesis." Their analysis of this Russian child-rearing practice and Mead's emphasis on child-rearing practices and national character formation was poorly received by anthropologists, some derisively calling it "diaperology." The idea that this work contributed to military intelligence is absurd.

What of Foerstel's and Gilliam's claims that Mead acted as an agent of U.S. imperial interests? In a 1970 *Redbook* article, Mead provided a clear statement of her position on the Vietnam War as well as U.S. imperialism and militarism. It is worth noting that Mead penned this strong statement against American militarism and favoring the autonomy of colonized peoples *before* the 1971 Thailand controversy.

> One great difficulty we have in thinking [Vietnam] through is that in our preoccupation with Vietnam we have come to treat the situation as unique. We would do better to place our intervention in Vietnam within the context of American engagements in the world in the past 25 years. . . . Among our successful efforts we can list the Truman Doctrine, the Marshall Plan, the Berlin airlift and the many programs, of which the Peace Corps is only one, in which we have been constructively involved in efforts to move toward a peaceful world. Among those that have been ill-advised, we must list the U-2 espionage episode, the Bay of Pigs, the invasion of the Dominican Republic and, above all, the military escalation of our intervention in Vietnam. Looking at the present situation within this larger

> context, we should be able to recognize . . . the war in Vietnam has been from the beginning a gross mistake. It is significant that all our errors have been military and our successes have been economic and social.[61]

With regard to Mead's views on race, Foerstel and Gilliam selectively use other scholars' writings. In support of their contention that Mead's record was questionable, they quote George Stocking's statement that Mead's culture and personality school "took over the sphere of determinism that had been governed by race." However, they edit out Stocking's next sentence: "But the political implications were very different," because it was now possible to argue that those who had been "denied full participation in modern society on the grounds of 'race' were agreed to be equipotentially full participants, if only the barriers of prejudice and discrimination could be removed."[62] When Foerstel and Gilliam consult the historian Carleton Mabee's assessments of Mead's record on race, they cite his article noting that Mead took no stand on World War II internment of Japanese Americans, but they ignore two other articles in which Mabee demonstrates Mead's overwhelmingly positive record on racial issues, including her postwar era work on human rights, her decision to boycott segregated colleges by refusing to speak at them, and her efforts to create a model integrated school in New York City—the Downtown Community School—which used scholarships and busing to achieve integration of students, staff, and trustees well before Brown vs. Board of Education.[63]

In this instance and elsewhere, Mabee points to the coherence between Mead's knowledge as an anthropologist and her citizenship praxis—her conviction that skin color had nothing to do with intellectual capabilities and her willingness to act on that conviction in a practical way. Furthermore, when circumstances required it, Mead accepted civil disobedience as a means to bring necessary change in race relations.[64]

Nor was Mead soft on sociobiology, as di Leonardo and others claim. In contending that Mead fought "fiercely (and successfully) against a 1976 American Anthropological Association resolution to condemn sociobiology's inherent racism and sexism," di Leonardo omits information within the very sources she cites. She cites evidence selectively from Jane Howard's biography of Mead, failing to include Howard's explanations of why Mead, who represented the majority opinion on the subject, opposed the motion. Indeed, Mead thought that the motion was "'the antithesis of free inquiry." Di Leonardo also omits Howard's interview with the eminent sociobiologist Edward O. Wilson in which he told Howard that he feared accepting a dinner invitation from Mead because "the grand matriarch" might scold him for "reintroducing genetics, social Darwinism, and biological determinism into anthropology." When they eventually met for dinner, Wilson recalled, she "mentioned the dangers of allowing the sociobiological approach to be subverted to racism, but she was more anxious by far to find a common ground and to establish her own priority in analyzing genetic factors." He was referring to Mead's idea that "certain genetically based personality types predominate in a particular society and confer their distinctive traits."[65] In other words, in speaking with Wilson, Mead was attempting to reconcile the biological and cultural viewpoints.

Mead and other Boasians understood that anthropology must embrace both human culture and biology. According to Howard's interview with Wilson, she repeated that axiom to him, just as she had affirmed it in *Sex and Temperament* and in *Male and Female.* She chose to emphasize culture during the 1930s when Hitler's genocidal policies were a threat. After his defeat, she felt freer to examine the potential influences of reproductive differences between men and women. In *Male and Female* she did just that.

Di Leonardo and others ignore the historical record on Mead's confrontation with colleagues in the American Association for the Advancement of Science concerning the psychologist Arthur R. Jenson, whose conclusion that Afro-Americans "as a race were less intelligent than white Americans" had gained national attention. He argued that it was reasonable to conclude that genetics explained the difference. When the Psychology Division of the Association (of which Mead had recently served as president) wished to make Jensen a fellow, Mead vigorously opposed the nomination. She stormed out of the meeting and immediately engaged the attention of news reporters at the scene. She insisted, Mabee reports, "that parts of Jensen's work were absolutely unjustified" and "that it might be dangerous for race relations for a long time to come if the AAAS made him a fellow." Privately, Mabee reports, Mead went further, "arguing that as long as I.Q. scores remained crude and biased, studies of genetic influences on intelligence were so likely to lead to dangerous distortions that such studies should not be conducted at all."[66] These few examples of Mead's racial politics demonstrate the consistency of her position, not her inconsistency. She was on the one hand a defender of free inquiry, and on the other, unalterably opposed to using science, especially flawed science, in the service of racial discrimination.

Problems also exist in interpretations of Mead's feminism. Di Leonardo, for example, refers to her "grotesquely anti-feminist pronouncements of the 1940s and 1950s." She believes that Mead opportunistically attached herself to Second Wave feminism after its proven attractiveness.[67] Indeed, Mead was not a take-it-to-the-streets feminist. She is supposed to have quipped: "I am not a feminist; I am an anthropologist." By this I think that she meant to say that she had gained her knowledge of women from her science and that her contribution to women's equality would be made through the area of her expertise, not in the political sphere. A quick perusal of the index of Mead's bibliography of published works in scientific and popular journals reveals about five hundred articles and books listed under the topic "women," many written before the Second Wave feminism of the 1960s and 70s had appeared.[68] A 1946 essay written for *Fortune* magazine, for example, anticipates Betty Friedan's *The Feminine Mystique* and disproves charges that Mead encouraged "Momism," idealized traditional femininity, failed to acknowledge class differences among women, and had no interest in gender stratification.[69]

Mead noted that "women in our society complain of the lack of stimulation, of the loneliness, of the dullness of staying at home." Mead suggested that much of the drudgery of household labor such as laundry and food preparation could be moved to work sites outside the home. She agreed with women's demands that as the "production side" of home life vanished, men should share in the "human" side of home activities such as bringing children to school and shopping "for which women have no special biological fitness." She continued in language that anticipates the Marxist feminist argument of the 1970s: "Elimination of the semi-voluntary slavery to housekeeping that we now impose upon married women in the United States should open the way for an equally significant set of inventions in that key spot of our civilization, the home, where not 'things' but human beings are produced and developed." In the same essay, Mead called attention to class differences among American women, some of whom, she reminded readers, worked out of necessity. She insisted that women were not "a special class of biologically determined defectives" and that they should have the same professional opportunities as men.[70]

Whether postmodern, neo-Marxist, or some blending of the two, much of the criticism of Margaret Mead's politics emanates from a condemnation of American anthropology in general. Thus according to Laura Nader, Mead's positions on race and on the Thai controversy are demonstrations of Mead's (and anthropology's) unwillingness to confront powerful institutions in American society. Consistent with this criticism, Peter Worsley claims that for Mead the revolutionary was a "deviant" or a "misfit" and that she couldn't tolerate the idea that "radical politics" could transform the "economic, political, and legal structure of society."[71]

In his thoughtful article based on a close reading of Mead's papers, Mabee understands Mead differently. While Mead herself was not a leftist, she did associate with leftists and, as Mabee tells us, very likely found herself subjected to FBI investigations as a result. According to Mabee, Mead "believed" that leftists "had a useful role to play in a democratic society in pressing for reform. . . . Indeed, in the context of the 1940s, she herself could be considered a radical in the sense that anyone who would be willing to struggle to make a school significantly interracial was a radical." But, Mabee continues, she believed that "united front" organizations, "even those against racial discrimination or nuclear war, could be manipulated by leftists and she therefore chose to avoid them and advised Ruth Benedict to do the same."[72]

Why does all of this matter? First, consider the thoughts of the philosopher of science I. C. Jarvie. After noting the "dogmatism and confusion" of current anthropology and proposing a return to the history of anthropology as a means to achieve self-understanding, he writes:

> It is time for anthropologists to do genuine history of anthropology, to cease treating their own predecessors the way they accused those predecessors of treating savages, namely as backward, stupid, etc. . . . to cease, in other words, transmitting oral myths about anthropological history which smack of the armchair on the veranda and to buckle down to some fieldwork, with its hardships and rewards, in the history of anthropology itself. The problems and debates of our anthropological ancestors deserve our concern and respect, partly as a matter of rights and partly because we have a lot to learn from them. What anthropology is about cannot be understood without its history, and the more flippantly its history is dismissed as an embarrassment, the poorer self-understanding will be, the more the signs of crisis will grow.[73]

Moreover, intended or not, their position in anthropology's indigenous wars places the Mead critics and anthropology itself in a compromised political position. The attacks on Mead and anthropology produce consequences in the culture wars swirling outside the discipline. Rewriting anthropology's history according to their own lights, the postmodernists obscure the progressive originating purposes of anthropology (perhaps most notably its insistence on the "sameness of humanity"),[74] deny its scientific efficacy, and thereby make anthropology vulnerable to a neo-conservative constituency only too delighted to see the discipline of anthropology (and its progressive values) rendered illegitimate as a science. More than one commentator has observed that ancestor-bashing, anti-Enlightenment, and anti-anthropology views enable the arguments of conservatives. Anthropologist Nicholas Thomas noted that "the potential ramifications in [sex education and child-rearing] of Derek Freeman's so-called refutation of [Mead's] work have not been overlooked by such bodies as the Moral Majority." David Hollinger reported that conservative Christians use Foucault and Kuhn to argue that the objectivity of science is questionable and that therefore their own "orthodox version of the biblical *episteme*" is rendered "cognitively legitimate once again."[75]

In their rush to judgment, ancestor-bashers blind themselves not only to Mead's historical situation but to their own. They create a Mead who is a symbolic figure of their own imaginations, not the complicated historical person that she was. Perhaps it is time to make Mead whole, to re-imagine her by re-establishing, alongside her symbolic persona, the human actor set in historical time and the changing, complicated individual and thinker that she was. She was socially engaged; she did grow and change in response to criticism; and she was prescient about the future of her world and her discipline. Mead and many of her peers—whatever their shortcomings, however naïve or "liberal"—understood that their specialized knowledge could matter. And many regarded themselves as public intellectuals with a social responsibility.

This is forgotten, unacknowledged, or not understood by younger generations of academic professionals. Their projects—a modern scholasticism or a Marxism that would be unrecognizable to Marx, set in an academic environment far removed from public culture and often written in a language incomprehensible to anyone but themselves—are detached from the public stage upon which Boas, Mead, Benedict, and their colleagues walked. Margaret Mead was a productive social scientist and a public presence. Giants are easy targets. But the attack on Mead displays a disturbing intellectual violence and destructiveness, even a scholarly dishonesty, that calls into question the motives and political objectives of its perpetrators.

The neo-conservative critique of relativism, multiculturalism—and indeed anthropology itself—benefits from anthropology's self-condemnation, from its thoughtless attacks upon its Founding Mother, even from its sloppy historical practices.

Now, as my tale of disciplinary suicide and matricide comes to a close, I wonder: Will the Mead-bashers be joining their vilified ancestors if the good ship "Anthropology" sinks?

Notes

Books by Ruth Benedict Cited in Notes

The Concept of the Guardian Spirit in North America, in *Memoirs of the American Anthropological Association.* Washington, D.C.: American Anthropological Association, 1923.

Patterns of Culture. Boston: Houghton Mifflin, 1934, 1959, 1989. New York: New American Library, 1946; Mentor Books, 1950.

An Introduction to Zuni Mythology. New York: Columbia University Press, 1935, AMS Press, 1969.

Race: Science and Politics. New York: Viking Press, 1940, 1959, 1964.

Race and Racism. London: George Routledge & Sons, 1942; Routledge & Kegan Paul, 1983.

The Chrysanthemum and the Sword: Patterns of Japanese Culture. Boston: Houghton Mifflin, 1946. Cleveland, Ohio: Meridian Books, 1967. New York: New American Library, 1974.

For a full bibliography of Ruth Benedict's work, including her many articles, see Margaret Mead, *Ruth Benedict* (New York: Columbia University Press, 1974). Mead reprinted many of Benedict's important articles in *An Anthropologist at Work: The Writings of Ruth Benedict* (Boston: Houghton-Mifflin, 1959) and in *Ruth Benedict.*

Books by Margaret Mead Cited in Notes

Coming of Age in Samoa. New York: William Morrow, 1928, 1961, 1964, 1973; Blue Ribbon Books, 1936; New American Library, 1949; Modern Library 1953; HarperCollins, 2001.

Changing Culture of an Indian Tribe. New York: Columbia University Press, 1932; Capricorn Books, 1966.

Sex and Temperament in Three Primitive Societies. New York: William Morrow, 1935, 1963, 1980; New American Library, 1950; HarperCollins, 2001.

Cooperation and Competition Among Primitive Peoples. New York: McGraw-Hill, 1937. Boston: Beacon, 1961.

From the South Seas: Studies of Adolescence and Sex in Primitive Societies, ed. by Mead. (Contains *Coming of Age in Samoa*; *Growing Up in New Guinea*, and *Sex and Temperament in Three Primitive Societies.*) New York: William Morrow, 1939.

Balinese Character, with Gregory Bateson. New York: New York Academy of Sciences, 1942.

And Keep Your Powder Dry: An Anthropologist Looks at America. New York: William Morrow, 1942, 1965; Berghahn Books, 2000.

Male and Female: A Study of the Sexes in a Changing World. New York: William Morrow, 1949; Dell, 1968.
Soviet Attitudes Toward Authority: An Interdisciplinary Approach to Problems of Soviet Character. New York: McGraw-Hill, 1951.
Cultural Patterns and Technological Change. Paris: UNESCO, 1953.
New Lives for Old: Cultural Transformations—Manus, 1928–1953. New York: William Morrow, 1956, 1966, 1975; New American Library, 1961.
An Anthropologist at Work: Writings of Ruth Benedict. Boston: Houghton Mifflin, 1959. New York: Atherton; 1966, Avon, 1973. Westport, Conn.: Greenwood Press, 1977.
The Golden Age of American Anthropology, ed. with Ruth Bunzel. New York: George Braziller, 1960.
American Women: The Report of the President's Commission on the Status of Women and other Publications of the Commission, ed. with Frances Bagley. New York: Charles Scribner's Sons, 1965.
The Wagon and the Star: A Study of American Community Initiative, with Muriel Brown. St. Paul, Minn.: Curriculum Resources, 1966.
A Rap on Race, with James Baldwin. Boston: J. B. Lippincott, 1971.
Ruth Benedict. New York: Columbia University Press, 1974, forthcoming, 2005.
Blackberry Winter: My Earlier Years. New York: William Morrow, 1972; Pocket Books, 1975. Gloucester, Mass.: Peter Smith, 1989.
Twentieth Century Faith: Hope and Survival. New York: Harper & Row, 1972.
A Way of Seeing, with Rhoda Metraux. New York: William Morrow, 1974.
World Enough: Rethinking the Future. Boston: Little, Brown, 1975.
Letters From the Field, 1925–1975. New York: Harper & Row, 1977.
Some Personal Views. New York: Walker, 1979.

For a complete bibliography of the works of Mead, including her many articles, see Jane Gordan, *Margaret Mead: The Complete Bibliography, 1925–1975.* The Hague: Mouton, 1976. Gordan compiled this work under Mead's direction.

Introduction

1. Mead and Benedict are notably absent from Russell Jacoby, *The Last Intellectuals: American Culture in the Age of Academe* (New York: Basic Books, 1987).

2. Margaret M. Caffrey, *Ruth Benedict: Stranger in This Land* (Austin: University of Texas Press, 1989), 180–81; Mead, *Blackberry Winter: My Earlier Years* (New York: William Morrow, 1972), 313, 124; Ruth Benedict to Margaret Mead, August 25, 1925, in Margaret Mead, *An Anthropologist at Work: Writings of Ruth Benedict* (Boston: Houghton Mifflin, 1959), 301; Judith Schachter Modell, *Ruth Benedict: Patterns of Life* (Philadelphia: University of Pennsylvania Press, 1983), 146–57.

3. Lois W. Banner, *Intertwined Lives: Margaret Mead, Ruth Benedict, and Their Circle* (New York: Alfred A. Knopf, 2003).

4. Ruth Benedict, *Race: Science and Politics* (New York: Viking Press, 1940); Ruth Benedict, *Race and Racism* (London: George Routledge & Sons, 1942), vii, 152.

5. Ruth Benedict, *The Chrysanthemum and the Sword: Patterns of Japanese Culture* (Boston: Houghton Mifflin, 1946).

6. Mary Catherine Bateson, *With a Daughter's Eye: A Memoir of Margaret Mead and Gregory Bateson* (1984; New York: Washington Square Press, 1985).

7. Mead to Papa Franz, July 16, 1930, Fieldwork, N119; Mead to Dr. Clark Wissler, August 25, 1930, Publications, I6; Mead to Professor Malinowski, August 9, 1930, Fieldwork, N119, Margaret Mead Papers, Library of Congress, Washington, D.C., hereafter Mead/LC; Mead, *Blackberry Winter,* 220–23; Margaret Mead, *The Changing Culture of an Indian Tribe* (New York: Columbia University Press, 1932), 134; Mead to Clark Wissler, July 2, 1930, August 6, 1930, Publications, I6, Mead/LC; Ruth Benedict, "The White Man and the Indian," New York *Herald Tribune,* November 13, 1932.

8. Mead, *Anthropologist at Work,* 91–93; Mead, *Blackberry Winter,* 122–24, 150, 333; Bateson, *With a Daughter's Eye,* 119, 240, 248, 288; Jacoby, *The Last Intellectuals,* 17, 16.

9. Margaret Mead, *Male and Female: A Study of the Sexes in a Changing World* (1949; New York: Dell, 1968), 355, 354, 49, 215, 345, 359.

10. Mead, *New Lives for Old: Cultural Transformation—Manus 1928–1953* (New York: William Morrow, 1956), 4–8, 168, 173–79, 362–451, 186; Mead, *Letters from the Field: 1925–1975* (New York: Harper & Row, 1977), 260–61; Mead to Peter Worsley, July 21, 1958, *New Lives for Old,* Publications, I-78, Mead/LC; Peter M. Worsley, "Margaret Mead: Science or Science Fiction?" *Science and Society* 21 (Spring 1957); Margaret Mead, *Soviet Attitudes Toward Authority: An Interdisciplinary Approach to Problems of the Soviet Character* (New York: McGraw-Hill, 1951), frontispiece, 2; Peter Worsley, "Foreword," and Angela Gilliam and Lenora Foerstel, "Margaret Mead's Contradictory Legacy," in *Confronting the Margaret Mead Legacy: Scholarship, Empire, and the South Pacific,* ed. Lenora Foerstel and Angela Gilliam (Philadelphia: Temple University Press, 1992), xi–xviii, 101–56; Virginia Yans-McLaughlin, "Science, Democracy, and Ethics: Mobilizing Culture and Personality for World War II," in *Malinowski, Rivers, Benedict and Others: Essays on Culture and Personality,* ed. George W. Stocking Jr. (Madison: University of Wisconsin Press, 1986), 184–217, 204.

11. Mead to Marie Eichelberger, March 10, 1955; Mead, "To Those I Love" [written in March 1955], Marie Eichelberger Folder, 1922–65, Special Correspondence, R-7, Mead/LC; Yans-McLaughlin, "Science, Democracy, and Ethics," 190–91; Mead, *Blackberry Winter,* 313, 124, 333; Bateson, *With a Daughter's Eye,* 143; Joan Gordan, ed., *Margaret Mead: The Complete Bibliography, 1925–1975* (The Hague-Paris: Mouton, 1976); Margaret Mead, *Culture and Commitment: A Study of the Generation Gap* (Garden City, N.Y.: Doubleday, 1970); Margaret Mead and James Baldwin, *A Rap on Race* (Boston: J. B. Lippincott, 1971); Margaret Mead, *Cultural Patterns and Technical Change* (Paris: UNESCO, 1953); Margaret Mead, *World Enough: Rethinking the Future* (Boston: Little, Brown, 1975); Margaret Mead, *Some Personal Views* (New York: Walker, 1979); Margaret Mead and Muriel Brown, *The Wagon and the Star: A Study of American Community Initiative* (St. Paul, Minn.: Curriculum Resources, 1966).

12. Betty Friedan, *The Feminine Mystique* (1963; New York: Dell, 1974), 128–30; *American Women: The Report of the President's Commission on the Status of Women and Other Publications of the Commission,* ed. Margaret Mead and Frances Bagley Kaplan (New York: Charles Scribner's Sons, 1965); Gordan, *Margaret Mead;* Mead, *Twentieth Century Faith: Hope and Survival* (New York: Harper & Row, 1972), 84–86, 94–95, 113, 118–19; Mead, *World Enough,* xxvi–xxvii, 34, 90–94, 100–103.

1. Woven Lives, Raveled Texts

1. Mead, *Blackberry Winter,* 313, 124; Bateson, *With a Daughter's Eye,* 136.

2. Marjorie Fulton Freeman to Mead, September 18, 1948, Ruth Fulton Benedict Papers, Vassar College, hereafter RFB/VC; Mead, *Anthropologist at Work,* 68; Mead, *Blackberry Winter,* 124; Bateson, *With a Daughter's Eye,* 142–49.

3. Bateson, *With a Daughter's Eye,* 137–39, 118–19, 147, 150; Eve Kosofsky Sedgwick, "Epistemology of the Closet," *Raritan* 4 (Spring 1988): 39–69.

4. Janet E. Halley, "The Politics of the Closet: Towards Equal Protection for Gay, Lesbian and Bisexual Identity," in *Reclaiming Sodom,* ed. Jonathan Goldberg (New York: Routledge, 1994), 157; Eve Kosofsky Sedgwick, "Across Gender, Across Sexuality: Willa Cather and Others," 536–49, in *A Cultural Studies Reader: History, Theory, Practice,* ed. Jessica Munns and Gita Rajan (London: Longman, 1995); Bateson, *With a Daughter's Eye,* 144; Benedict, Letters from Mead, Mead, Special Correspondence, S1, Mead/LC. See also Hilary Lapsley, *Margaret Mead and Ruth Benedict: The Kinship of Women* (Amherst: University of Massachusetts Press, 1999).

5. Bateson, *With a Daughter's Eye,* 137; Regna Darnell, *Edward Sapir: Linguist, Anthropologist, Humanist* (Berkeley: University of California Press, 1990), 171–88; Mead, *Anthropologist at Work,* 91, 93.

6. Judith Schachter Modell, *Ruth Benedict: Patterns of Life* (Philadelphia: University of Pennsylvania Press, 1983), 146–57; Margaret M. Caffrey, *Ruth Benedict: Stranger in This Land* (Austin: University of Texas Press, 1989), 180–81; Mead, Articles, Books and Other Writings, "Life History, 1935," Addition III, Publications and Other Writings, S9, Mead/LC; Benedict to Mead, August 25, 1925, in Mead, *Anthropologist at Work,* 301; Mead, *Anthropologist at Work,* 4, 206–7, 548.

7. Mead, *Anthropologist at Work,* 167, 83, 195; Bateson, *With a Daughter's Eye,* 143, 149; Mead to Benedict, Paris, July 15, 1926; Benedict to Mead, October 16, 1931, Special Correspondence, S5, Mead /LC; Poems, 1927, Ruth Benedict, Publications & Other Writings, Q15, Mead/LC; Bateson, *With a Daughter's Eye,* 152; Halley, "The Politics of the Closet," 166.

8. Mead, *Anthropologist at Work,* 206, 207, 211; Caffrey, *Ruth Benedict,* 188; Mead, *Blackberry Winter,* 129; Mead, *Ruth Benedict,* 34, 35; Benedict, "Configurations of Culture," *American Anthropologist* 34 (January–March 1932): 4.

9. Mead, *Anthropologist at Work,* 206–9; Benedict, "Configurations of Culture," 1; Mead, *Ruth Benedict,* 34, 35; Mead, *Social Organization of Manu'a,* Bishop Museum Bulletin LXXVI, Honolulu, 1930; Caffrey, *Ruth Benedict,* 200–202; Mead to Benedict, September 3, 1928, en Route to St. Louis, Special Correspondence, Mead/LC.

10. Edward Sapir, "Observations of the Sex Problem in America," *American Journal of Psychiatry* 8 (1928): 523, 525, 528, 529; Richard Handler, "Vigorous Male and Aspiring Female: Poetry, Personality, and Culture in Edward Sapir and Ruth Benedict," in Stocking, *Malinowski, Rivers, Benedict and Others,* 144–47; Edward Sapir to Ruth Benedict, April 29, 1929, in Mead, *Anthropologist at Work,* 195, 207–12.

11. Benedict, "Configurations of Culture," 24, 25, 26; Darnell, *Edward Sapir,* 143; Victor Barnouw, "Ruth Benedict: Apollonian and Dionysian," *University of Toronto Quarterly* 18 (1949): 241–53.

12. Benedict, "Anthropology and the Abnormal," *Journal of General Psychology* 10 (1934): 59–80, reprinted in Mead, *Anthropologist at Work,* 262–83; Barnouw, "Ruth Benedict: Apollonian and Dionysian."

13. Benedict to Mead, October 16, 1932, Special Correspondence, Mead/LC.

14. Benedict to Mead in New Guinea, July 20, 1932; Benedict to Reo Fortune, August 2, 1932; Benedict to Mead, October 9, 1932, in Mead, *Anthropologist at Work,* 320, 323, 319, 206, 207, 202, 208; Benedict, *Patterns of Culture,* 2nd ed. (Boston: Houghton Mifflin, 1958), 7–8, 1, 265, 278, 245; Handler, "Vigorous Male and Aspiring Female," 147.

15. Mead, *Blackberry Winter,* 226, 229, 237–38, 254, 255, 258.

16. Mead, *Sex and Temperament,* (1935) reprinted in Mead, *From the South Seas: Studies of Adolescence and Sex in Primitive Societies* (New York: William Morrow, 1939), vi, viii, 322, 318; Benedict, [undated] Journal Fragment, in Mead, *Anthropologist at Work,* 145; Mead, *Blackberry Winter,* 259; George W. Stocking Jr., "Essays on Culture and Personality," in Stocking, *Malinowski, Rivers, Benedict,* 5.

17. William C. Manson, "Abram Kardiner and the Neo-Freudian Alternative in Culture and Personality," in Stocking, *Malinowski, Rivers, Benedict,* 78; Stocking, "Essays in Culture and Personality," 5; Margaret Mead, ed., *Cooperation and Competition among Primitive Peoples* (New York: McGraw Hill, 1937); Mead, *Sex and Temperament,* 241–53; Marvin Harris, *Rise of Anthropological Theory: A History of Theories of Culture* (London: Routledge & Kegan Paul, 1968), 433–35, 447–48; Jane Howard, *Margaret Mead: A Life* (New York: Simon and Schuster, 1984), 217.

18. Howard, *Margaret Mead,* 188–89, 213, 269, 290, 293, 366–67; Bateson, *With a Daughter's Eye,* 95, 113; Sedgwick, "Across Gender, Across Sexuality," 547; Mead, *Sex and Temperament,* 270; Benedict, "Sex in Primitive Society," *American Journal of Orthopsychiatry* 9 (1939): 570–75; Jennifer Terry, *An American Obsession: Science, Medicine, and Homosexuality in Modern Society* (Chicago: University of Chicago Press, 1999), 284–86, 164–66.

19. Mead, *Anthropologist at Work,* 346, 350, 353, 364, 347; Ruth Benedict, *Race and Racism,* 162; Bateson, *With a Daughter's Eye,* 113; Yans-McLaughlin, "Science, Democracy, and Ethics," 190–91, 207, 193, 198, 206; Margaret Mead, *And Keep Your Powder Dry: An Anthropologist Looks at America* (New York: William Morrow, 1942), 21, 24, 204, 206, 217; Caffrey, *Ruth Benedict,* 310, 319, 321.

20. Benedict to Robert Hashima, August 15, 1945, in Peter T. Suzuki, "Overlooked Aspects of *The Chrysanthemum and the Sword,*" *Dialetical Anthropology* 24 (1999): 225; Paul Boyer, "Justifications, Rationalizations, Evasions: Hiroshima, Nagasaki, and the American Conscience," in Paul Boyer, *By the Bomb's Early Light: American Thought and Culture at the Dawn of the Atomic Age* (New York: Pantheon, 1985), 184–85; Caffrey, *Ruth Benedict,* 322; Benedict, *Chrysanthemum and the Sword,* 13, 15, 184, 188, 315–16; Christopher Shannon, "A World Made Safe for Differences: Ruth Benedict's *The Chrysanthemum and the Sword,*" *American Quarterly* 47 (1995): 660.

21. Benedict, "The Past and the Future," Review of John Hersey's *Hiroshima, The Nation,* December 7, 1946, 656–57; Boyer, *By the Bomb's Early Light,* 204–10.

22. Benedict, "Recognition of Cultural Diversities in the Postwar World," *Annals of the American Academy of Political and Social Sciences* 228 (July 1943): 101–7. Benedict, *Chrysanthemum and the Sword,* 314–17; Benedict, "Answering a Few Questions," Fourth Lecture at UNESCO Seminar on Childhood Education in Podebrady, Czechoslovakia, 21 July-25 August 1948, RFB/VC; John F. Embree, "A Note on Ethnocentrism in Anthropology," Letters to the Editor, *American Anthropologist* 52 (July-September 1950): 430–32; Arthur Schlesinger, quoted in Terry, *An American Obsession,* 326; *New York Daily News,* quoted in Terry, *An American Obsession,* 326.

23. Arthur Schlesinger, *The Vital Center: The Politics of Freedom* (1950; London: 1970), 1; Margaret Mead, *Male and Female,* 355, 354, 49, 215, 345, 359.

24. Mead, *Male and Female,* as quoted on back jacket copy, 45, 79; Mead, *New Lives for Old,* 4, 7, 12, 13, 168; Mead, *Male and Female,* 45, 79, 417–18, 288, 254, n.2, 372, n.15, 418; Mead, Introduction to 1949 edition, Introduction to 1967 edition, *Male and Female,* 254; Angela Gilliam and Lenora Foerstel, eds., *Confronting the Margaret Mead Legacy: Scholarship, Empire, and the South Pacific* (Philadelphia: Temple University Press, 1992), 117–29, and Yans-McLaughlin, "Mobilizing Culture and Personality for World War II," 204; Mead, *Blackberry Winter,* 220; Betty Friedan, *The Feminine Mystique* (1963; New York: Dell, 1974), 127–28.

25. Mead, *Soviet Attitudes toward Authority: An Interdisciplinary Approach to Problems of the Soviet Character* (New York: McGraw-Hill, 1951), frontispiece, 2; Jules Henry, "National Character and War," "Letters to the Editor," *American Anthropologist* 53 (1951): 134–35; John F. Embree, "National Character and War," Letters to the Editor, *American Anthropologist* 53 (1951): 134–35.

26. Mead, *New Lives for Old,* 4, 6–8, 168, 178, 179, 173, 362, 370, 371, 405, 403, 404, 442–57, 186; Mead, *Letters from the Field,* 260–61.

27. Mead, *New Lives for Old,* 137–39, 118–19, 147, 135–36; Mead, *Anthropologist at Work,* 3.

28. Ibid., xvi–xvii, 3, 56.

29. Ibid., 519, "The Eucharist," xxi, 4, 56; Bateson, *With a Daughter's Eye,* 150, 152, 141.

30. Benedict to Mead in New Guinea, July 20, 1932; Jaime de Angulo, Berkeley, California, to Benedict, May 19, 1925; Mead to Benedict, March 21, 1933, in Mead, *Anthropologist at Work,* 319, 201–8, 296–98, 334; Micaela di Leonardo, *Exotics at Home: Anthropologies, Others, American Modernity* (Chicago: University of Chicago, 1998), 145–93, 334–66, and Foerstel and Gilliam, *Confronting the Margaret Mead Legacy.*

31. Mead, *Anthropologist at Work,* 425; Mead, *Blackberry Winter,* 329.

32. Mead, *Anthropologist at Work,* 425, 493–94; Mead, *Blackberry Winter,* 329; Sedgwick, "Across Gender, Across Sexuality," 538–39.

33. Rhoda Metraux, "Foreword," in Mead, *Anthropologist at Work* (New York: Atherton Press, 1966), iv, v; Rhoda Metraux, Letters to Mead, 1948–1978, Mead to Rhoda Metraux, 1948–1978, Special Correspondence, Mead/LC; Bateson, *With a Daughter's Eye,* 150–52, 142, 136; Mead, *Blackberry Winter,* 313–14.

34. Mead, *Anthropologist at Work,* 519; Mead, *Ruth Benedict,* 55, 58, 64, 75; Bateson, *With a Daughter's Eye,* 144.

35. "Ruth Fulton Benedict: A Memorial," 1949, as quoted in Lapsley, *Margaret Mead and Ruth Benedict,* 303.

2. "The Bo-Cu Plant"

1. Elmer Verner McCollum, "Biographical Memoir of Stanley Rossiter Benedict, 1884–1936," National Academy of Sciences, *Biographical Memoirs* (Washington, D.C.: National Academy of Sciences, 1952), 27, 155–77.

2. Bram Dijkstra, *Idols of Perversity: Fantasies of Feminine Evil in Fin-de-Siècle Culture* (New York: Oxford University Press, 1986); Elaine Showalter, *Sexual Anarchy: Gender and Culture at the Fin de Siècle* (New York: Viking, 1990); Nina Auerbach, *Woman and the Demon: The Life of a Victorian Myth* (Cambridge, Mass.: Harvard Uni-

versity Press, 1990); Richard Dellamora, *Masculine Desire: The Sexual Politics of Victorian Aestheticism* (Chapel Hill: University of North Carolina Press, 1990).

3. Mead, *Ruth Benedict,* 18; John Keegan, *The First World War* (London: Hutchinson, 1998).

4. Mead, *Anthropologist at Work,* 142, 194; Benedict, untitled mss., unpublished paper file, RFB/VC; Gloria Bowles, *Louise Bogan's Aesthetic of Limitation* (Bloomington: Indiana University Press, 1987), 20–23.

5. Beatrice Forbes Robertson-Hale, *What Women Want: An Interpretation of the Women's Movement* (New York: Frederic A. Stokes, 1914), 30; Benedict, "Mary Wollstonecraft"; Mead, *Anthropologist at Work,* 491–519; Janet Todd, *Mary Wollstonecraft: A Revolutionary Life* (New York: Columbia University Press, 2000); Lois W. Banner, "The Protestant Crusade: Religious Missions, Benevolence, and Reform in the United States, 1790–1840," Ph.D. diss., Columbia University, 1970, 322.

6. Mary Wollstonecraft file, RFB/VC; *Anthropologist at Work,* 146.

7. Benedict, "The Vision in Plains Culture," *American Anthropologist* 24 (1922): 1–23; Benedict, "The Concept of the Guardian Spirit in North America," *Memoirs of the American Anthropological Association* (Washington, D.C.: American Anthropological Association, 1923); Benedict, *Patterns of Culture,* 266–67.

8. Mead, *Anthropologist at Work,* 44–48.

9. Helene Silverberg, *Gender and American Social Science: The Formative Years* (Princeton, N.J.: Princeton University Press, 1998); Dorothy Ross, *The Origins of American Social Science* (Cambridge: Cambridge University Press, 1991).

10. Abraham Kardiner and Edward Preble, *They Studied Man* (New York: World, 1961), 204; Abram Maslow to Benedict, December 30, 1939, March 6, 1940, RFB/VC ; Mead, *Blackberry Winter* draft, "Marriage and Graduate School," 9, Mead/LC; Mead to Martha Ramsey Mead, March 11, 1923, Mead/LC; Robert H. Lowie, *Robert H. Lowie, Ethnologist: A Personal Record* (Berkeley: University of California Press, 1959), 135; Robert E. Lowie, *The History of Ethnological Theory* (New York: Farrar & Rinehart, 1937), 134. Nancy J. Parezo, "Anthropology: The Welcoming Science," in *Hidden Scholars: Women Anthropologists and the Native American Southwest,* ed. Nancy J. Parezo (Albuquerque: University of New Mexico Press, 1983), 3–37; Lowie, *Lowie,* 83; Regna Darnell, *Edward Sapir: Linguist, Anthropologist, Humanist* (Berkeley: University of California Press, 1990).

11. Mead, "Introduction," in *The Golden Age of American Anthropology,* ed. Mead and Ruth Bunzel (New York: George Braziller, 1960), 5–6; Lowie, *Lowie,* 83; Darnell, *Edward Sapir;* Michael Rogin, *Blackface, White Noise: Jewish Immigrants in the Hollywood Melting Pot* (Berkeley: University of California Press, 1996), 68–69; Sander Gilman, *The Jew's Body* (New York: Routledge, 1991), 81–82.

12. Alfred Kroeber, "The Superorganic," *American Anthropologist* 19 (1917): 201. On Kroeber, see Julian H. Steward, *Alfred Kroeber* (New York: Columbia University Press, 1973); and Theodora Kroeber, *Alfred Kroeber: A Personal Configuration* (Berkeley: University of California Press, 1970), 263–64; Eric Wolf, "Kroeber," in *Totems and Teachers: Perspectives on the History of Anthropology,* ed. Sydel Silverman (New York: Columbia University Press, 1981), 67–97; Clyde Kuckhohn, "The Influence of Psychiatry on Anthropology in America During the Past One Hundred Years," in *Personal Character and Cultural Milieu,* ed. Douglas Haring (Syracuse, N.Y.: Syracuse University

Press, 1956), 490–91; Alfred Kroeber, "On the Principle of Order in Civilization as Exemplified by Changes of Fashion," *American Anthropologist* 21 (1919): 235–63; Benedict to Mead, November 6, 1928, S-5; Benedict to Mead, April 2, 1932, Special Correspondence, Mead/LC, S-5.

13. Alexander Goldenweiser, *Early Civilization: An Introduction to Anthropology* (New York: Alfred A. Knopf, 1922); Charlotte Perkins Gilman, *The Man-Made World, or Our Androcentric Culture* (New York: Charlton, 1911); Ruth Bunzel, "Alexander Goldenweiser," in *Golden Age of American Anthropology,* ed. Mead and Bunzel, 508; and William H. Fanton, "Sapir as Museologist," ed. William Cowan et al., *New Perspectives in Language, Culture, and Personality: Proceedings of the Edward Sapir Centenary Conference* (Amsterdam: John Benjamins, 1986), 229; Mead, mss., "After Reading An Anthropologist at Work Through" file, Ruth Fulton Benedict Biography for Columbia University Press, Mead/LC.; Alexander Goldenweiser, "Sex and Primitive Society," in *Sex in Civilization,* ed. V. F. Calverton and S. D. Schmalhausen (Garden City, N.Y.: Garden City Publishing, 1929), 53; Goldenweiser, "Men and Women as Creators," in *Our Changing Morality,* ed. Freda Kerchway (New York: Albert and Charles Boni, 1927), 129–43; Goldenweiser, *Anthropology: An Introduction to Primitive Culture* (New York: F. S. Crofts, 1946), 142.

14. Mead, *Benedict,* 25; Mrs. Henry Cowell to Margaret, February 23, 1973, Jane Howard Papers, Columbia University, hereafter Howard/CU; Stanley Diamond, "Paul Radin," in Silverman, ed., *Totems and Teachers,* 67–97.

15. Robert S. Lowie, with Leta S. Hollingworth, "Science and Feminism," *Scientific Monthly* (1919): 277–84; Lowie, "Religion," in *The Making of Man: An Outline of Anthropology,* ed. V. F. Calverton (New York: Random House, 1931), 744–57; Alfred Kroeber, "Heredity, Environment, and Civilization: Factors Controlling Human Behavior as Illustrated by the Natives of the Southwestern United States," *Scientific American Supplement* 86, October 5, 1918, 211–12. Sabine Lang, *Men as Women, Women as Men: Changing Gender in Native American Cultures* (Austin: University of Texas Press, 1998), 30–31; Robert H. Lowie to Risa Lowie, March 1, 1918; Robert H. Lowie to Paul Radin, October 2, 1920; Robert H. Lowie Papers, Bancroft Library, University of California, Berkeley.

16. Alfred Louis Kroeber, "Reflections on Edward Sapir: Scholar and Man," in *Edward Sapir: Appraisals of His Life and Work,* ed. Konrad Koener (Amsterdam: John Benjamins, 1984), 131; Robert McMillan, interview with Ruth Bunzel, in McMillan, "The Study of Anthropology, 1931–37, at Columbia University and the University of Chicago," Ph.D. diss., New York University, 1986, 92; Edward Sapir, "The Woman's Man," *New Republic,* September 16, 1916, 167; Toni Flores, "The Poetry of Edward Sapir," *Dialectical Anthropology* 11 (1986): 159. Richard Handler, "The Dainty and the Hungry Man: Literature and Anthropology in the Work of Edward Sapir," *History of Anthropology* 1 (1983): 208–31; Edward Sapir, "Observations on the Sex Problem in America," *American Journal of Psychiatry* 8 (1928): 519–34; and "The Discipline of Sex," *American Mercury* 16 (1929): 413–29; Edward Sapir to Benedict, April 29, 1929, Mead/LC, S-15; Edward Sapir to Leslie White, June 30, 1926, Leslie White Papers, University of Michigan, quoted by Deslie Deacon, *Elsie Clews Parsons: Inventing Modern Life* (Chicago: University of Chicago Press, 1997), 260.

17. Mead to Benedict, December 2, 1932, Mead/LC, S-4. Jean Houston interview with Mead, Mead/LC; Mead to Benedict, December 2, 1932, Mead/LC, S-4; Benedict

to Mead, April 27, 1936, Mead/LC, S-5; Mead to Benedict, May 2, 1937, Mead/LC, S-4. In 1929 Sapir attacked Mead by name as incompetent in a review of Boas's *Anthropology and Modern Life* in *New Republic* 57 (1929): 279; Léonie Adams to Edmund Wilson, October 5, 1930, Edmund Wilson Papers, Beinecke Library, Yale University.

18. Mead to Benedict, April 2, 1932, Mead/LC, S-5; Mead, *Benedict,* 38. See *Blackberry Winter* draft, "The Years Between Field Trips," Mead/LC.

19. Robert Lynd, "Ruth Benedict," in *Ruth Fulton Benedict: A Memorial* (New York: Viking Fund, 1949), 23; Sidney W. Mintz, "Ruth Benedict," in Silverman, ed., *Totems and Teachers,* 61; Alfred L. Kroeber, "Franz Boas: The Man," in *Franz Boas Memoir Series,* ed. Kroeber et al. (Washington, D.C.: American Anthropological Association, 1943), 5–26; Helen Codere, "The Amiable Side of Kwakiutl Life: The Potlatch and the Play Potlatch," *American Anthropologist* 58 (April 1956): 334–51.

20. Benedict, *Patterns of Culture,* 227.

21. Richard Chase, "Ruth Benedict: The Woman as Anthropologist," *Columbia University Forum* 2 (Spring 1959): 19–22; Benedict, *An Introduction to Zuni Mythology* (New York: Columbia University Press, 1935), I:xxi; Benedict to Ruth Landes, January 26, 1937, RFB/VC.

22. Benedict to Mead, May 15, 1932, Mead/LC, S-4; Barnouw, "Ruth Benedict: Apollonian and Dionysian," 242; *Theorizing Masculinities,* ed. Harry Brod and Michael Kaufman (Thousand Oaks, Ca.: Sage, 1994).

23. Barnouw, "Ruth Benedict: Apollonian and Dionysian," 242; Betsey Erkilla, *The Wicked Sisters: Women Poets, Literary History and Discord* (New York: Oxford University Press, 1992), 3–16. Benedict, "Anthropology and the Humanities," *American Anthropologist* 50 (Oct.-Dec. 1948): 585–93; "Ruth Fulton Benedict (1878–1948)," *Journal of American Folk-lore* 62 (Oct.-Dec. 1949): 345. On Wylie and Buck, see Laura Johnson Wylie, *Studies in the Evolution of English Criticism* (Boston: Ginn, 1903), and *Toward a Feminist Rhetoric: The Writing of Gertrude Buck,* ed. JoAnn Campbell (Pittsburgh, Pa.: University of Pittsburgh Press, 1996).

24. Benedict, "Male Dominance in Thai Culture," in *About Bateson: Essays on Gregory Bateson,* ed. John Brockman (New York: E. P. Dutton, 1977), 215–21.

25. Mead, *Anthropologist at Work,* 109.

26. Lois W. Banner, *Intertwined Lives: Margaret Mead, Ruth Benedict, and Their Circle* (New York: Alfred A. Knopf, 2003), chaps. 1 and 4; Mead, *Anthropologist at Work,* 131.

27. Ellen Key, *The Woman Movement,* trans. Mamah Bouton Borthwick (Chicago: Ralph Fletcher Seymour, 1911), 108; Edward Carpenter, *Love's Coming of Age* (Chicago: Charles H. Kerr, 1910); Havelock Ellis, *Little Essays on Love and Virtue* (New York: George H. Doran, 1922); Beatrice Hinkle, *The Re-Creating of the Individual: A Study of Psychological Types and Their Relation to Personality* (New York: Dodd, Mead, 1932); Diane Hoeveler, *Romantic Androgyny: The Women Within* (University Park: Pennsylvania State University Press); Joanne Meyerowitz, *How Sex Changed: A History of Transsexuality in the U.S.* (Cambridge, Mass.: Harvard University Press, 2002); George Chauncey, "From Sexual Inversion to Homosexuality: The Changing Medical Conceptualization of Female 'Deviance,'" in *Passion and Power: Sexuality in History,* ed. Kathy Peiss and Christina Simmons (Philadelphia: Temple University Press, 1989), 87–117.

28. Virginia Woolf, *A Room of One's Own* (New York: Harcourt Brace, 1929), 102.

29. Esther S. Goldfrank, *Notes on an Undirected Life: As One Anthropologist Tells*

It (Flushing, N.Y.: Queens College Press, 1978), 222; Elizabeth Stassinos, "Ruthlessly: Ruth Benedict's Pseudonyms and the Art of Science Writ Large," Ph.D. diss., University of Virginia, 1998; Joseph Conrad, *The Nigger of the "Narcissus"* (New York: W.W. Norton, 1979), 3, 9, 15.

30. Mead to Benedict, September 20, 1927, Mead/LC, S-3; Benedict to Mead, November 20, 1925, Mead/LC S-4.

31. Dream Research File, Mead/LC, A-3; Mead, *Anthropologist at Work,* 106; Benedict to Mead, August 3, 1938, Mead/LC, S-5.

32. Mead, *Anthropologist at Work,* 106; RFB to MM, August 3, 1938, Mead/LC, S-5.

33. Judith Schachter Modell, *Ruth Benedict: Patterns of a Life* (Philadelphia: University of Pennsylvania Press, 1983), 164; Ruth Landes, "A Woman Anthropologist in Brazil," in *Women in the Field: Anthropological Experience,* ed. Peggy Golde (Chicago: Aldine, 1970), 120.

34. Benedict, "Speech before Seminar, Committee on the Study of Adolescents," February 12, 1937, typescript mss., RFB/VC; Benedict, "Sex in Primitive Societies," *American Journal of Orthopsychiatry* 9 (1939): 570–73; Benedict to Mead, August 3, 1938, Mead/LC, S-5; Benedict to Dr. Oliver Cope, February 27, 1943, RFB/VC. The article to which she referred is "We Can't Afford Race Prejudice," *Frontiers of Democracy,* October 9, 1942, 2.

35. T. George Harris, "About Ruth Benedict and her Lost Manuscript," *Psychology Today* 4 (1970): 51–52.

36. Richard Handler, "Ruth Benedict and the Modernist Sensibility," in *Modernist Anthropology: From Fieldwork to Text,* ed. Marc Manganaro (Princeton, N.J.: Princeton University Press, 1990), 180.

37. Adelin Linton and Charles Wagley, *Ralph Linton* (New York: Columbia University Press, 1971), 10, 48–49; Benedict to Mead, August 3, 1938, Mead/LC, S-5; Jane Howard interview with Robert L. Suggs, July 23, 1979, Howard/CU; Mintz, "Benedict," in Silverman, ed., *Totems and Teachers,* 156–57, 161.

38. Abram Kardiner oral interview, Columbia University Oral History Project; Abram Kardiner, *Sex and Morality* (Indianapolis: Bobbs-Merrill, 1954), 83; Kardiner, *My Analysis with Freud: Reminiscences* (New York: Norton, 1977).

3. Margaret Mead, the Samoan Girl and the Flapper

1. "Scientist Goes on Jungle Flapper Hunt," *New York Sun Times,* 8 November 1925, Clippings and Articles, 1926–28, Mead/LC, L3; Micaela di Leonardo, *Exotics at Home: Anthropologies, Others, American Modernity* (Chicago: University of Chicago, 1998), 18.

2. Lawrence W. Levine, *The Unpredictable Past: Explorations in American Cultural History* (New York: Oxford University Press, 1993), 191.

3. Kenneth Yellis, "Prosperity's Child: Some Thoughts on the Flapper," *American Quarterly* 21 (1969): 49; Frederick Lewis Allen, *Only Yesterday: The 1920s in America* (1931; New York: Harper & Row, 1986), 73–101.

4. John D'Emilio and Estelle Freedman, *Intimate Matters: A History of Sexuality in America,* 2nd ed. (New York: Harper & Row, 1997), 188–94; Carol Smith-Rosenberg, *Disorderly Conduct: Visions of Gender in Victorian America* (New York: Oxford University Press, 1985), 247–52.

5. Levine, *Unpredictable Past,* 201; Allen, *Only Yesterday,* 82.

6. Sinclair Lewis, *Main Street* (1920; New York: Harcourt, Brace & World, 1948),

1; Francis Scott Fitzgerald, *This Side of Paradise* (1920; Cambridge: Cambridge University Press, 1995), 6, 183.

7. Franz Boas to Mead, July 14, 1925, Special Correspondence, Mead/LC, B2.

8. Mead to Franz Boas, January 5, 1926, Pago Pago [Tau, Man'ua] Samoa Mead fieldtrip 1925–26, Mead/LC, N1; Mead to Franz Boas, January 16, 1926, Pago Pago [Tau, Man'ua] B/B61 Boas Papers Film 1263 Reel 15, American Philosophical Society Library, Philadelphia, hereafter Boas/APS.

9. Franz Boas to Mead, February 15, 1926, New York, Samoa, Mead fieldtrip 1925–26, Mead/LC, N1; Mead to Franz Boas, March 19, 1926, Tau, Manu'a, B/B61 Boas Papers, Film 1263, Reel 15, Boas/APS.

10. Mead to E. S. C. Handy, March 15, 1926, New York, Publications and Other Writings File, Mead/LC, I4.

11. Mead to Emily Fogg Mead, March 15, 1928, New York, Family Papers, Mead/LC, A7; Dennis Porter, "Anthropological Tales: Unprofessional Thoughts on the Mead/Freeman Controversy," *Notebooks in Cultural Anthropology* 1 (1984): 15–37; Lutkehaus, "Margaret Mead and the 'Wind Rustling in the Palm Trees' School of Ethnography," in *Women Writing Culture,* ed. Ruth Behar and Deborah Gordon (Berkeley: University of California Press, 1995).

12. "American Girl to Study Cannibals," *Boston Post,* May 24,1928; "Will Shows Fear of Cannibals," *New York Sun,* August 22, 1928; Clippings and Articles 1928–1930, Mead/LC, L3.; Mead to Emily Fogg Mead, May 19, 1928, Family Papers, Mead/LC, A7; William Morrow to Mead, June 20, 1928, Publications and Other Writings File, Mead/LC, 12.

13. William Morrow & Co., Order Form-Coming of Age in Samoa, Publications and Other Writings, Mead/LC, Box I2; "Samoa Is the Place for Women," *New York Sun,* January 23, 1929; Frederick O'Brien, "Where Neuroses Cease from Troubling and Complexes Are at Rest," *World,* October 21, 1928; "But I Came of Age in Samoa," cartoon, *Esquire,* December 1928, Clippings and Articles 1928–1930, Mead/LC, L3.

14. Robert Lowie, "Review of *Coming of Age in Samoa;* Edward Sapir, "The Discipline of Sex," *American Mercury,* 1929, 13–20, reprinted in *Selected Writings in Language, Culture and Personality,* ed. David Mandelbaum (Berkeley: University of California Press, 1949).

15. H. L. Mencken, "Adolescence," *American Mercury,* November, 1929, 379–80. Clippings and Articles 1928–1930, Mead/LC, L3.

16. Emily Fogg Mead to Mead, Philadelphia, July 19, 1933, Family Papers, Mead/LC, A7.

17. Lutkehaus, "Margaret Mead and the 'Wind Rustling in the Palm Trees' School," 203; Mead, *Coming of Age in Samoa,* 195, 151.

18. Mead, "Samoan Children at Work and Play," *Natural History* 28 (1928): 26–36; Mead, "The Role of the Individual in Samoan Culture," *Journal of the Royal Anthropological Institute* 55 (July-December 1928): 485; Mead, "South Sea Hints on Bringing up Children," *Parents' Magazine* 4 (1929): 20–22, 49–52; Michel Foucault, *The Care of the Self, The History of Sexuality,* vol. 3 (Harmondsworth: Penguin, 1988); Charles Taylor, *Sources of the Self: The Making of Modern Identity* (Cambridge: Cambridge University Press, 1989); and John Chynoweth Burnham, *Psychoanalysis and American Medicine, 1894–1918: Medicine, Society and Culture* (New York: International Universities Press, 1967), 89.

19. Carolyn Steedman, *Strange Dislocations: Childhood and the Idea of Human Interiority, 1780–1930* (London: Virago, 1995), 12, 9.

20. "Books on the Table: Samoan Adolescence," *The Argonaut,* January 19, 1929, 12. Clippings and Articles 1928–1930, Mead/LC, L3.

21. Untitled book review, *Philadelphia Record,* August 15, 1928; Untitled book review, *Brooklyn Eagle,* August 22, 1929, Clippings and Articles 1928–1930, Mead/LC, L3.

22. Caspar Hunt, "The Younger Generation in Samoa," *Travel,* April 1929; "Utopian Marriages, for the Woman," New York *Telegram,* January 1930, Clippings and Articles 1928–1930, Mead/LC, L3; Mead, "The Role of the Individual in Samoan Culture," 495. Freda Kirchwey, "This Week: Sex in the South Seas," *Nation,* October 18, 1928, 427; Lowie, Review of *Coming of Age in Samoa*; Derek Freeman, *Margaret Mead and Samoa: The Making and Unmaking of an Anthropological Myth* (Cambridge, Mass: Harvard University Press, 1983); di Leonardo, *Exotics at Home,* 172; Mead, "Americanization in Samoa," *American Mercury,* March 16, 1929, 268.

23. "Adolescence Not a Dangerous Period of Life in Samoa," *St. Louis Globe Democrat,* December 29, 1929; O'Brien, "Where Neuroses Cease," Clippings and Articles 1928–1930, Mead/LC, L3; Mead, "South Sea Hints on Bringing up Children," 22; Mead, *Coming of Age,* 199.

24. Mead, *Coming of Age in Samoa,* 200.

25. Hunt, "The Younger Generation in Samoa"; Aldous Huxley, "The Problem of Faith," *Harper's Magazine,* January 1933, Clippings and Articles 1928–1930, File L3, Mead/LC.

26. Review of *Coming of Age, Psychoanalytic Review* (1929): 115–16, Clippings and Articles 1928–1930, Mead/LC, L3.

27. Kirchwey, "This Week: Sex in the South Seas," 427.

28. Mead, "Adolescence in Primitive and Modern Society," in *The New Generation: The Intimate Problems of Parents and Children,* ed. V. F. Calverton and S. D. Schmalhausen (New York: Macauley, 1930), 185.

29. Levine, *The Unpredictable Past,* 203.

30. O'Brien, "Where Neuroses Cease from Troubling"; Nels Anderson, "In the Light of Samoa," *Survey,* January 15, 1929, Clippings and Articles 1928–30, Mead/LC, L3.

Part II. Erasures and Incursions

1. Mead, *Sex and Temperament in Three Primitive Societies,* 320, xvi, xxii, xvii, 161, 176, 223, 233, 236, 245, 253, 255, 256; see also Richard C. Thurnwald, Review of *Sex and Temperament, American Anthropologist* 38 (October-December 1936) 664; Richard Handler, "Vigorous Male and Aspiring Female: Poetry, Personality, and Culture in Edward Sapir and Ruth Benedict," in Stocking, *Malinowski, Rivers, Benedict and Others*; Mead and Reo Fortune, 1934, Box R-4, Mead/LC.

2. Mead, *Blackberry Winter,* 226–27; Mead, Review of *Patterns of Culture, Nation,* 12 December 1934; Benedict, *Patterns of Culture,* 180; Mead, "A New Preface" to Benedict, *Patterns of Culture,* vi–xii, x, 9, 10, 278, 20, 248, 10, 12, 189; *Oxford English Dictionary,* 2nd ed. (London: Clarendon, 1989), 8:76–76; Benedict, *Race: Science and Politics;* Benedict, *Race and Racism,* viii, 38, 150, 111, 152.

3. Mead, *Male and Female,* 355, 343; Betty Friedan, *The Feminine Mystique* (New York: Dell, 1974), 117–41.

4. Coming of Age, but Not in Samoa

This chapter is a revised version of Louise M. Newman, "Coming of Age, but Not in Samoa: Reflections on Margaret Mead's Legacy for Western Liberal Feminism," *American Quarterly* 48 (1996): 233–72.

1. Mead, "Preface," *Coming of Age in Samoa* (New York: William Morrow, 1973).

2. Stephen O. Murray, "On Boasians and Margaret Mead: Reply to Freeman," *Current Anthropology* 32 (August-October 1991): 488.

3. See Betty Friedan, *The Feminine Mystique* (1963; New York: Dell, 1974), 145, 135; Alice Rossi, *The Feminist Papers: From Adams to de Beauvoir* (New York: Columbia University Press, 1973), xx; Rosalind Rosenberg, *Beyond Separate Spheres: Intellectual Roots of Modern Feminism* (New Haven: Yale University Press, 1982), 208; *Woman, Culture and Society,* ed. Michelle Zimbalist Rosaldo and Louise Lamphere (Stanford, Calif.: Stanford University Press, 1974), 18.

4. See George W. Stocking Jr., *Victorian Anthropology* (New York: Free Press, 1987); and Stocking, *Race, Culture, and Evolution: Essays in the History of Anthropology* (New York: Free Press, 1968).

5. Mead, *Blackberry Winter,* 156.

6. William I. Thomas, "A Difference in the Metabolism of the Sexes," *American Journal of Sociology* 3 (July 1897): 41; Nicholas Thomas, *Colonialism's Culture: Anthropology, Travel and Government* (Princeton, N.J.: Princeton University Press, 1994), 102; Gail Bederman, "'Civilization,' the Decline of Middle-Class Manliness, and Ida B. Wells' Antilynching Campaign, 1892–94," *Radical History Review* 52 (Winter 1992): 9.

7. Mead, *Sex and Temperament in Three Primitive Societies,* 4th ed. (New York: HarperCollins, 2001), 289.

8. Todd Gernes suggested the term *cultural comparativism* in a personal communication.

9. Mead, *Coming of Age in Samoa,* 4 (emphasis added).

10. Charlotte Perkins Gilman, *The Home: Its Work and Influence* (New York: McClure Phillips, 1905); Elsie Clews Parsons, *The Family* (New York: Putnam, 1906); Gilman, *The Man-Made World;* Mary Roberts Coolidge, *Why Women Are So* (New York: Henry Holt, 1912); and Parsons "Facing Race Suicide," *Masses* 6 (June 1915): 15.

11. Dorothy Ross, *G. Stanley Hall: The Psychologist as Prophet* (Chicago: University of Chicago Press, 1972); Cynthia Eagle Russet, *Sexual Science: The Victorian Construction of Womanhood* (Cambridge, Mass.: Harvard University Press, 1989), 57; Mead, *Coming of Age in Samoa,* 5, 223.

12. Mead, *Coming of Age in Samoa,* 223, 5, 201.

13. Ibid., 206.

14. Thomas, *Colonialism's Culture,* 100.

15. Mead, *Coming of Age in Samoa,* 236–37.

16. Mead, *Sex and Temperament,* 261–62.

17. Ibid., 292.

18. Freda Kirchwey's review, "Sex in the South Seas," *Nation,* October 24, 1928, 427; Mead, Preface to *Sex and Temperament.*

19. Hortense Powdermaker, Review of *Sex and Temperament in Three Primitive*

Societies, Annals of the American Academy of Political and Social Science 181 (September 1935): 221–22; Mead, Preface to *Sex and Temperament.*

20. Mead, *Coming of Age,* 248; Jeannette Mirsky, Review of *Sex and Temperament in Three Primitive Societies, Survey* 71 (October 1935): 315; Mead, Preface to *Sex and Temperament.*

21. Patricia R. Hill, *The World Their Household: The American Woman's Foreign Mission Movement and Cultural Transformation, 1870–1920* (Ann Arbor: University of Michigan Press, 1985), 36–40, 213–22; Barbara Welter, "She Hath Done What She Could," *American Quarterly* 30 (Winter 1978): 627, n.12; Lester Ward, "Our Better Halves," *Forum* 6 (November 1888): 275; Otis T. Mason, "Woman's Share in Primitive Culture," *American Antiquarian* (January 11, 1889): 3–13; Otis T. Mason, *Woman's Share in Primitive Culture* (New York: Appleton, 1894, 1898); Lester Ward, *The Psychic Factors of Civilization* (New York: Ginn, 1893).

22. M. R. [Clarissa Chapman] Armstrong, "Sketches of Mission Life, No. IV," *Southern Workman* 10 (April 1881): 44.

23. Joan Mark, *A Stranger in Her Native Land: Alice Fletcher and the American Indians* (Lincoln: University of Nebraska Press, 1988); Frederick E. Hoxie and Joan T. Mark, "Introduction," in *With the Nez Perces: Alice Fletcher in the Field, 1889–92,* by E. Jane Gay (Lincoln: University of Nebraska Press, 1981), xiv–vi.

24. Jeanne Madeline Moore, "Bebe Bwana," *American History Illustrated,* 21 October 1986, 37; May French-Sheldon to Henry Morton Stanley, Henry Morton Stanley to May French-Sheldon, May French-Sheldon papers, Manuscripts, Library of Congress, Washington, D.C.; hereafter, MFS/LC.

25. "With Gayest Parisian Clothes She Traveled Alone through African Jungles," *Evening Sun,* 15 February 1915; T. J. Boisseau, "'They Called me *Bebe Bwana*': A Cultural Study of an Imperial Feminist," *Signs* 21 (Autumn 1995): 116–46; French-Sheldon, *Sultan to Sultan: Adventures among the Masai and Other Tribes of East Africa* (Boston: Arena, 1892), 66, 380, 381; Obituary, French-Sheldon, London *Evening News,* 1 February 1936; Obituary, *West London Observer,* 21 February 1936; Obituary, *New York Times,* 11 February 1936; Clippings from a newspaper in Covina, California, 27 September 1923; San Francisco *Chronicle,* 21 February 1924; Scrapbook, Container 7, MFS/LC.

26. Frances Drewry McMullen, "'Going Native' for Science," *Woman's Journal,* 15 July 1930, 8; *American Magazine,* September 1934, 42; "Women Explorers Held Equal to Men," *New York Times,* 14 March 1934, 9.

27. Mead, *Male and Female,* 39.

28. Jane Howard, *Margaret Mead: A Life* (New York: Simon and Schuster, 1984), 398; Mead, Letter from Alitoa, January 15, 1932, in Mead, *Letters from the Field,* 103; Howard, *Margaret Mead,* 398; Mead, "The Comparative Study of Cultures and the Purposive Cultivation of Democratic Values, 1941–1949," in Mead, *Perspectives on a Troubled Decade: Science, Philosophy and Religion, 1939–1949* (New York: Harper, 1950), 91.

29. Mead and Baldwin, *A Rap on Race,* 28.

30. Ibid.

31. Mead, "Race Majority—Race Minority," in *The People in Your Life: Psychiatry and Personal Relations by Ten Leading Authorities,* ed. Margaret M. Hughes (1951; New York: Books for Libraries Press, 1971), 132.

32. Mead, "Americanization in Samoa"; Mead, "Human Differences and World

Order," in *World Order: Its Intellectual and Cultural Foundations,* ed. Ernest Johnson (New York: Harper, 1945), 42–43.

33. See *Margaret Mead and Samoa;* Derek Freeman, *The Fateful Hoaxing of Margaret Mead: A Historical Analysis of her Samoan Research* (Boulder, Colo.: Westview, 1999); Jean Bethke Elshtain, "Coming of Age in America: Why the Attack on Margaret Mead?" *The Progressive,* October 1983, 33–35.

34. *Gender at the Crossroads of Knowledge: Feminist Anthropology in the Postmodern Era,* ed. Micaela di Leonardo (Berkeley: University of California Press, 1991), 27, 10–11.

5. "A World Made Safe for Differences"

This chapter is a revised version of Christopher Shannon, "A World Made Safe for Differences: Ruth Benedict's *The Chrysanthemum and the Sword,*" *American Quarterly* 47 (1995): 659–80.

1. Christopher Lasch, "The Cultural Cold War: A Short History of the Congress for Cultural Freedom," in Lasch, *The Agony of the American Left* (New York: Knopf, 1969); Peter Coleman, *The Liberal Conspiracy: The Congress for Cultural Freedom and the Struggle for the Mind of Postwar Europe* (New York: Free Press, 1989).

2. Benedict, *Chrysanthemum and the Sword,* 15.

3. "Foreword" in Benedict, *Patterns of Culture,* viii.

4. Herbert Marcuse, "Repressive Tolerance," in Marcuse, Barrington Moore Jr., and Robert Wolff, *A Critique of Pure Tolerance* (Boston: Beacon Press, 1969), 115, 84.

5. R. Jeffrey Lustig, *Corporate Liberalism: The Origins of Modern American Political Theory, 1890–1920* (Berkeley: University of California Press, 1982).

6. Richard Handler, "Boasian Anthropology and the Critique of American Culture," *American Quarterly* 42 (June, 1990): 266–68; Virginia Yans-McLaughlin, "Science, Democracy, and Ethics: Mobilizing Culture and Personality for World War II," in *Malinowski, Rivers, Benedict and Others: Essays on Culture and Personality,* ed. George Stocking Jr. (Madison: University of Wisconsin Press, 1986), 210.

7. Richard Handler, "Vigorous Male and Aspiring Female: Poetry, Personality, and Culture in Edward Sapir and Ruth Benedict," in Stocking, *Malinowski, Rivers, Benedict and Others,* 152; Yans-McLaughlin, "Science, Democracy, and Ethics," 204, 207, 212.

8. Marcuse, "Repressive Tolerance," 86, 105.

9. George W. Stocking Jr., "Franz Boas and the Culture Concept in Historical Perspective," in Stocking, *Race, Culture, and Evolution.*

10. Mead, *Ruth Benedict,* 57; Margaret M. Caffrey, *Ruth Benedict: Stranger in This Land* (Austin: University of Texas Press, 1989), 315–16.

11. Judith Schachter Modell, *Ruth Benedict: Patterns of Life* (Philadelphia: University of Pennsylvania Press, 1983), 268–70; Caffrey, *Ruth Benedict,* 321.

12. Mead, *And Keep Your Powder Dry,* 8–9, 241, 247; John Dower, *War without Mercy: Race and Power in the Pacific War* (New York: Pantheon, 1986), 119.

13. Dower, "Primitives, Children, Madmen," in Dower, *War without Mercy;* Benedict, *Chrysanthemum and the Sword,* 13, 15, 16.

14. Benedict, *Chrysanthemum and the Sword,* 14–15.

15. Ibid., 14.

16. Ibid., 98, 177.

17. Ibid., 20, 21, 23, 28, 146–47, 223.

18. Mead, *And Keep Your Powder Dry.*

19. Dower, *War without Mercy,* 19, 128, 304.

20. Benedict, *Chrysanthemum and the Sword,* 79–80; hereafter page numbers in parentheses in the text will refer to this text.

21. David Hollinger, "Ethnic Diversity, Cosmopolitanism, and the Emergence of the American Liberal Intelligentsia," in Hollinger, *In the American Province: Studies in the History and Historiography of Ideas* (Bloomington: Indiana University Press, 1985), 56–73; Lasch, *The True and Only Heaven: Progress and Its Critics* (New York: Norton, 1991), esp. chap. 10, "The Politics of the Civilized Minority."

6. "White Maternity, Rape, and the Sexual Exile in *A Rap on Race*"

1. Louise M. Newman, "Coming of Age, but Not in Samoa: Reflections on Margaret Mead's Legacy for Western Liberal Feminism," *American Quarterly* 48 (1996): 234, 235.

2. Micaela di Leonardo, *Exotics at Home: Anthropologies, Others, American Modernity* (Chicago: University of Chicago, 1998), 164–65.

3. Hilary Lapsley, *Margaret Mead and Ruth Benedict: The Kinship of Women* (Amherst: University of Massachusetts Press, 1999).

4. Jean Walton, "A People of Her Own: Margaret Mead," in Walton, *Fair Sex, Savage Dreams: Race, Psychoanalysis, Sexual Difference* (Durham, S.C.: Duke University Press, 2001).

5. Baldwin and Mead, *A Rap on Race,* 16–17.

6. *Women in the Field: Anthropological Experience,* ed. Peggy Golde (Chicago: Aldine, 1970), 294; Baldwin and Mead, *A Rap on Race,* 14–15.

7. Baldwin and Mead, *A Rap on Race,* 17.

8. Ibid., 18–20.

9. Ibid., 214–15.

10. Ibid., 215.

11. Ibid., 15.

12. Note the conflation of "ordinary" with "white" in this passage.

13. Joan Riviere, "Womanliness as a Masquerade," in *Formations of Fantasy,* ed. Victor Burgin, James Donald, and Cora Kaplan (London: Methuen, 1986), 37.

14. Riviere, "Womanliness," 10, Frantz Fanon, *Black Skin, White Masks,* trans. Charles Lam Markmann (New York: Grove Weidenfeld, 1967), 179.

15. Fanon, *Black Skin, White Masks,* 201, 200.

16. Baldwin and Mead, *A Rap on Race,* 107; Mead, *Letters from the Field,* 19.

17. Baldwin and Mead, A Rap on Race, 11.

18. Ibid., 202, 216.

Part III. Imperial Visions

1. Micaela di Leonardo, *Exotics at Home: Anthropologies, Others, American Modernity* (Chicago: University of Chicago, 1998), 165, 187; Lenora Foerstel and Angela Gilliam, eds., *Confronting the Margaret Mead Legacy: Scholarship, Empire, and the South Pacific* (Philadelphia: Temple University Press, 1992).

2. Di Leonardo, *Exotics at Home* 261, 227, 189; Eric Wolf, "American Anthropolo-

gists and American Power," in *Reinventing Anthropology,* ed. Dell Hymes (New York: Vintage Books, 1974), 251–63, 257; Peter M. Worsley, "Margaret Mead: Science or Science Fiction?" *Science and Society* 21 (Spring 1957): 122–34, 128.

3. Worsley, "Margaret Mead," 124.

7. Of Feys and Culture Planners

1. Mead to Geoffrey Gorer, October 1, 1936, Mead/LC, N5:4.

2. Thorstein Veblen, *The Engineers and the Price System* (New York: B. W. Heubsch, 1921), 138; Jane Howard, *Margaret Mead: A Life* (New York: Simon and Schuster, 1984), 26; Anonymous, *History and Purpose of Technocracy* (Savannah, Ohio: 1984), 8; Mead, *Blackberry Winter,* 186–88.

3. Rodney Needham, *Exemplars* (Berkeley: University of California Press, 1985). On the squares, see Sullivan, "A Four-Fold Humanity: Margaret Mead and Psychological Types," *Journal of the History of the Behavioral Sciences* 40 (Spring 2004): 183–206.

4. See Mead, *Blackberry Winter,* 217; Bateson, *With a Daughter's Eye* (1984 ed.), 132; "Summary Statement of the Problem of Personality and Culture," Tchambuli, 1933, Mead/LC, N102:2.

5. Mead, *Sex and Temperament,* xiv; "Summary Statement of the Problem of Personality and Culture."

6. William McDougall, *An Introduction to Social Psychology* (London: Methuen, 1908); McDougall, *An Outline of Psychology* (London: Methuen, 1923); McDougall, *An Outline of Abnormal Psychology* (London: Methuen 1933), 442; Mead/LC N102:2.

7. Untitled document, Mead/LC, S11:8.

8. John Dewey, *Human Nature and Conduct* (1922; New York: Modern Library, 1930), 38.

9. Mead/LC, N5:1; Mead/LC, N6:4.

10. Untitled diagram, Mead/LC, S11; Mead to Bateson, January 12, 1935, Mead/LC, S 1:6.

11. "Summary Statement of the Problem of Personality and Culture"; Bateson, *With a Daughter's Eye,* 132.

12. Statement of the Squares Hypothesis, May 16, 1937, N12:2.; Benedict's notes on Mead's lecture course on this subject of 1935, Mead/LC, O40:7; Marcel Mauss, "A Category of the Human Mind: The Notion of Person; the Notion of Self," in *The Category of the Person: Anthropology, Philosophy, History,* ed. Michael Carrithers, Steven Collins, and Steven Lukes (Cambridge: Cambridge University Press, 1985), 1–25; Clifford Geertz, "Person, Time and Conduct in Bali," in Geertz, *The Interpretation of Cultures* (New York: Basic Books, 1973), 360– 411; Thomas J. Csordas, "Introduction," in *Embodiment and Experience: The Existential Ground of Culture and Self,* ed. Thomas J. Csordas (Cambridge: Cambridge University Press, 1994), 8; see Mead, *Coming of Age in Samoa,* 158.

13. "Research Meditations," December 28, 1936, Mead/LC, N12:2.

14. Mead to Helen Lynd, February 6, 1937, Mead/LC, N5:5; "Summary Statement of the Problem of Personality and Culture," Mead/LC; Ian Hacking, *Rewriting the Soul: Multiple Personality and the Sciences of Memory* (Princeton, N.J.: Princeton University Press, 1995); Louis A. Sass, *Madness and Modernism: Insanity in the Light of Modern Art, Literature, and Thought* (Cambridge, Mass: Harvard University Press, 1994).

15. Eugen Bleuler, *Dementia Praecox or the Group of Schizophrenias,* trans. Joseph Zinkin (1911; New York: International University Press, 1952); Bleuler, *The Theory of Schizophrenic Negativism,* trans. William A. White (New York: Nervous and Mental Disease Publishing Co., 1912); Morton Prince, *The Dissociation of a Personality* (New York: Greenwood 1908); Mead to Gregory Bateson, February 27, 1935, Mead/LC S1:6; Mead, *Blackberry Winter,* 220; Ernst Kretschmer, *Physique and Character: An Investigation of the Nature of Constitution and the Theory of Temperament,* trans. W. J. Sprott (1922; New York: Harcourt, Brace 1925); Bateson, *Naven: A Survey of the Problems Suggested by a Composite Picture of a New Guinea Tribe drawn from Three Points of View* (1936; Stanford, Calif.: Stanford University Press, 1958), 160.

16. Mead to John Dollard, September 23, 1936, Mead/LC, N5:4; Gerald Sullivan, *Bali as It Might Have Been Known: Margaret Mead, Gregory Bateson, Wolfgang Weck, Schizophrenia and Human Agency,* Ph.D. diss., University of Virginia, 1998, 275–308; Bateson and Mead, *Balinese Character* (New York: New York Academy of Sciences, 1942), 25, 131; Clifford Geertz, "Deep Play: Notes on the Balinese Cockfight," in Geertz, *The Interpretation of Cultures* (New York; Basic Books, 1973), 417ff, 436; Unni Wikan, *Managing Turbulent Hearts: A Balinese Formula for Living* (Chicago: University of Chicago Press, 1990).

17. Mead to Helen Lynd, February 6, 1937, Mead/LC, N5:5; "Statement of the Squares Hypothesis"; "Summary Statement of the Problem of Personality and Culture"; Boas, *The Mind of Primitive Man* (New York: Macmillan, 1911); "The Problem of Race," in *The Making of Man: An Outline of Anthropology,* ed. V. F. Calverton (New York: Modern Library, 1931), 113–41; Franz Boas, "Race," *Encyclopaedia of the Social Sciences* (New York: Macmillan), 13:25–36; Mead to Lynd, February 6, 1937, Mead/LC, N5:5; Mead/LC, N5:5; Wikan, *Managing Turbulent Hearts,* xv; untitled document, Mead/LC, S11:8.

18. Mead, *From the South Seas,* xvii.

19. Gerald Sullivan, *Margaret Mead, Gregory Bateson, and Highland Bali: Fieldwork Photographs of Bayung Gedé, 1936–1939* (Chicago: University of Chicago Press 1999), 23; Benedict, *Patterns of Culture,* 237, 46; Mead to Jeannette Mirsky, April 2, 1937, Mead/LC, N5:5.

20. Mead to Dollard, September 23, 1936, Mead/LC, N5:4. See also Bateson and Mead, *Balinese Character,* 255; Mead to Gorer, August 20, 1936, N5:4, Mead to Dollard, September 23, 1936, Mead/LC, N5:4.

21. Mead, *Blackberry Winter,* 230; Report to Clark Wissler, N5:6, Mead to Bernard Mishkin, September 4, 1937, N5:5, Mead to Gorer, August 20, 1936, Mead/LC, N5:4.

22. See Thomas Hobbes, *Leviathan,* ed. C. B. MacPherson (1651; Harmondsworth: Penguin, 1968) 262; John Locke, *The Second Treatise of Government* (1690; Indianapolis: Hackett, 1980); John Dewey, *Human Nature and Conduct,* 10, 303; Margaret M. Caffrey, *Ruth Benedict: Stranger in This Land* (Austin: University of Texas Press, 1989), 186; Howard, *Margaret Mead,* 48; Mead, *Blackberry Winter,* 113; John Patrick Diggins, *The Promise of Pragmatism: Modernism and the Crisis of Knowledge and Authority* (Chicago: University of Chicago Press, 1994).

23. Mead to Erich Fromm, March and June 1937, N5:5, Mead to Dollard, August 23, 1937, Mead/LC, N5:5.

24. Mead to Gorer, August 16, 1937, Mead/LC, N5:5.

25. Mead to Gorer, June 6, 1937, N5:5; untitled diagram, Mead/LC, S11:8; Benedict, *Patterns of Culture,* 46.

26. Gorer, *Bali and Angkor, or Looking at Life and Death* (London: Michael Joseph, 1936) xx; see Aristotle, *Nichomachean Ethics,* trans. Terence Irwin (Indianapolis: Hackett [n.d.]1985).

27. Mead to Robert S. Hale, August 25, 1937, Mead/LC, N5:5; Thorstein Veblen, *The Theory of the Leisure Class: An Economic Study in the Evolution of Institutions* (New York: Macmillan, 1899); "Research Considerations," July 17/36, Mead/LC, N12:2 (Mary Wolfskill and Patricia Francis of the Library of Congress's Manuscript Division verified this point). Bateson, *With a Daughter's Eye,* 132–39; Mead, *Blackberry Winter,* 217–20. Hilary Lapsley, *Margaret Mead and Ruth Benedict: The Kinship of Women* (Amherst: University of Massachusetts Press, 1999), 223; Mead to Robert S. Hale, August 25, 1937, Mead/LC, N5:5.

28. Mead to Fromm, March 28, 1937, Mead/LC, N5:5.

29. "Research Suggestions"; Mead to Gorer, October 1, 1936, Mead/LC, N5:4.

8. The Lady of the Chrysanthemum

1. Yamada Entai, "My Personal History," *Nihon Keizai Shinbun,* December 24, 1992. Most probably the female visitor was Helen Mears, a journalist who visited Japan doing research for her book. Helen Mears, *Mirror for Americans: Japan* (Boston: Houghton Mifflin, 1948).

2. Mears, *Mirror for Americans.*

3. Ibid.

4. Etsu Inagaki Sugimoto, *A Daughter of a Samurai* (New York: Doubleday, Page, 1925); Etsuko Sugimoto, "Bushi no musume no mita America," *Fujin no Tomo-sha,* January 1940.

5. Ruth Benedict to Dr. M. F. Ashley Montagu, May 28, 1943; Benedict to Dr. Bingham Dai, May 28, 1943; Office of War Information to Benedict, June 28, 1943; Benedict to Esther, September 11, 1943; Bjarne Braatoy, Memorandum to Ruth Benedict, January 22, 1944, RFB/VC.

6. *Science Illustrated,* August 1948; Benedict, Report 25: *Japanese Behavior Patterns,* Office of War Information, Washington, D.C., September 15, 1945, RFB/VC; Pauline Kent, "Benedict no Nihon Kenkyu," *Nihonjin no Kodo Pattern* (Tokyo: NHK Books, 1997), 190–93.

7. Pauline Kent, "Ruth Benedict's Original Wartime Study of the Japanese," *International Journal of Japanese Sociology* 3 (1994) 81–97, 82–84; Robert Hashima, "Rusu Benedektuo joshi no Tsuioku" [Reminiscences about Professor Ruth Benedict], *Minzokugaku Kenkyu* 15:1 (1949): 68–69, republished in Peter T. Suzuki, "Ruth Benedict, Robert Hashima, and *The Chrysanthemum and the Sword,*" *Interdisciplinary Anthropology* 3 (1985): 58–62.

8. Mead, "Provisional Analytical Summary of Institute of Pacific Relations Conference on Japanese Character Structure," Relations Conference on Japanese Character Structure, December 16–17, 1944, RFB/VC; Taiheiyo Mondai Chosa-Kai (I.P.R.) no Kenkyu, Kenkyu series 33, Waseda Daigaku Shaki Kagaku Kenkyusho, 1994.

9. Mead, "Provisional Analytical Summary." Quotations in the following paragraphs are from this source.

10. Office of War Information to Benedict, January 1, 1945; Kent, "Benedict no Nihon Kenkyu"; Hashima, "Rusu Benedekuto Joshi no Tsuioku."

11. Benedict, *Japanese Behavior Patterns,* 40–44.

12. Ibid., 40–44; Benedict, "Anthropology and the Abnormal," *Journal of General Psychology* 10(1934): 59–80; Mead, *Anthropologist at Work,* 280.

13. Benedict, *Japanese Behavior Patterns,* 56.

14. Ibid., 32–33.

15. Hashima, "Rusu Benekekuto Joshi no Tsuioku"; Benedict to Ferris Greenslet, Houghton Mifflin, October 22, 1945, RFB/VC.

16. Benedict to Greenslet, November 14, 1945; Greenslet to Benedict, November 16, 1945; Benedict to Greenslet, December 26, 1945; Greenslet to Benedict April 23, 1946; Greenslet to Benedict, April 25, 1946, RFB/VC.

17. Greenslet to Benedict, July 3, 1946; Greenslet to Benedict, July 15, 1946, RFB/VC.

18. Benedict, *Chrysanthemum and the Sword,* 2.

19. Ibid., 295; Benedict, untitled manuscript, RFB/VC.

20. Mead, *Ruth Benedict;* Caffrey, *Ruth Benedict,* translated into Japanese by Nanako Fukui (Kansai University Press, 1993).

9. Ruth Benedict's Obituary for Japanese Culture

This chapter is a revised version of part two of Douglas Lummis, *Uchinaru Gaikoku* (Tokyo: Jiji Tsushinsha, 1981), published in English as *A New Look at* "The Chrysanthemum and the Sword" (Tokyo: Shohakusha, 1982).

1. See Tsurumi Kazuko, "Kiku to Katana: Amerikajin no Mita Nihonteki Dotockukan [*The Chrysanthemum and the Sword:* Japanese morals as seen by an American], *Shiso* (April 1947): 221–24; Tokushu: Rusu Bendikuto Kiku to Katana No Ataerumono, *Minzokugaku Kenkyu* [Japanese Journal of Ethnology] 14 (1949) [Special issue: Proposals from Ruth Benedict's *The Chrysanthemum and the Sword*], esp. the critiques by Watsuji Tetsuro and Yanagita Kunio; C. Douglas Lummis, trans., "Yanagita Kunio's Critique of *The Chrysanthemum and the Sword,*" *Kokusai Kankei Kenkyu* (Tsuda College) 24 (1998): 125–40; J. W. Bennett and M. Nagai, "The Japanese Critique of Methodology of Benedict's *Chrysanthemum and the Sword,*" *American Anthropologist* 55 (1953): 404–11; and Sakuta Keiichi, "Haji no bunka saiko" [A reconsideration of shame cultures], *Shiso no kagaku* 4 (1964).

2. See Freeman, *Margaret Mead and Samoa;* Martin Orans, *Not Even Wrong—Margaret Mead, Derek Freeman, and the Samoans* (Novato, Calif.: Chandler and Sharp, 1996); Benedict, *Patterns of Culture;* Clifford Geertz "Us/Not-Us: Benedict's Travels," in Geertz, *Works and Lives: The Anthropologist as Author* (Stanford, Calif.: Stanford University Press, 1988), 128.

3. Benedict, *Chrysanthemum and the Sword,* 5.

4. Benedict, "German Defeatism at the Beginning of the Fifth Winter of War," Office of War Information, 1943, Box 99, Folder 99.4, RFB/VC, 1, 6; Benedict, *Chrysanthemum and the Sword,* 315.

5. Ibid., 299–300.

6. Lummis, "The Beauty of the World of Death," in Lummis, *A New Look,* 11–28; Benedict, "Anthropology and the Humanities," in Mead, *Anthropologist at Work,* 460; Mead, "A New Preface," in Benedict, *Patterns of Culture* (1959), ix.

7. Mead, *Anthropologist at Work,* 301; Benedict, "The Story of My Life," in Mead, *Anthropologist at Work,* 100.

8. Benedict, "The Story of My Life," 98–99.

9. Benedict, "Resurgam," in Mead, *Anthropologist at Work,* 194.

10. Ibid., xviii; Benedict, *Patterns of Culture,* 21–22.

11. Mead, *Anthropologist at Work,* 5.

12. Ibid., 202, 206.

13. Ibid., 207, 293–303.

14. Ruth Bunzel, "Introduction to Zuni Ceremonialism," *Forty-Seventh Annual Report of the Bureau of American Ethnology to the Secretary of the Smithsonian Institution, 1929–1930* (Washington, D.C.: United States Government Printing Office, 1932), 494, 494fn.

15. Ibid., 293.

16. Ibid., 136.

17. Benedict, "Love That Is Water," in Mead, *Anthropologist at Work,* 474.

18. Benedict, "Countermand," in Mead, *Anthropologist at Work,* 476.

19. Benedict, "Preference," in Mead, *Anthropologist at Work,* 177.

20. Mead, *Anthropologist at Work,* 144.

21. Benedict, *Chrysanthemum and the Sword,* 294, 224; Pauline Kent, "Ruth Benedict's Original Wartime Study of the Japanese," *International Journal of Japanese Sociology* 3 (1994): 92.

22. *The Invention of Tradition,* ed. Eric Hobsbawm and Terence Ranger (Cambridge: Cambridge University Press, 1983). See Lummis, "A Critique of Ruth Benedict's Picture of Japan," in Lummis, *A New Look.*

23. Lummis, "Yanagita Kunio's Critique," 131, 132.

24. C. Douglas Lummis, Robert Hashima Interview, October 16, 1996, Box 104, Folder 4, RFB/VC. Quotations from Hashima in the following paragraphs are from this interview.

25. Benedict, *Chrysanthemum and the Sword,* 296, 295.

26. Ibid., 294, 295.

27. Ibid., 59.

10. The Parable of Manus

1. H. G. Barnett, *Anthropology in Administration* (Evanston, Ill: Row, Peterson, 1956), 49; Mead, "Introduction for the Apollo Edition, 1967," vii–xi, in Mead, *Male and Female* (1967 ed.), viii.

2. Mead, *New Lives,* 43, 498; Mead, *Letters from the Field,* 260, 379, 332, 262.

3. Rodney Needham, Review of *New Lives, American Anthropologist* 59 (June 1957): 262–64; Mead, *New Lives,* 502; Mead, *Letters from the Field,* 320.

4. Mead, *New Lives,* 458.

5. Bateson, *With a Daughter's Eye* (1994 ed.), 175.

6. Mead, *New Lives,* 8, 17, 10.

7. Mead, *New Lives,* 4; Mead, "Problems of the Atomic Age," *The Survey* 85 (July 7, 1949): 385; Mead, *New Lives,* 8; Mead, "Toward More Vivid Utopias," *Science* 126 (November 1957): 957–61.

8. Mead, "Toward More Vivid Utopias," in Mead, *New Lives,* 17.

9. Mead, *New Lives,* 12, 25–27, 255, 291, 37–42, 260–61, 429–33, 250, 341, 379, 405.

10. Michel-Rolph Trouillot, "Anthropology and the Savage Slot: The Poetics and Politics of Otherness," in *Recapturing Anthropology: Working in the Present,* ed. Richard G. Fox (Santa Fe, N.M.: School of American Research Press, 1991), 27; Mead, *New Lives,* xxi; Mead, *Blackberry Winter,* 57; Mead, *New Lives,* 24.

11. Mead, *New Lives,* 388–98, 37–42, 405; Mead, "Preface—1965," in *New Lives,* xi–xv; Mead, *And Keep Your Powder Dry* (1965 ed.), xii, 269–70, 276, 282; Mead, "The Years Between—1943–1965," 263–311.

12. Mead, "Toward More Vivid Utopias," 960; Mead, *New Lives,* 5.

13. Mead, *New Lives,* 277, 7–8; Paul Boyer, *By the Bomb's Early Light: American Thought and Culture at the Dawn of the Atomic Age* (New York: Pantheon, 1985), 33–45.

14. Mead, *New Lives,* 432; see also, 37–42, 260–61, 429–33; 454–55.

15. Richard J. Kozicki, "The United Nations and Colonialism," in *The Idea of Colonialism,* ed. Robert Strausz-Hope and Harry W. Hazard (New York: Frederick A. Praeger, Inc., 1958), 17–18, 42, 362–63, 384; see also Stefan T. Possony, "Colonial Problems in Perspective," 17–43; and Erasmus H. Kloman Jr., "Colonialism and Western policy," 366–82 in *The Idea of Colonialism,* ed. Straus-Hope and Hazard.

16. Lenora Foerstel, "Margaret Mead from a Cultural-Historical Perspective," in Foerstel and Angela Gilliam, eds., *Confronting the Margaret Mead Legacy: Scholarship, Empire, and the South Pacific* (Philadelphia: Temple University Press, 1992), 67, 71; Gilliam, "Leaving a Record for Others: An Interview with Nahau Rooney," 42; Gilliam and Foerstel, "Margaret Mead's Contradictory Legacy," in *Confronting the Margaret Mead Legacy,* 143, 144.

17. Mead, *New Lives,* 75, 100, 165, 168–69, 170, 177, 183, 202, 204–06, 217, 290, 310–12, 448–49, 525; Mead, "Cultural Factors in Community Education Programs," in *Education and Culture: Anthropological Approaches,* ed. George D. Spindler (New York: Holt, Rinehart, and Winston, 1963), 501–3.

18. Theodore Schwartz, quoted in Howard, *Margaret Mead,* 283; Mead, *Culture and Commitment,* 89.

19. Mead, *New Lives,* 165.

20. Ibid., 170, 173.

21. Charles Rowley, *The New Guinea Villager: The Impact of Colonial Rule on Primitive Society and Economy* (New York: Frederick A. Praeger, 1966), 166–67.

22. Mead, *New Lives,* 167, 168, 173, 175.

23. Ibid., 173.

24. Ibid., 168, 170, 178.

25. W. E. Tomasetti, *Australia and the United Nations: New Guinea Trusteeship Issues from 1946 to 1966, New Guinea Research Bulletin No. 36* (Canberra, Australia: Australian National University, 1970), 46.

26. Mead, *New Lives,* 287, 451.

27. Ibid., 442–43.

28. Lola Romanucci-Ross, *Mead's Other Manus: Phenomenology of the Encounter* (South Hadley, Mass.: Bergin and Garvey, 1985), 202–7.

29. Ibid, 361.

30. Ibid., 407, 410, 409, 328.

31. Elaine Tyler May, *Homeward Bound: American Families in the Cold War Era* (New York: Basic Books, 1988), 88–89, 146, 96–97.

32. Mead, *New Lives*, 444–45.

33. Peter M. Worsley, "Margaret Mead: Science or Science Fiction?" *Science and Society* 21 (Spring 1957): 126; Mead, *New Lives*, 23.

Part IV. Echoes and Reverberations

1. Micaela di Leonardo, *Exotics at Home: Anthropologies, Others, American Modernity* (Chicago: University of Chicago, 1998), 184.

2. Edwin McDowell, "New Samoa Book Challenges Margaret Mead's Conclusions," *New York Times*, January 31, 1983; Derek Freeman, *Margaret Mead and Samoa: The Making and Unmaking of an Anthropological Myth* (Cambridge, Mass: Harvard University Press, 1983); Derek Freeman, *The Fateful Hoaxing of Margaret Mead: A Historical Analysis of her Samoan Research* (Boulder, Colo.: Westview, 1999).

3. As an initial introduction, see J. W. Bennett and M. Nagai, "The Japanese Critique of Methodology of Benedict's *Chrysanthemum and the Sword*," *American Anthropologist* 55 (1953): 404–11.

4. Di Leonardo, *Exotics at Home*, 186, 363.

11. Imagining the South Seas

This chapter is a revised version of Sharon Tiffany, "Imagining the South Seas: Thoughts on *Coming of Age* and the Sexual Politics of Paradise," *Pacific Studies* 24, September/December 2001, used with permission of the editor.

Epigraph: Lelei Lelaulu, "Letter to the Editor: The Real Samoa," *Newsweek*, February 28, 1983, 8.

1. Edwin McDowell, "New Samoa Book Challenges Margaret Mead's Conclusions," *New York Times*, January 1983, Mead/LC, A1, C21; Mead, *Coming of Age;* Derek Freeman, *The Fateful Hoaxing of Margaret Mead: A Historical Analysis of her Samoan Research* (Boulder, Colo.: Westview, 1999), 161; Freeman, *Margaret Mead and Samoa: The Making and Unmaking of an Anthropological Myth* (Cambridge, Mass: Harvard University Press, 1983), 228, 288; Freeman, "Paradigms in Collision: The Far-Reaching Controversy over the Samoan Researches of Margaret Mead and Its Significance for the Human Sciences," *Academic Questions* 5 (Summer 1992): 23–33; Freeman, "Preface Addendum, 1984," *Margaret Mead and Samoa*, xvii–xx; Freeman, "Foreword: Mead's Samoa as of 1996," in Freeman, *Margaret Mead and the Heretic: The Making and Unmaking of an Anthropological Myth* (Ringwood, Australia: Penguin, 1996), vi–xiv; Freeman, "Was Margaret Mead Misled or Did She Mislead on Samoa? *Current Anthropology* 41 (August-October 2000): 609–14; E. E. Evans-Pritchard, *Social Anthropology and Other Essays* (New York: Macmillan, 1962), 96; Peter M. Worsley, "Margaret Mead: Science or Science Fiction?" *Science and Society* 21 (Spring 1957); John Leo, "Bursting the South Sea Bubble: Anthropologist Attacks Margaret Mead's Research in Samoa," *Time*, February 14, 1983, 68–70; Richard Bernstein, "Samoa: A Paradise Lost?" *New York Times Magazine*, April 24, 1983, 48–50, 54–60, 66–67; Sharon Begley, John Carey, and Carl Robinson, "In Search of the Real Samoa," *Newsweek*, February 14, 1983, 56; James P. Sterba, "Tropical Storm: New Book Debunking Margaret Mead Dispels Tranquility in Samoa," *Wall Street Journal*, April 14, 1983, 1, 16; Jane Howard, "Angry Storm over the South Seas of Margaret Mead," *Smithsonian Magazine*, April 14, 1983, 66–75. For additional documentation see Tiffany, "Imagining the South Seas."

2. James E. Côté, *Adolescent Storm and Stress: An Evaluation of the Mead-Freeman Controversy* (Hillsdale, N.J.: Lawrence Erlbaum Associates, 1994); Richard Feinberg, "Margaret Mead and Samoa: *Coming of Age* in Fact and Fiction," *American Anthropologist* 90 (1988): 656–63; Nicole J. Grant, "From Margaret Mead's Field Notes: What Counted as 'Sex' in Samoa?" *American Anthropologist* 97 (1995): 678–82; Lowell D. Holmes, *Quest for the Real Samoa;* Jeannette Marie Mageo, "*Malosi:* A Psychological Exploration of Mead's and Freeman's Work and of Samoan Aggression," *Pacific Studies* 11 (1988): 25–65; Martin Orans, *Not Even Wrong;* Martin Orans, "Hoaxing, Polemics, and Science," *Current Anthropology* 41 (August-October 2000): 615–16; Paul Shankman, "Sex, Lies, and Anthropologists: Margaret Mead, Derek Freeman, and Samoa," in *Research Frontiers in Anthropology,* ed. Melvin Ember (New York: Prentice-Hall, 1994), 111–26; Paul Shankman, "The History of Samoan Sexual Conduct and the Mead-Freeman Controversy," *American Anthropologist* 98 (1996): 555–67; Lutkehaus, "Margaret Mead and the 'Wind-Rustling- in-the-Palm-Trees School' of Ethnography," in *Women Writing Culture,* ed. Ruth Behar and Deborah Gordon (Berkeley: University of California Press, 1995), 186–206. For additional documentation, see Tiffany, "Imagining the South Seas."

3. Bronislaw Malinowski, *Sex and Repression in Savage Society* (London: Kegan Paul, 1927); Havelock Ellis, "Preface," Bronislaw Malinowski, *The Sexual Life of Savages in North- Western Melanesia* (1929; New York: Harcourt, Brace & World, Harvest Books, 1962), vii–xiii; letters from Havelock Ellis to William Morrow, August 26, 1928, and Bronislaw Malinowski to William Morrow, August 22, 1928, Mead/LC, S9.

4. Nancy F. Cott, "The Modern Woman of the 1920s, American Style," in *Toward a Cultural Identity in the Twentieth Century,* ed. Francise Thébaud, vol. 5, *A History of Women in the West* (Cambridge: Harvard University Press, 1994), 76–91; Rayna Rapp and Ellen Ross, "The 1920s: Feminism, Consumerism, and Political Backlash in the United States," in *Women in Culture and Politics: A Century of Change,* ed. Judith Friedlander (Bloomington: Indiana University Press, 1986), 52–61.

5. William Morrow to Mead, January 11, 1929; William Morrow to Mead, November 4, 1928, January 11m 1929, Mead/LC, I2; Mead, *Growing Up in New Guinea: A Comparative Study of Primitive Education* (New York: William Morrow, 1930); Mead, *Sex and Temperament;* Mead, *Anthropologists and What They Do,* 122–26.

6. Mead, *Anthropologists and What They Do,* 125, 123.

7. Freeman, "There's Tricks i' th' World: An Historical Analysis of the Samoan Researches of Margaret Mead," *Visual Anthropology Review* 7 (Spring 1991): 118.

8. Sarah Graham-Brown, *Images of Women: The Portrayal of Women in Photography of the Middle East 1860–1950* (New York: Columbia University Press, 1988); and Michael Sturma, *South Sea Maidens: Western Fantasy and Sexual Politics in the South Pacific* (Westport, Conn.: Greenwood Press, 2002).

9. Freeman, *Margaret Mead and Samoa,* 70, 287; Freeman, "There's Tricks i' th' World," 117–18.

10. Mead, "The Arts in Bali," *Yale Review* 30 (December 1940): 336–37; Mead, "1925–1939: 'Lest One Good Custom Should Corrupt the World,'" in Mead, *From the South Seas,* x; Mead, *Anthropologist At Work,* 206–9, 547, n. 24.

11. Louis D. Froelick, "*Moana of the South Seas:* One of Seven Immortals in a Radiant Land of Morning Light," *Asia* 25 (1925): 389–92, 450; Alison Nordström, "Pho-

tography of Samoa: Production, Dissemination, and Use," in Blanton, *Picturing Paradise: Colonial Photography of Samoa, 1875–1925* (Daytona Beach, Fla.: Daytona Beach Community College, 1995), 11–40.

12. John Berger, *Ways of Seeing* (London: British Broadcasting Corporation and Penguin Books, 1972), 45–47; Catherine A. Lutz and Jane L. Collins, *Reading National Geographic* (Chicago: University of Chicago Press, 1993), 187–216.

13. Jane C. Desmond, *Staging Tourism: Bodies on Display from Waikaikai to Sea World* (Chicago: University of Chicago Press, 1999), 11–12; Lutz and Collier, *Reading National Geographic*, 197–201; Alison Devine Nordström, "Early Photography in Samoa: Marketing Stereotypes of Paradise," *History of Photography* 15 (Winter 1991): 273–74; Alison Devine Nordström, "Wood Nymphs and Patriots: Depictions of Samoans in *The National Geographic Magazine*," *Visual Sociology* 7 (1992): 49–59.

14. Lutz and Collins, *Reading National Geographic*, 198; Haunani-Kay Trask, *From a Native Daughter: Colonialism and Sovereignty in Hawai'i*, rev. ed. (Honolulu: Center for Hawaiian Studies, University of Hawai'i Press, 1999), 136–37.

15. Barbara E. Adams, William Morrow and Company, to Mead, April 3, 1967; Karen Graff on behalf of Mead to Ross Claiborne, June 12, 1967, Mead/LC, I2; John D'Emilio and Estelle Freedman, *Intimate Matters: A History of Sexuality in America*, 2nd ed. (New York: Harper & Row, 1988), 301–25.

16. Mead, "The Arts in Bali," 336–37.

17. James A. Michener, *The World Is My Home: A Memoir* (New York: Random House, 1992), 279; John P. Hayes, *James A. Michener: A Biography* (Indianapolis: Bobbs-Merrill, 1984), 74–77; James A. Michener, *Tales of the South Pacific* (New York: Macmillan, 1947), 2, 147–48, 155.

18. Stanley Green, *The Rodgers and Hammerstein Story* (New York: John Day, 1963), 136, 174; Philip D. Beidler, "*South Pacific* and American Remembering; or, 'Josh, We're Going to Buy This Son of a Bitch!'" *Journal of American Studies* 27 (1993): 207–22; Adrian Vickers, *Bali: A Paradise Created* (Berkeley, Calif.: Periplus Editions, 1989); Stanley Green, *The World of Musical Comedy*, 4th ed. (New York: Da Capo Press, 1980), 217–18; John P. Hayes, *James A. Michener*, 77; Margaret Jolly, "From Point Venus to Bali Ha'i: Eroticism and Exoticism in Representations of the Pacific," in *Sites of Desire / Economies of Pleasure: Sexualities in Asia and the Pacific*, ed. Lenore Manderson and Margaret Jolly (Chicago: University of Chicago Press, 1997), 112; Peter Mesenhöller, "Kulturen zwischen Paradies und Hölle: Die Fotografie als Mittler zwischen den Welten," in *Der Geraubte Schatten: Die Photographie als ethnographisches Dokument*, ed. Thomas Theye (München: Münchner Stadtmuseum, Verlag C. J. Bucher, 1989), 350–79. [Peter Mesenhöller, "Cultures Between Paradise and Hell: The Photograph as Mediator Between Worlds," in *The Great Station: Photography as Ethnographic Document*, ed. Thomas Theye (Munich: Munich State Museum, C. J. Bucher, Publisher, 1989), 350–79.]

19. Robert Minton, "Challenging Margaret Mead," *Boston Globe Magazine*, July 17, 1983, 8–9, 14, 16, 20–22, 26.

20. Edmund Leach, "The Shangri-La That Never Was," *New Society*, March 24, 1983, 477–78; George E. Marcus, "One Man's Mead," *New York Times Book Review*, March 27, 1983, 3, 22, 24.

21. Derek Freeman, "Foreword: Mead's Samoa as of 1996," vi–xiv; Peter Mon-

aghan, "Fantasy Island: Review of *Heretic,* by David Williamson," *Lingua Franca,* July-August 1996, 7–8; Helen Thomson, "An Intellectual and Nostalgic Cabaret: Review of *Heretic,* by David Williamson," *The Age* (Sydney, Australia), July 22, 1996, B7.

22. Ruth Behar, "Introduction: Out of Exile," in *Women Writing Culture,* 1–29; Freeman, "There's Tricks i' th' World," 118.

23. Adam Kuper, "Love under the Palm Trees," *The Times Higher Education Supplement,* April 29, 1983, 15.

12. Symbolic Subordination and the Representation of Power in *Margaret Mead and Samoa*

This chapter is a revised version of Angela Gilliam, "Symbolic Subordination and the Representation of Power in *Margaret Mead and Samoa,*" *Visual Anthropology Review* 9 (Spring 1993): 105–15.

1. Frank Heimans, *Margaret Mead and Samoa* (Evanston, Ill., Altschul Group, 1988); Mead, *Coming of Age;* Derek Freeman, *Margaret Mead and Samoa: The Making and Unmaking of an Anthropological Myth* (Cambridge, Mass: Harvard University Press, 1983); Jeannette Marie Mageo, "Malosi: A Psychological Exploration of Mead's and Freeman's Work and of Samoan Aggression," *Pacific Studies* 11 (March 1988): 27; Marianne Hester, *Lewd Women and Wicked Witches: A Study of the Dynamics of Male Domination* (New York: Routledge, 1992), 198.

2. Fox Butterfield, "Study Cites Role of Biological and Genetic Factors in Violence," *New York Times,* November 13, 1992; Philip J. Hilts, "U.S. Puts a Halt to Talks Tying Genes to Crime," *New York Times,* September 5, 1992; David L. Wheeler, "U. of Md. Conference That Critics Charge Might Foster Racism Loses NIH Support," *Chronicle of Higher Education,* September 2, 1992. Derek Freeman, "Inductivism and the Test of Truth," *Canberra Anthropology* 6:2: 101–92.

3. James Clifford, "On Ethnographic Allegory," in *Writing Culture: The Poetics and Politics of Ethnography,* ed. James Clifford and George Marcus (Berkeley: University of California, 1986), 102, 103; Robert F. Berkhofer Jr., *The White Man's Indian: Images of the American Indian from Columbus to the Present* (New York: Vintage Books, 1978).

4. Angela Gilliam and Lenora Foerstel, "Margaret Mead's Contradictory Legacy," in *Confronting the Margaret Mead Legacy: Scholarship, Empire, and the South Pacific,* ed. Foerstel and Gilliam (Philadelphia: Temple University Press, 1992), 55–73.

5. Derek Freeman, "There's Tricks i' th' World: An Historical Analysis of the Samoan Researches of Margaret Mead," *Visual Anthropology Review* 7 (Spring 1991): 114; Adam Kuper, "Coming of Age in Anthropology?" *Nature* 338 (April 1989): 453–55; Lowell Holmes, *Quest for the Real Samoa: The Mead-Freeman Controversy and Beyond* (South Hadley, Mass.: Bergin & Garvey, 1987), 175; Freeman, *Margaret Mead and Samoa,* 225, 278, 301; Freeman, "Inductivism and the Test of Truth."

6. Epeli Hau'ofa, "Anthropology and Pacific Islanders," *Oceania* 45 (1975): R Crocombe, "Anthropology, Anthropologies, and Pacific Islanders," *Oceania* 47 (1975): 65–73.

7. Marvin Harris, *Rise of Anthropological Theory: A History of Theories of Culture* (London: Routledge & Kegan Paul, 1968); Eleanor Leacock, "Anthropologists in Search of a Culture: Margaret Mead, Derek Freeman, and All the Rest of Us," in Foerstel and Gilliam, *Confronting the Margaret Mead Legacy,* 5; George W. Stocking Jr., *Victorian Anthropology* (New York: Free Press, 1987), 327.

8. Clifford, "On Ethnographic Allegory," 102; Gilliam and Foerstel, "Margaret Mead's Contradictory Legacy," 143.

9. Freeman, "There's Tricks i' th' World," 114; Freeman, *Margaret Mead and Samoa,* 114–20, 278, 301; David M. Schneider, "The Coming of a Sage to Samoa," *Natural History* 6 (June 1983): 5–10.

10. Albert Wendt, "Three Faces of Samoa: Mead's, Freeman's and Wendt's," *Pacific Islands Monthly,* April 1983, 69; Tuaopepe Felix S. Wendt, "Book Review Forum: Derek Freeman, *Margaret Mead and Samoa: The Making and Unmaking of an Anthropological Myth,*" *Pacific Studies* 7 (Spring 1984): 91–140, 98–99.

11. Anne Beth Fischel, "The Politics of Inscription in Documentary Film and Photography," Ph.D. dissertation, University of Massachusetts, 1992.

12. Catherine A. Lutz and Jane L. Collins, *Reading National Geographic* (Chicago: University of Chicago Press, 1993), 119–85, 188–92.

13. Wendt, "Book Review," 99, 98.

14. Lelei Lelaulu to Gilliam, personal communication, April 29, 1986.

15. Jane Gaines, "White Privilege and Looking Relations: Race and Gender in Feminist Film Theory," in *Issues in Feminist Film Criticism,* ed. Patricia Erens (Bloomington: Indiana University Press, 1990), 200; E. Ann Kaplan, *Women and Film: Both Sides of the Camera* (New York: Methuen, 1983); Laura Mulvey, "Visual Pleasure and Narrative Cinema," in *Issues in Feminist Film Criticism,* ed. Patricia Erens (Bloomington: Indiana University Press, 1990), 33; Peter M. Worsley, "Margaret Mead: Science or Science Fiction?" *Science and Society* 21 (Spring 1957): 122–134; Robert Stam, Robert Burgoyne, and Sandy Flitterman-Lewis, *New Vocabularies in Film Semiotics: Structuralism, Post-Structuralism, and Beyond* (New York: Routledge, 1992), 174.

16. Fischel, "The Politics of Inscription," 123.

17. Christian Hansen, Catherine Needham, and Bill Nichols, "Pornography, Ethnography, and the Discourses of Power," in *Representing Reality,* ed. Bill Nichols (Bloomington: Indiana University Press, 1991), 207; Diane Mei Lin Mark, "The Reel Hawaii," in *Moving the Image: Independent Asian Pacific American Media Arts,* ed. Robert Leong (Los Angeles: UCLA Asian American Studies Center and Visual Communications, 1991), 109.

18. Wendt, "Three Faces of Samoa," 69; Pascal Bruckner, *The Tears of the White Man: Compassion as Contempt* (New York: Free Press, 1983), 114; Gilliam and Foerstel, "Margaret Mead's Contradictory Legacy," 110–12, 142–44.

19. Clyde Taylor to Angela Gilliam, personal communication, July 22, 1992.

20. Freeman, "There's Tricks i' th' World"; Stam, personal communication to Gilliam, July 24, 1992; Pauline Marie Rosenau, *Post-Modernism and the Social Sciences: Insights, Inroads, and Intrusions* (Princeton, N.J.: Princeton University Press, 1992), 119; Gilliam, "Papua New Guinea and the Geopolitics of Knowledge Production," in Foerstel and Gilliam, *Confronting the Margaret Mead Legacy,* 267–98, 271–75; Glenn Jordan, "On Ethnography in an Intertextual Situation: Reading Narratives or Deconstructing Discourse?" in *Decolonizing Anthropology: Moving Further Toward an Anthropology of Liberation,* ed. Faye Harrison (Washington, D.C.: Association of Black Anthropologists/American Anthropological Association, 1991), 59.

21. Clyde Taylor, "We Don't Need Another Hero: Anti-Theses on Aesthetics," in *Blackframes: Critical Perspectives on Black Independent Cinema,* ed. Mbye B. Cham and

Claire Andrade-Watkins (Boston: MIT Press, 1990); Adam Kuper, "Coming of Age in Anthropology?" *Visual Anthropology Review* 6 (Spring 1990): 22–25.

22. Trinh T. Minh-ha, *Woman, Native, Other* (Bloomington: Indiana University Press, 1989); J. Stephen Lansing, "The Decolonization of Ethnographic Film," *Visual Anthropology Review* (Spring 1990): 13.

23. Mariana Torgovnick, *Gone Primitive: Savage Intellects, Modern Lives* (Chicago: University of Chicago Press, 1990), 242.

24. Leacock, "Anthropologists in Search of a Culture," 26.

13. Misconceived Configurations of Ruth Benedict

This chapter is a revised version of Pauline Kent, "Misconceived Configurations of Ruth Benedict," *Japan Review* 7 (1996): 33–60. Used with permission of the editor of *Japan Review*, the International Research Center for Japanese Studies, Kyoto.

1. Ruth Benedict, *Chrysanthemum and the Sword* (Boston: Houghton Mifflin, 1946); Douglas Lummis, *A New Look at* "The Chrysanthemum and the Sword," ed. Masayuki Ikeda (Tokyo: Shohakusha, 1982), 2.

2. Margaret Mead, *Anthropologist at Work;* Judith Schachter Modell, *Ruth Benedict: Patterns of Life* (Philadelphia: University of Pennsylvania Press, 1983); Margaret M. Caffrey, *Ruth Benedict: Stranger in This Land* (Austin: University of Texas Press, 1989); Hilary Lapsley, *Margaret Mead and Ruth Benedict: The Kinship of Women* (Amherst: University of Massachusetts Press, 1999).

3. C. Douglas Lummis, *Nihon bunka he no bohimei: Kiku to katana saiko,* [An Obituary for Japanese Culture, A Reassessment of "The Chrysanthemum and the Sword"], trans. Kaji Etsuko, *Shiso no kagaku* 6, part 1, 104 (March 1979): 86–99; part 2, 103 (April 1979): part 3, 105 (1979): 85–99; Lummis, *Uchi naru gaikoku Kiku to Katana saiko* [A Foreign Country Within: A Reconsideration of "The Chrysanthemum and the Sword"], trans. Kaji Etsuko (Tokyo: Jiji Tsushinsha, 1981); Lummis, *A New Look at* "The Chrysanthemum and the Sword." For praise of Lummis, see Ikeda Masayuki in Ikeda Masayuki and C. Douglas Lummis, *Nihonjinron no shinso: hikaku bunka no otoshiana to kanosei* [The depths of Nihonjinron: The pitfalls and potential of comparative culture] (Tokyo: Haru Shobo, 1985); Soeda Yoshiya, *Nihonbunka shiron: Benedikuto Kiku to Katana wo yomu* [Essays on Japanese culture: Reading Benedict's *The Chrysanthemum and the Sword*] (Tokyo: Shinyosha, 1993; Nishikawa Nagao, "Chikyu jidai no minzoku / bunka riron" [An ethnic / cultural theory for a global era], *Ritsumeikan kokusai kenkyu* 6 (1994): 158–76.

4. Ross Mouer and Yoshio Sugimoto, *Images of Japanese Society: A Study in the Social Construction of Reality* (London: Kegan Paul International, 1986), 62.

5. Tsurumi Kazuko, "Kiku to katana: Amerikajin no mita Nihonteki dotokukan" [*The Chrysanthemum and the Sword:* Japanese morals as seen by an American] *Shiso* (April, 1947): 221–24; *Tokushu: Rusu Benedikuto Kiku to katana no ataerumono* [*Special Issue: Proposals from Ruth Benedict's* The Chrysanthemum and the Sword] *Minzokugaku Kenkyu* [*Japanese Journal of Ethnology*] 14 (1949); Sakuta Keiichi, "Haji no bunka saiko" [A reconsideration of shame culture], *Shiso no kagaku* 4 (1964): 2–11; Shimada Hiromi, "Haji no bunka toshite no Nihon: Kiku to Katana e no hanpatsu to juyo" [Japan as a shame culture: Resistance and acceptance towards *The Chrysanthemum and the Sword*], *Nihon toiu moso* [Delusion: Japan] (Tokyo: Nihonhyoron-

sha, 1994), 40–53; Soeda, *Nihonbunka shiron;* Lapsley, *Margaret Mead and Ruth Benedict,* 232, quoted from Margaret Mead in *Ruth Benedict: A Memorial,* 25–26. See Pauline Kent, "An Appendix to *The Chrysanthemum and the Sword:* A Bibliography," *Japan Review* 6 (1995): 107–25.

6. Lummis, *A New Look,* 2, 11–28, 59–60, 76–78

7. Ibid.

8. Benedict, "Anthropology and the Humanities," *American Anthropologist* 50 (1948): 585; Lummis, *A New Look,* 14; Mead, *Anthropologist at Work,* 460.

9. Marshall Hyatt, *Franz Boas, Social Activist* (New York: Greenwood Press, 1990), x; Arnold Krupat, *Ethnocriticism: Ethnography, History, Literature* (Berkeley: University of California Press, 1992), 72.

10. Krupat, *Ethnocriticism,* 65, 67.

11. Ibid., 60, 73.

12. Mead, *Anthropologist at Work.*

13. Ibid., 97–112; Lummis, *A New Look,* 23.

14. Benedict, "Story of My Life," Mead, *Anthropologist at Work.;* Lummis, *A New Look,* 22, 23., 27, 59, 55; Caffrey, *Ruth Benedict;* Nakano Takashi, "Rekishiteki genjitsu no saikosei" [The reconstruction of historical facts], in *Raifu hisutori no shakaigaku* [The sociology of life history], ed. Nakano Takashi and Sakurai Atushi (Tokyo: Kobundo, 1995), 191–218, 201–7. See Helene Bowen Raddeker, "The Past Through Telescopic Sights—Reading the Prison-Life-Story of Kaneko Fumiko" *Japan Forum* 7 (1995): 156–69, 157–8.

15. Caffrey, *Ruth Benedict,* 1–5; Lapsley, *Margaret Mead and Ruth Benedict,* 261–62.

16. Caffrey, *Ruth Benedict,* vii, viii.

17. Ibid., 13.

18. Marvin Harris, *Rise of Anthropological Theory: A History of Theories of Culture* (London: Routledge & Kegan Paul, 1968), 313; Lummis, *A New Look,* 49–50.

19. Lummis, *A New Look,* 65; Kato Shuichi, ed., *Haabat Noman: Hito to Gyoseki* [*Herbert Norman: The man and his works*] (Tokyo: Iwanami Shoten, 2002). Nishi Yoshiyuki, *Shin* "Kiku to Katana" *no yomikata* [A New Reading of *The Chrysanthemum and the Sword*], (Tokyo: PHP Books, 1983); *Watsuji Tetsuro's Collected Works,* 20 Vols. (Tokyo: Iwanami Shoten, 1961–78). Soeda, *Nihonbunka shiron,* 77–78.

20. Mead, *Anthropologist at Work,* xix.

21. Caffrey, *Ruth Benedict,* 400, 401.

22. Lois Banner, "Gender and the Making of Margaret Mead as an Intellectual," American Anthropology Association Annual Meeting, November 29, 2001, Washington, D.C.; Lapsley, *Margaret Mead and Ruth Benedict,* 27, 262; Mead, *Blackberry Winter,* 117–18, 83–96.

23. Mead, *Anthropologist at Work,* 55; Modell, *Ruth Benedict,* 282.

24. Ibid., 304.

25. Ibid., 259–61.

26. Ibid., 289.

27. Caffrey, *Ruth Benedict,* 9–12.

28. Abraham H. Maslow and John J. Honigmann, "Synergy: Some Notes of Ruth Benedict, *American Anthropologist* 72 (1970): 320–33; and Papers I, III, and VI of The Anna Shaw Lectures, 1941, RFB/VC, 326.

29. Ruth Benedict, "Anthropology and the Social Bases of Morale," Anna Shaw Lecture VI, 17 March 1941, RBF/VC.

30. Clifford Geertz, "Us/Not-Us: Benedict's Travels," in Geertz, *Works and Lives: The Anthropologist as* Author (Stanford, Calif.: Stanford University Press, 1988), 120–21, 116; Geoffrey Gorer, "Themes in Japanese Culture," in *Personal Character and Cultural Milieu,* ed. Douglas Haring (New York: Syracuse University Press, 1948), 273–90; Weston La Barre, "Some Observations on Character Structure in the Orient: The Japanese," *Psychiatry,* 8 (1945): 319–42; John Embree, *The Japanese, War Background Studies,* no. 7 (Washington, D.C.: Smithsonian Institution, 1943); John Embree, "A Note on Ethnocentrism in Anthropology," *American Anthropologist* 52 (1950): 430–32; John Dower, *War Without Mercy: Race and Power in the Pacific War* (New York: Pantheon Books, 1986), 118–46.

31. Adrian Pinnington, "Yoshimitsu, Benedict, Endo: Guilt, Shame and the Postwar Idea of Japan," *Japan Forum* 13 (2001): 91–105.

Part V. Re-Thinking Benedict and Mead

1. Roy A. Rappaport, "Desecrating the Holy Woman: Derek Freeman's Attack on Margaret Mead," *American Scholar* 55 (1986): 319; Mead, *Blackberry Winter,* 333.

2. Mead, *Anthropologist at Work,* 143; Judith Schachter Modell, *Ruth Benedict: Patterns of Life* (Philadelphia: University of Pennsylvania Press, 1983), 103

14. Margaret Mead: Anthropology's Liminal Figure

1. Mead, *Blackberry Winter* draft, "Vicissitudes of Public Lives," February 2, 1971, Mead/LC.

2. Victor Turner, *The Ritual Process* (1969; Ithaca, N.Y.: Cornell University Press, 1977); Bateson, *With a Daughter's Eye;* Jane Howard, *Margaret Mead: A Life* (New York: Simon and Schuster, 1984); Hilary Lapsley, *Margaret Mead and Ruth Benedict: The Kinship of Women* (Amherst: University of Massachusetts Press, 1999); Lois W. Banner, *Intertwined Lives: Margaret Mead, Ruth Benedict, and Their Circle* (New York: Alfred A. Knopf, 2003).

3. Marvin Harris, *Rise of Anthropological Theory: A History of Theories of Culture* (London: Routledge & Kegan Paul, 1968); Stanley Diamond, *In Search of the Primitive: A Critique of Civilization* (New Brunswick, N.J.: Rutgers University Press, 1974), 110–11.

4. Russell Jacoby, *The Last Intellectuals: American Culture in the Age of Academe* (New York: Basic Books, 1987), 5.

5. Geoffrey Gorer and J. Rickman, *The People of Great Russia* (London: Cresset, 1949); William Beeman, "Margaret Mead, Cultural Studies and International Understanding," in *The Study of Culture at a Distance,* ed. Mead and Rhoda Metraux (Chicago 1949; New York and Oxford: Berghahn, 2000), xiv–xxxi.

6. Clifford Geertz, *Works and Lives: The Anthropologist as Author* (Stanford, Calif.: Stanford University Press, 1988), 111.

7. Lutkehaus, "Margaret Mead and the 'Wind-Rustling-in-the-Palm-Trees' School of Ethnology," in *Women Writing Culture,* ed. Ruth Behar and Deborah Gordon (Berkeley: University of California Press, 1995), 186–206.

8. Peter M. Worsley, "Margaret Mead: Science or Science Fiction?" *Science and Society* 21 (Spring 1957); Geertz, *Works and Lives,* 102–27.

9. Mead and Rhoda Metraux, *A Way of Seeing* (1970; New York: William Morrow, 1974); Alice B. Kehoe, Review of Mead and Metraux, *A Way of Seeing, American Sociological Review* (1971): 538–39.

10. Paul Bohannan, review of Mead and Metraux, *A Way of Seeing, American Ethnologist* (1980): 198.

11. Mead, "Visual Anthropology in a Discipline of Words," in *Principles of Visual Anthropology,* 2nd ed., ed. Paul Hockings (Berlin and New York: Mouton de Gruyter, 1995), 3–10; Ira Jacknis, "Margaret Mead and Gregory Bateson in Bali: Their Use of Photography and Film," *Cultural Anthropology* 3 (May 1988): 160–77; Andrew Lakoff, "Freezing Time: Margaret Mead's Diagnostic Photography," *Visual Anthropology Review* 12 (Spring 1996): 1–18; Alison Griffith, *Wondrous Differences: Cinema, Anthropology, and Turn-of-the-Century Visual Culture* (New York: Columbia University Press, 2001), xx; Harris, *Rise of Anthropological Theory,* 417.

12. Lenora Foerstel and Angela Gilliam, eds., *Confronting the Margaret Mead Legacy: Scholarship, Empire, and the South Pacific* (Philadelphia: Temple University Press, 1992); Eric Watkins, *Anthropology Goes to War* (Madison: University of Wisconsin Press, 1992); *New York Times,* December 28, 1971; Mead, "Being an Anthropologist in the 1970s," *Blackberry Winter* draft, Mead, LC.

13. Nancy Lutkehaus, "She Was Very Cambridge: Camilla Wedgwood and the History of Women in British Anthropology," *American Ethnologist* 13 (November 1986): 776–99; Robert Levine, n.d., "An Anthropology of Childhood: Re-Examining Margaret Mead's Approach to Child Development," paper presented in The Legacy of Margaret Mead lecture series, Brown University, 2001–2, 1; Lawrence A. Hirschfeld, "Why Don't Anthropologists Like Children?" *American Anthropologist* 104 (June 2002): 611–27; Worsley, "Margaret Mead: Science or Science Fiction?" 123.

14. Mead, "The Vicissitudes of Public Lives." I will deal with celebrity and its dialectical relationship with the intellectual and the professor in my forthcoming book, *Margaret Mead and the Media.* See also Daniel Boorstin, *The Image: A Guide to Pseudo-Events in America* (1961; New York: Vintage Books, 1992); Leo Braudy, *The Frenzy of Renown: Fame and Its History* (New York: Oxford, 1986); Neal Gabler, *Life the Movie: How Entertainment Conquered Reality* (1998; New York: Vintage Books, 2000); Richard Schickel, *Intimate Strangers* (New York: Doubleday, 1985); Jacoby, *The Last Intellectuals;* Bruce Robbins, *Secular Vocations: Intellectuals, Professionals, Culture* (New York: Verso, 1993); *Intellectuals: Aesthetics, Politics, Academics,* ed. Bruce Robbins (Minneapolis: University of Minnesota Press, 1990); Edward W. Said, *Representations of the Intellectual: The 1993 Reith Lectures* (New York: Vintage Books, 1994); Dana Polan, "Professors," *Discourse* 16 (fall 1993): 28–49; Polan, "The Professor of History," in *The Persistence of History: Film and the Historians,* ed. Vivian Sobchack (New York: Routledge, 1996), 236–56; and Andrew Ross, *No Respect: Intellectuals and Popular Culture* (New York: Routledge, 1989.)

15. Robert Foster, "Margaret Mead's *Redbook* Project: A Problem in the Sociology of Culture," unpublished paper, 1982; Jeremy MacClancy, Introduction, in *Popularizing Anthropology,* ed. Jeremy MacClancy and Chris McDonagh (London: Routledge, 1996), 1–57.

16. Claude Lévi-Strauss, *Totemism,* trans. Rodney Needham (Boston: Beacon, 1963); Polan, "Professors," 80–81.

17. Derek Freeman, *The Fateful Hoaxing of Margaret Mead: A Historical Analysis of her Samoan Research* (Boulder, Colo.: Westview, 1999); Paul Shankman, "Mead's Other Samoa," paper presented at American Anthropological Association, Washington D.C., November 29, 2001.

18. Richard Hofstadter, *Anti-Intellectualism in American Life* (New York: Knopf, 1963); *Looking High and Low: Art and Cultural Identity*, ed. Brenda Jo Bright and Liza Bakewell (Tucson: University of Arizona Press, 1995); Lawrence W. Levine, *Highbrow/Lowbrow: The Emergence of Cultural Hierarchy in America* (Cambridge, Mass.: Harvard University Press, 1988); Kirk Varnadoe and Adam Gropnik, *High and Low: Modern Art and Popular Culture* (New York: Museum of Modern Art, 1990).

19. Levine, *Highbrow/Lowbrow*, 223.

20. Andreas Huyssen, "Mass Culture as Woman: Modernism's Other," in *Studies in Entertainment: Critical Approaches to Mass Culture*, ed. Tania Modleski (Bloomington: Indiana University Press, 1986), 188–208.

21. Henrika Kuklik, *The Savage Within: The Social History of British Anthropology, 1885–1945* (Cambridge: Cambridge University Press, 1991).

22. Donna Haraway, *Primate Visions: Gender, Race, and Nature in the World of Modern Science* (London: Routledge, 1989), 26–58; Roy Grinker, *In the Arms of Africa: The Life of Colin M. Turnbull* (New York: St. Martin's Press, 2000).

23. Ross, *No Respect: Intellectuals and Popular Culture*, 102–134.

24. Rena Lederman, "Unchosen Grounds: Cultivating Subfield Affinities for a Public Presence," in *Unwrapping the Sacred Bundle: Reconfiguring the Discipline of Anthropology*, ed. D. Segal and S. Yanagisako (Durham: Duke University Press, 2004); C. P. Snow, *The Two Cultures* (1959; Cambridge: Cambridge University Press, 1993).

25. Harris, *Rise of Anthropological Theory;* Napoleon Chagnon, *Yanomamo*, 4th ed. (Fort Worth, Tex.: Harcourt Brace Jovanovich, 1992) and Brian Ferguson, *Yanomami Warfare: A Political History* (Santa Fe, N.M.: School of American Research Press, 1995).

26. Worsley, "Margaret Mead: Science or Science Fiction," 125; Micaela di Leonardo, *Exotics at Home: Anthropologies, Others, American Modernity* (Chicago: University of Chicago, 1998).

27. *Blackberry Winter* draft, Mead/LC.

28. Di Leonardo, *Exotics at Home.*

29. Polan, "Professors," 28.

30. MacClancy, *Popularizing Anthropology*, 2–3.

31. Martin Jay, "The Academic Woman as Performance Artist," *Salmagundi* 98–99 (1993): 28–34.

32. Lutkehaus, "Margaret Mead," 202.

33. Daniel Boorstin, *The Image: A Guide to Pseudo-Events in America* (New York: Vintage, 1992).

34. Nancy Lutkehaus, "At the Museum: Margaret Mead, American Icon," *Natural History* 110 (December 2001): 14–15. See unpublished Letter to the Editor of *Natural History* magazine, December 2001.

15. "It is besides a pleasant English word"

1. Benedict to Ferris Greenslet, Houghton Mifflin, November 1933, Correspondence, Series I, RFB/VC.

2. Benedict, *Patterns of Culture,* 46, 48.

3. George W. Stocking Jr., *Race, Culture and Evolution: Essays in the History of Anthropology* (New York: Free Press, 1967); Mead, *Anthropologist at Work.*

4. George W. Stocking Jr., ed., *Malinowski, Rivers, Benedict and Others: Essays on Culture and Personality* (Madison: University of Wisconsin Press, 1986).

5. Margaret M. Caffrey, *Ruth Benedict: Stranger in This Land* (Austin: University of Texas Press, 1989); Judith Schachter Modell, *Ruth Benedict: Patterns of Life* (Philadelphia: University of Pennsylvania Press, 1983); Mead, *Anthropologist at Work.*

6. Edward Sapir to Benedict, June 14, 1925, in Mead, *Anthropologist at Work,* 180.

7. Modell, *Ruth Benedict,* 172.

8. Benedict, "Fragment," December 1930, RFB/VC; Benedict, "The Story of My Life," 1935, in Mead, *Anthropologist at Work,* 97–112, 99.

9. Benedict to Ferris Greenslet, Houghton Mifflin, November 1933, Correspondence, Series I, RFB/VC.

10. Benedict to Franz Boas, June 13, 1933, in Mead, *Anthropologist at Work,* 410–11.

11. Sapir to Benedict, June 14, 1925, in Mead, *Anthropologist at Work,* 180.

12. Benedict, "The Science of Custom," *Century Magazine* 117 (1929): 649, 648.

13. Benedict, "Psychological Types in the Cultures of the Southwest," *Proceedings of the Twenty-third International Congress of Americanists, September 1928* (New York: 1930), 572– 81; Mead, *Anthropologist at Work,* 249, 253, 261.

14. Benedict, "Configurations of Culture in North America," *American Anthropologist* 34 (1932): 1–17; Mead, *Ruth Benedict,* 104.

15. Ibid., 83.

16. Benedict to Mead, August 10, 1932, in Mead, *Anthropologist at Work,* 322; Benedict, Copy for *Patterns of Culture,* 1934, Correspondence, Series I, RFB/VC.

17. Benedict to Mead, October 9, 1932, in *Anthropologist at Work,* 323–24, 323.

18. Benedict to Mead, January 20, 1932, in Mead, *Anthropologist at Work,* 318.

19. Virginia Woolf, 1922, *The Letters of Virginia Woolf* (London: Hogarth, 1976), 175; Leon Edel, *Bloomsbury: A House of Lions* (New York: Lippincott, 1979), 159; Modell, *Ruth Benedict,* 25–26.

20. Roger Fry, "Art and Life" (1917), in Fry, *Vision and Design* (1920; Harmondsworth: Penguin, 1937), 19; Fry, *Transformations: Critical and Speculative Essays* (London: Chatto and Windus, 1926).

21. Benedict, *Patterns of Culture,* 48, 49.

22. Wilhelm Worringer, *Form in Gothic* (London: Tiranti, 1927), 11, 88.

23. Alfred Kroeber, "Review of *Patterns of Culture,*" *American* Anthropologist 37 (2935): 689.

24. Kroeber, *Anthropology* (London: George G. Harrop, 1923), 54.

25. Roger Fry, "The Artist's Vision" (1919), in Fry, *Vision and Design,* 49, 50; Benedict, *Patterns of Culture,* 229; Benedict, *Chrysanthemum and the Sword.*

26. Roger Fry, *Cézanne: A Study of His Development* (London: Hogarth Press, 1927), 18, 22.

27. Virginia Woolf, "Mr. Bennett and Mrs. Brown," in *The Essays of Virginia Woof,* ed. Andrew McNeillie (London: Hogarth Press, 1988), 3:384–92.

28. Roger Fry, 1926, Preface to *Transformations*; Fry, "Art and Life," 19.

29. Fry, "An Essay in Aesthetics" (1909), in Fry, *Vision and Design,* 22–40, 33.

30. Fry, "Art and Science" (1919), in Fry, *Vision and Design,* 73.
31. Sapir to Benedict, June 14, 1925, in Mead, *Anthropologist at Work,* 180.
32. Roger Fry, "El Greco" (1920), in Fry, *Vision and Design,* 170.
33. Benedict, *Patterns of Culture,* 258.
34. Ibid., 243.

16. On the Political Anatomy of Mead-bashing, or Re-thinking Margaret Mead

I wish to acknowledge the support of National Science Foundation Grant SBR 9515-387 for assistance in developing this chapter. I also thank Herbert S. Lewis for sharing his knowledge of anthropology past and present with me, which has substantially increased my understanding of the issues considered in this essay, though he bears no responsibility for my errors.

1. Leach, "Acting in Inverted Commas," *Times Literary Supplement,* October 4, 1974, quoted in I. C. Jarvie, *Thinking About Society: Theory and Practice* (Dordrecht, Holland: D. Reidel, 1986), 197.

2. Freeman's *Margaret Mead and Samoa* formulated a Popperian critique of Mead's Samoan fieldwork that led to a more general critique of American cultural anthropology. See *The Samoa Reader: Anthropologists Take Stock,* ed. Hiram Caton (Lanham, Md.: University Press of America, 1990). I am aware that there are as many postmodernisms as there are postmodernists. Nonetheless, postmodernists constitute an academic "school" sharing common concerns. See Pauline Marie Rosenau, *Postmodernism and the Social Sciences: Insights, Inroads, and Intrusions* (Princeton, N.J.: Princeton University Press, 1992).

3. For criticisms of the critics, see, for example, Nancy McDowell, "The Oceanic Ethnography of Margaret Mead, " *American Anthropologist* 82 (1980): 278–302; James Côté, *Adolescent Storm and Stress: An Evaluation of the Mead-Freeman Controversy* (Hillsdale, N.J.: Lawrence Erlbaum, 1994); Nicole J. Grant, "From Margaret Mead's Footnotes: What Counted as 'Sex' in Samoa," *American Anthropologist* 97 (1995): 678–82; and Martin Orens, *Not Even Wrong: Margaret Mead, Derek Freeman, and the Samoans* (Novato, Calif.: Chandler & Sharp, 1996), 96–97.

4. Lawrence E. Klein, "Enlightenment as Conversation," in *What's Left of the Enlightenment: A Postmodern Question* (Stanford, Calif.: Stanford University Press, 2001), 163, 148–49. On the oppositional stance of postmodernists, see also Herbert Klein, "Righting Wrongs (*sic*) and Wronging (*sic*) Wright," *Anthropology Today* 14 (April 1988): 24.

5. Klein, "Enlightenment," 148, 156.

6. See my film *Margaret Mead: An Observer Observed,* Filmmaker's Library, 1997, for this point in the context of Mead's 1930s Bali and New Guinea fieldwork. The Margaret Mead archives at the Library of Congress, Washington, D.C., contain perhaps the best demonstration of her awareness of the anthropologist's subjectivity. See also Gregory Bateson, *Naven* (Stanford, Calif.: Stanford University Press, 1936). Mead did extensive fieldwork with Bateson in New Guinea and Bali in the 1930s. William Beeman's Introduction to a new edition of *The Study of Culture at a Distance,* ed. Mead and Rhoda Metraux (New York: Berghahn Books, 2000), points to postmodern tendencies in this book.

7. On the pragmatist tradition, see *Rorty and His Critics,* ed. Robert Brandon (Ox-

ford: Blackwell, 2000); and Rorty, *The Revival of Pragmatism: Essays on Social Thought, Law, and Culture* (Durham, N.C.: Duke University Press, 1998).

8. David A. Hollinger, "Science as a Weapon in *Kulturkampf* in the United States During and After World War II," in his *Science, Jews, and Secular Culture: Studies in Mid-Twentieth-Century American Intellectual History* (Princeton, N.J.: Princeton University Press, 1996), 171, 166.

9. This is a major theme of Foerstel and Gilliam, *Confronting the Margaret Mead Legacy: Scholarship, Empire, and the South Pacific* (Philadelphia: Temple University Press, 1992), which will be discussed extensively below. On the complexity of the relationship between scholars and "natives," see Sue-Ellen Jacobs, "Unraveling the Threads of Oppression: Notes on the Development of Ethical and Humanistic Anthropology," in *Crisis in Anthropology: View from Spring Hill 1980,* ed. E. Adamson Hoebel et al. (New York: Garland, 1982), 379–404; and George W. Stocking, "Anthropology in Crisis? A View from Between the Generations," ibid., 407–19.

10. Madan Sarup, *An Introductory Guide to Post-Structuralism and Postmodernism* (Athens: University of Georgia Press), quoting Terry Eagleton, *Literary Theory* (Oxford: Basil Blackwell 1983), 142.

11. Russell Jacoby, *The End of Utopia: Politics and Culture in an Age of Apathy* (New York: Basic Books, 1999), 13, 136, contains a critique of intellectual discourse for neglecting a social agenda. On the ethics of contemporary anthropology, see Philip Carl Salzman, "On Reflexivity," *American Anthropologist* 104 (1992): 805–13.

12. Rorty, "Response to Jacques Bouveresse," in Grandon, *Rorty and His Critics,* 153.

13. Habermas, quoted in Rosenau, *Postmodernism,* 123. See also Rosenau on right-wing politics, 122–23, 159–64.

14. See Hollinger, *Science, Jews, and Secular Culture,* chaps. 5, 6, 8; and Herbert S. Lewis, "The Passion of Franz Boas," *American Anthropologist* 103 (2001): 447–67.

15. Herbert Lewis, "The Misrepresentation of Anthropology and Its Consequences," *American Anthropologist* 100 (1998): 716–23. Rosenau, *Postmodernism,* notes German influences, particularly Nietzsche, on postmodernism.

16. See a discussion of Bauman in Stephen T. Leonard, Introduction to *Intellectuals and Public Life: Between Radicalism and Reform,* ed. Leon Fink, Stephen T. Leonard, and Donald M. Reid (Ithaca, N.Y.: Cornell University Press, 1996), 17.

17. See Dell Hymes, *Reinventing Anthropology* (New York: Pantheon Books, 1969), especially Eric R. Wolf, "American Anthropologists and American Society," 251–63, and Laura Nader, "Up the Anthropologist—Perspectives Gained from Studying Up," 284–311.

18. Micaela di Leonardo, *Exotics at Home: Anthropologies, Others, American Modernity* (Chicago: University of Chicago, 1998).

19. Wolf, "American Anthropologists"; di Leonardo, 261, 362f.

20. Norman Markowitz, personal communication to author, January 14, 2003. See also Markowitz, *The Rise and Fall of the People's Century: Henry A. Wallace and American Liberalism, 1941–1948* (New York: Free Press, 1973).

21. Di Leonardo, *Exotics at Home,* 20, expresses her debt to Wolf. Other works by these Marxists include: Wolf, *Peasant Wars of the Twentieth Century* (New York: Harper & Row, 1969); and Sidney Mintz, *Worker in the Cane: A Puerto Rican Life History* (New Haven, Conn.: Yale University Press, 1960).

22. Eleanor Leacock, "Anthropologists in Search of a Culture: Margaret Mead, Derek Freeman, and all the Rest of Us," in Foerstel and Gilliam, *Confronting the Margaret Mead Legacy,* 12, 19. Mead was interested in social structural issues as well as culture. See, for example, her *Social Organization of Manu'a.*

23. Marvin Harris, "Margaret and the Giant Killer," in Caton, *Samoa Reader,* 236. For her typical acknowledgment of scholars working outside her cultural anthropology, see Mead, "Some Cultural Anthropological Responses to Technical Assistance Experience," *Information sur Les Sciences Sociale* IX (December 1970): 49–59.

24. Judith Buber Agassi, "Epistemological and Methodological Concerns of Feminist Social Scientists," in *Critical Rationalism, The Social Sciences, and the Humanities: Essays for Joseph Agassi,* ed. I. C. Jarvie II (Dordrecht: Kluwer, 1995), 153–63, ably discusses some of the issues.

25. Di Leonardo, *Exotics at Home,* 95, 260, 190–93.

26. My discussion of Newman is based on both her essay and her book *White Women's Rights: The Origins of Racial Feminism in America* (New York: Oxford University Press, 1999). Newman's fn. 31, p. 266, in her essay claims that Mead believed in the superiority of Western culture but offers no proof. She quotes a reviewer of one of Mead's books, but the quote does not address Newman's claim. In her essay, Newman cites one archival source, a letter from Mead's published collection of her field letters (see 269, fn. 55).

27. Newman, "Coming of Age, But Not in Samoa: Reflections on Margaret Mead's Legacy for Western Liberal Feminism," *American Quarterly* 48 (1996): 183 (reprinted in this book as chapter 4).

28. Ibid., 226, fn. 50.

29. Russell Jacoby, *Social Amnesia: A Critique of Contemporary Psychology from Adler to Lang* (Boston: Beacon Press, 1975), 4, 101. See also Sarup, *Postmodernism,* 143ff.

30. On Rorty, see Brandon, *Rorty and His Critics,* 35.

31. See George W. Stocking, "Anthropology in Crisis?" in *Crisis in Anthropology,* ed. Hoebel, 407–19.

32. Hollinger, "Science," 160–61. See also Virginia Yans-McLaughlin, "Science, Democracy, and Ethics: Mobilizing Culture and Personality for World War II," in *Malinowski, Rivers, Benedict and Others: Essays on Culture and Personality,* ed. George W. Stocking Jr. (Madison: University of Wisconsin Press, 1986).

33. James T. Kloppenberg, "Pragmatism: An Old Name for Some New Ways of Thinking?" in *The Revival of Pragmatism,* ed. Morris Dickstein (New York: William Morrow, 1992), 262.

34. Ibid., 84.

35. Comment by Deborah Gewertz at a planning meeting for an exhibit on Mead, "Human Nature and the Power of Culture," Library of Congress, September 18, 2000.

36. Hollinger, "Science," 158, 157, 161.

37. David Hollinger, ms. of "Cultural Relativism," for *Cambridge History of Science,* Vol. 7: *Social Science,* n.p.

38. Winthrop Sargeant, "It's All Anthropology," *New Yorker,* December 30, 1961, 31.

39. Foerstel, quoted in Jane Howard, *Margaret Mead: A Life* (New York: Simon and Schuster, 1984), 435.

40. Quoted in Rae Goodale, *The Visible Scientists* (Boston: Little, Brown, 1975), 148–49.

41. Sargeant, "Anthropology," 31.

42. Di Leonardo, *Exotics at Home,* 17.

43. Foerstel and Gilliam, *Confronting the Margaret Mead Legacy,* 67.

44. Selections in Forestel and Gilliam written by New Guineans offer negative evaluations of anthropology. See Gilliam's interview with Nahau Rooney, 31–53; Warilea Iamo, "The Stigma of New Guinea: Reflections of Anthropology and Anthropologists," 75–79; John D. Waiko, "Tugata: Culture, Identity, and Commitment," 233–66. New Guinea filmmaker Leoni Kanawa was more positive to me in the mid-1990s. Without my requesting it, she volunteered that she had become appreciative of Mead's success in documenting and therefore preserving her ancestral history.

45. Lola Romanucci-Ross, *Mead's Other Manus: Phenomenology of the Encounter* (South Hadley, Mass.: Bergen and Garvey, 1985), 164.

46. Peter Worsley accepts their erroneous description of Mead's "uncritical support" of the Vietnam War in Foerstel and Gilliam, 145, and in his Introduction to the book. Di Leonardo, *Exotics at Home,* several times cites Foerstel and Gilliam for her characterization of Mead as a Cold Warrior (e.g., 239). See also Laura Nader, "The Phantom Factor: Impact of the Cold War on Anthropology," in Noam Chomsky et al., *The Cold War and the University: Toward an Intellectual History of the Postwar Years* (New York: Free Press, 1997), 126–27.

47. Worsley, Introduction to *Confronting the Margaret Mead Legacy,* ed. Foerstel and Gilliam, xvii.

48. Ibid.

49. On Mead during World War II, see Yans-McLaughlin, "Science, Democracy, and Ethics." The Mead collection at the Library of Congress amply documents Mead's opposition to the Vietnam War. See File K75, "Draft, Sept. 22. 1971, in support of Cease Fire in Vietnam." See also Mead interview transcript from CBS News Special Report, "Vietnam: A War That Is Finished," April 29, 1975, in which Mead states that the Vietnam War attacked Americans' "ethical existence." My film, *Margaret Mead: An Observer Observed,* shows Mead on camera advocating withdrawal from Vietnam to the Senate Foreign Relations Committee and opposition to nuclear proliferation.

50. Mead, *Social Organization of Manu'a,* xviii.

51. Mead, "Reply to Suggs and Carr," and "Comment on Peace and Disarmament Resolutions," by Suggs and Carr, *Fellow Newsletter: American Anthropological Association,* 3 (September 1962).

52. Eric Wakin, *Anthropology Goes to War: Professional Ethics and Counter Insurgence in Thailand,* Monograph 7 (Madison: University of Wisconsin Center for Southeast Asian Studies, 1992), 230, 233–34, discusses the controversy and its contest. See also E12, no. 1–3, Mead/LC; and Hedda M. van Ooijen, "American Anthropologists at War: The Vietnam Era," unpublished paper, in my possession.

53. Letter from D. L. Olmstead to Carleton Mabee, February 4, 1981, given to the author by Mabee.

54. Quoted by Foerstel and Gilliam, 138. For news coverage of the committee investigation, see Israel Shenker, "Anthropologists Clash over Their Colleagues Ethics on Thailand," *New York Times,* December 27, 1971. The Committee Report, which condemned anthropologists using stolen documents, is in E12, folder 2, Mead/LC.

55. Letter from David Olmstead to Ad Hoc Committee members, E12, folder 1, Mead/LC.

56. File E12, #2, Mead/LC. Letter from Herbert P. Phillips to James Gibbs, December 29, 1971, and letter to D. J. Blakeslee from Herbert P. Phillips, February 3, 1972.

57. A. David Rossin, nuclear research engineer for Commonwealth Edison in Illinois, to Carleton Mabee, July 22, 1980, given to the author by Mabee. Mabee's notes relating to his NSF grant SES 82047, citing *Science,* April 23, 1976, indicate that antinuclear groups would use the National Council of Churches position on plutonium.

58. "Are Shelters the Answer?" *New York Times Magazine,* November 26, 1961. In her forthcoming book on civil defense, *Homeland Security: the Myth of Civil Defense in the Nuclear Age,* Dee Garrison portrays Mead as strongly opposed to nuclear policy (personal communication to author, May 1, 2003).

59. Gene M. Lyons, *The Uneasy Partnership: Social Science and the Federal Government in the Twentieth Century* (New York: Russell Sage Foundation, 1969), 136–37, 162–63; Margaret M. Caffrey, *Ruth Benedict: Stranger in This Land* (Austin: University of Texas Press, 1989), 330ff.

60. Carleton Mabee, "Margaret Mead and a Pilot Experiment in Progressive and Interracial Education: The Downtown Community School," *New York History* (January 1984): 26.

61. "A Reasonable View of Vietnam," *Redbook,* February 1970, 48, 50.

62. George W. Stocking Jr., "Anthropology and the Science of the Irrational: Malinowski's Encounter with Freudian Psycho-Analysis," in Stocking, *Malinowski, Rivers, Benedict,* 5, quoted in Foerstel and Gilliam, 143–44.

63. Foerstel and Gilliam, 126, 149. They fail to cite two other articles by Mabee, both of which view Mead's racial politics as positive. See Mabee, "Mead's Approach to Controversial Public Issues: Racial Boycotts in the AAAS," *The Historian* 48 (February 1986): 191–208; and "Mead and a 'Pilot Experiment.'"

64. For Mead on Civil Disobedience, see "Margaret Mead Answers," *Redbook,* January 1969, 33, 35.

65. Di Leonardo, *Exotics at Home,* citing Howard, 398; see also Howard, *Margaret Mead,* quoting Wilson, 397–98.

66. Mabee, "Margaret Mead's Approach to Controversial Public Issues," 201–3.

67. Di Leonardo, *Exotics at Home,* 193, 258–59, 261.

68. See Joan Gordan, ed., *Margaret Mead: The Complete Bibliography, 1925–1975* (The Hague-Paris: Mouton, 1976).

69. Mead, "What Women Want," *Fortune* 34 (December 1946): 14ff.

70. Ibid.

71. Nader, "Phantom Factor," 126; Peter Worsley, "Margaret Mead: Science or Science Fiction?" *Science and Society* 21 (Spring 1957).

72. Mabee, "Margaret Mead and a Pilot Experiment," 193–94.

73. Jarvie, *Thinking About Society,* 193–94.

74. See Herbert Lewis, "The Misrepresentation of Anthropology and Its Consequences," *American Anthropologist* 100 (September *Science and Society* 21 (Spring 1957); 1998): 720.

75. Thomas, *Out of Time: History and Evolution in Anthropological Discourse* (Cambridge: Cambridge University Press, 1989), 15; Hollinger, "Science," 171.

Contributors

Lois W. Banner, Professor of History and Gender Studies, University of Southern California, is the author of *Intertwined Lives: Margaret Mead, Ruth Benedict, and Their Circle* (Knopf, 2003) and numerous books and articles in U.S. women's history.

Margaret M. Caffrey, Associate Professor, Department of History, University of Memphis, is the co-editor of a forthcoming collection of Mead letters in association with Pat Francis and is the author of *Ruth Benedict: Stranger in This Land* (University of Texas Press, 1989)

Nanako Fukui, Professor, Kansai University, Osaka, is the author of *Samayoeruhito* (Osaka: Kansai University, 1993) and *Nihonjin no Kodo Pattern* (Tokyo: NHK Books, 1997).

Angela Gilliam, an expert in Black feminist anthropology teaching at Evergreen State College, co-edited *Confronting the Margaret Mead Legacy: Scholarship, Empire and the South Pacific* (Temple University Press, 1992) with Leonora Foerstal. Recently she published "A Black Feminist Perspective on the Sexual Commodification of Women in the New Global Culture," in *Black Feminist Anthropology: Theory, Politics, Praxis and Poetics,* edited by Irma McClaurin (Rutgers University Press, 2001) and "Globalization, Identity, and Assaults on Equality in the United States: A Perspective for Brazil" in *Souls: A Critical Journal of Black Politics, Culture and Society* 5 (Spring 2003).

Dolores E. Janiewski, Associate Professor in U.S. History, Victoria University of Wellington, New Zealand, is the author of *Sisterhood Denied: Race, Gender, and Class in a New South Community* (Temple University Press, 1985) and articles in gender, colonialism, and indigenous history, including "Engendering the Invisible Empire: Imperialism, Feminism, and US Women's History," *Australian Feminist Studies* 16 (2001).

Pauline Kent, Associate Professor, Intercultural Communications Faculty, Ryukoku University, Kyoto, Japan, is the author of "The Consummate Cultural Relativist? Ruth Benedict's Approach to Japan in

Re-Mapping Japanese Culture, edited by Alison Tokita (Monash Asia Institute, 2000) and other articles, including "Misconceived Configurations of Ruth Benedict," *Japan Review* 7 (1996).

C. Douglas Lummis, retired from Tsuda College in Tokyo is now on the academic staff at Okinawa International University. He is the author of *Uchinaru Gaikoku: "Kiku to Katana" Saiko* [The Abroad Within: Rethinking "The Chrysanthemum and the Sword"] (Tokyo: Jiji Tsushin, 1981) reprinted by Chikuma Shobo in 1997, and *Radical Democracy* (Cornell University Press, 1996).

Nancy Lutkehaus, Associate Professor of Anthropology and Gender Studies, University of Southern California, has written *Margaret Mead and the Media: Anthropology and the Making of an American Icon,* to be published by Princeton University Press. She has also written the introduction to the reprint of *Blackberry Winter,* published by Kodansha Press, and "Margaret Mead and the 'Wind-Blowing-in-the-Palm-Trees' School of Ethnography," in Ruth Behar and Deborah Gordon (eds), *Women Writing Culture* (Berkeley: University of California Press, 1995).

Judith Schachter Modell, Professor of History and Art History, Carnegie Mellon University, wrote *Ruth Benedict, Patterns of a Life* (University of Pennsylvania Press, 1983). She subsequently published *A Sealed and Secret Kinship: Policies and Practices in American Adoption* (Berghahn Books, 2002). She is currently working on a biographical study of "A 100 Percent Hawaiian Man."

Maureen Molloy, Professor of Women's Studies at the University of Auckland, co-edited a forthcoming collection on the work of Slavoj Zizek and is preparing a book on Margaret Mead's early work and its reflection of American issues and mores.

Louise M. Newman, Associate Professor in U.S. Women's and Gender History at the University of Florida, has written *White Women's Rights: The Racial Origins of Feminism in the United States* (Oxford, 1999).

Christopher Shannon, Research Associate, Jacques Maritain Center, University of Notre Dame, is the author of *Conspicuous Criticism: Tradition, the Individual and Culture in American Social Thought from Veblen to Mills* (Johns Hopkins University Press, 1996) and *A World Made Safe for Differences: Cold War Intellectuals and the Politics of Identity* (Rowman and Littlefield, 2001).

Gerald Sullivan, Visiting Assistant Professor, Department of Anthropology, University of Notre Dame, has published *Margaret Mead, Gregory Bateson and Highland Bali: Fieldwork Photographs of Bayung Gede 1936–1939* (University of Chicago Press, 1999), "A Four-Fold Humanity: Margaret Mead and Psychological Types," *Journal of the History of the Behavioral Sciences* 40 (2004), and "The Individual in Culture or Cultures and Personalities without Embarrassment," *Journal of Pacific Studies.*

Sharon Tiffany, Professor of Anthropology and Women's Studies, University of Wisconsin, Whitewater, is the author of "Imagining the South Seas: Thoughts on Coming of Age and the Sexual Politics of Paradise," *Pacific Studies* 24 (2001) and books on gender and anthropology.

Jean Walton, Professor of English and Women's Studies, Chair of Department of English, University of Rhode Island, is the author of *Fair Sex, Savage Dreams: Race, Psychoanalysis, Sexual Difference* (Duke University Press, 2001).

Virginia Yans, Board of Governors Distinguished Service Professor, Department of History, Rutgers–the State University of New Jersey, has produced a PBS documentary, *Margaret Mead: An Observer Observed,* which is distributed by Filmmakers Library. She is the author of several articles on Mead and of works in immigration history.

Index